In Vitro Propagation and Secondary Metabolite Production from Medicinal Plants: Current Trends (Part 2)

Edited by

Mohammad Anis
Plant Biotechnology Laboratory
Department of Botany
Aligarh Muslim University
Aligarh-202002, India

&

Mehrun Nisha Khanam
Plant Biotechnology Laboratory
Department of Botany
Aligarh Muslim University
Aligarh-202002, India

University Centre for Research & Development
Chandigarh University
Mohali-140413
Punjab, India

In Vitro Propagation and Secondary Metabolite Production from Medicinal Plants: Current Trends

(Part 2)

Editors: Mohammad Anis and Mehrun Nisha Khanam

ISBN (Online): 978-981-5196-35-1

ISBN (Print): 978-981-5196-36-8

ISBN (Paperback): 978-981-5196-37-5

© 2024, Bentham Books imprint.

Published by Bentham Science Publishers Pte. Ltd. Singapore. All Rights Reserved.

First published in 2024.

BENTHAM SCIENCE PUBLISHERS LTD.
End User License Agreement (for non-institutional, personal use)

This is an agreement between you and Bentham Science Publishers Ltd. Please read this License Agreement carefully before using the book/echapter/ejournal ("**Work**"). Your use of the Work constitutes your agreement to the terms and conditions set forth in this License Agreement. If you do not agree to these terms and conditions then you should not use the Work.

Bentham Science Publishers agrees to grant you a non-exclusive, non-transferable limited license to use the Work subject to and in accordance with the following terms and conditions. This License Agreement is for non-library, personal use only. For a library / institutional / multi user license in respect of the Work, please contact: permission@benthamscience.net.

Usage Rules:

1. All rights reserved: The Work is the subject of copyright and Bentham Science Publishers either owns the Work (and the copyright in it) or is licensed to distribute the Work. You shall not copy, reproduce, modify, remove, delete, augment, add to, publish, transmit, sell, resell, create derivative works from, or in any way exploit the Work or make the Work available for others to do any of the same, in any form or by any means, in whole or in part, in each case without the prior written permission of Bentham Science Publishers, unless stated otherwise in this License Agreement.
2. You may download a copy of the Work on one occasion to one personal computer (including tablet, laptop, desktop, or other such devices). You may make one back-up copy of the Work to avoid losing it.
3. The unauthorised use or distribution of copyrighted or other proprietary content is illegal and could subject you to liability for substantial money damages. You will be liable for any damage resulting from your misuse of the Work or any violation of this License Agreement, including any infringement by you of copyrights or proprietary rights.

Disclaimer:

Bentham Science Publishers does not guarantee that the information in the Work is error-free, or warrant that it will meet your requirements or that access to the Work will be uninterrupted or error-free. The Work is provided "as is" without warranty of any kind, either express or implied or statutory, including, without limitation, implied warranties of merchantability and fitness for a particular purpose. The entire risk as to the results and performance of the Work is assumed by you. No responsibility is assumed by Bentham Science Publishers, its staff, editors and/or authors for any injury and/or damage to persons or property as a matter of products liability, negligence or otherwise, or from any use or operation of any methods, products instruction, advertisements or ideas contained in the Work.

Limitation of Liability:

In no event will Bentham Science Publishers, its staff, editors and/or authors, be liable for any damages, including, without limitation, special, incidental and/or consequential damages and/or damages for lost data and/or profits arising out of (whether directly or indirectly) the use or inability to use the Work. The entire liability of Bentham Science Publishers shall be limited to the amount actually paid by you for the Work.

General:

1. Any dispute or claim arising out of or in connection with this License Agreement or the Work (including non-contractual disputes or claims) will be governed by and construed in accordance with the laws of Singapore. Each party agrees that the courts of the state of Singapore shall have exclusive jurisdiction to settle any dispute or claim arising out of or in connection with this License Agreement or the Work (including non-contractual disputes or claims).
2. Your rights under this License Agreement will automatically terminate without notice and without the

need for a court order if at any point you breach any terms of this License Agreement. In no event will any delay or failure by Bentham Science Publishers in enforcing your compliance with this License Agreement constitute a waiver of any of its rights.
3. You acknowledge that you have read this License Agreement, and agree to be bound by its terms and conditions. To the extent that any other terms and conditions presented on any website of Bentham Science Publishers conflict with, or are inconsistent with, the terms and conditions set out in this License Agreement, you acknowledge that the terms and conditions set out in this License Agreement shall prevail.

BENTHAM SCIENCE PUBLISHERS PTE. Ltd.
80 Robinson Road #02-00
Singapore 068898
Singapore
Email: subscriptions@benthamscience.net

CONTENTS

PREFACE	i
LIST OF CONTRIBUTORS	iii
CHAPTER 1 BIOACTIVE COMPONENTS IN *SENNA ALATA* L. ROXB	1
Archana Pamulaparthi, Vamshi Ramana Prathap and *Ramaswamy Nanna*	
INTRODUCTION	1
Results and Discussion	5
Physicochemical Parameters	6
Phytochemical Analysis	8
DISCUSSION	11
CONCLUSION	12
REFERENCES	12
CHAPTER 2 PLANT TISSUE CULTURE: A POTENTIAL TOOL FOR THE PRODUCTION OF SECONDARY METABOLITES	15
Madhukar Garg, Soumi Datta and *Sayeed Ahmad*	
INTRODUCTION	16
SECONDARY METABOLITES	17
Primary *vs* Secondary Metabolites	17
The Biosynthesis of Secondary Metabolites	18
The Physiological Function of Secondary Metabolites	19
Classification of Secondary Metabolites	19
Alkaloids	19
Terpenoids	19
Steroids	20
Quinones	20
Phenylpropanoid	20
Secondary Metabolites and Plant Tissue Culture	20
Callus Cultures	21
Organ Cultures	23
Cell Suspension Cultures	24
Fungal Elicitation	25
Micropropagation	25
Strategies to Increase *in vitro* Synthesis of Secondary Metabolite	27
Conventional Strategies	28
Biotechnological Strategies (Metabolic Engineering, etc.)	35
Use of Nanoparticles for Secondary Metabolite (s) Enhancement	38
Large-scale Production of SMs by Bioreactors	38
CONCLUSION AND PERSPECTIVES	38
ABBREVIATIONS	39
REFERENCES	40
CHAPTER 3 IN VITRO PROPAGATION AND SECONDARY METABOLITE PRODUCTION FROM *WITHANIA SOMNIFERA* (L.) DUNAL	64
Praveen Nagella, Wudali Narashima Sudheer and *Akshatha Banadka*	
INTRODUCTION	65
BIOSYNTHESIS OF WITHANOLIDES	67
***IN VITRO* PROPAGATION STUDIES**	69
Direct Organogenesis	71
Indirect Organogenesis	72

Rooting of the Regenerated Plantlets	73
Acclimatization	73
***IN VITRO* PRODUCTION OF WITHANOLIDES**	74
Production of Withanolides from Cell/Callus Culture	75
Production of Withanolides from Shoot Culture	76
Production of Withanolides from Adventitious Root Culture	76
Production of Withanolides from Hairy Root Culture	77
ELICITATION STRATEGIES FOR IMPROVED PRODUCTION OF WITHANOLIDES	79
METABOLIC ENGINEERING FOR INCREASED PRODUCTION OF WITHANOLIDES	81
GENETIC TRANSFORMATION STUDIES IN *WITHANIA SOMNIFERA*	82
CONCLUSION AND PROSPECTS	84
REFERENCES	85

CHAPTER 4 *IN VITRO* PROPAGATION AND PHYTOCHEMICAL SCREENING OF SOME IMPORTANT MEDICINAL PLANTS OF NORTHERN INDIA-A REVIEW ... 92
Rafiq Lone, Shakir Ahmad Mochi, Younis Ahmad Hajam, Ibraq Khurshid and *Azra N. Kamili*

INTRODUCTION	93
Acorus calamus L	101
Alcea rosea L. (Malvaceae)	102
Hyoscyamus niger L.	103
Gymnema sylvestre R. Br.	104
Abies pindrow (Royle ex D.Don)	106
Heracleum candicans Wall.	108
Jurinea dolomiaea Boiss	109
Arnebia benthamii Wall. ex G. Don	110
Atropa acuminata Royle Ex Lindl.	111
Meconopsis aculeta Royle L.	111
SUMMARY AND CONCLUSION	113
REFERENCES	113

CHAPTER 5 PHYTOCHEMISTRY, ANTIOXIDANTS, ANTIMICROBIAL ACTIVITIES AND EDIBLE COATING APPLICATION OF *ALOE VERA* ... 119
Awad Y. Shala, Hayam M. Elmenofy, Eman Abd El-Hakim Eisa and *Jameel M. Al-Khayri*

INTRODUCTION	120
BIOACTIVE COMPONENTS	122
FACTORS AFFECTING PLANT CHEMICAL COMPOSITION	125
ANTIOXIDANT ACTIVITY OF *A. VERA*	127
ANTIMICROBIAL ACTIVITY	129
Antibacterial Activity	129
Antifungal Activity	132
***ALOE VERA* AS AN EDIBLE COATING FOR VEGETABLES AND FRUITS**	135
Effect of *A. vera* Coating on Microbial Decay and Physiological Disorders of Fruits	135
Microbial Activity	135
Physiological Disorders	137
Impact of *Aloe vera* Coatings on Fruit and Vegetables' Physical-chemical Qualities	138
Fruit Firmness	138
Soluble Solids Content and Acidity	139
Respiration and Ethylene Production	139
Sensory Attributes and Color	140
Total Phenolic Content and Antioxidant Activity	140

IN VITRO PROPAGATION RESPONSES OF *ALOE VERA*	142
Microstructural and Histochemical Changes in *Aloe vera* from *In Vitro* to *In Vivo*	146
CONCLUSION	148
ACKNOWLEDGMENTS	149
REFERENCES	149

CHAPTER 6 MICROPROPAGATION AND PHYTOCHEMICAL STUDIES ON *OROXYLUM INDICUM* (L) KURZ – A REVIEW 161
Samatha Talari and *Rama Swamy Nanna*

INTRODUCTION	161
In-vitro Seed Germination	164
Zygotic Embryo Culture	165
Clonal Propagation	165
Nodal Culture	165
Mericlone Technology	166
In vitro Rooting and Plantlet Establishment	167
Callus Induction	168
Organogenesis	169
Somatic Embryogenesis	169
Phytochemical Studies	171
Toxicological Activity	171
Antimicrobial Activity	172
Antioxidant Activity	173
Anti-inflammatory Activity	174
Hepatoprotective Activity	174
Nephroprotective Activity	174
Antidiabetic Activity	174
Immunomodulatory Activity	174
Anticancer Activity	175
Anti-angiogenic Activity	176
CONCLUSION	176
LIST OF ABBREVIATIONS	176
REFERENCES	176

CHAPTER 7 EXPLORING PLANT TISSUE CULTURE IN *OCIMUM BASILICUM* L. 180
Priyanka Chaudhary, Shivika Sharma and *Vikas Sharma*

INTRODUCTION	180
EVALUATION OF HEREDITARY STABILITY OF TISSUE CULTURE RAISED PLANTLETS *VIA* MOLECULAR MARKERS	182
SHOOT INDUCTION	183
In Vitro Rooting	184
ACCLIMATIZATION	185
SOMATIC EMBRYOGENESIS	185
PRODUCTION OF SECONDARY METABOLITES IN *O.BASILICUM*	186
ORGAN AND CALLUS CULTURE	186
CELL SUSPENSION CULTURE	187
IN VITRO PLANT CELL ELICITATION	188
CONCLUSION AND FUTURE PERSPECTIVES	188
REFERENCES	189

CHAPTER 8 PLANT TISSUE CULTURE: A PERPETUAL SOURCE FOR THE PRODUCTION OF THERAPEUTIC COMPOUNDS FROM RHUBARB 196

Shahzad A. Pandith and *Mohd. Ishfaq Khan*
INTRODUCTION	196
RHUBARB: A General Account	196
ETHNOBOTANY AND PHARMACOLOGY OF RHUBARB	200
PHYTOCONSTITUENTS FROM RHUBARB WITH THERAPEUTIC SIGNIFICANCE	207
Flavonoids	208
Anthraquinones	209
Stilbenoids	211
IN VITRO PROPAGATION OF RHUBARB: CHALLENGES AND PROSPECTS	214
SOMATIC EMBRYOGENESIS AS AN ALTERNATIVE TO *IN VITRO* MICROPROPAGATION OF RHUBARB	216
INITIATION AND ESTABLISHMENT OF CELL SUSPENSION CULTURES: A PROSPECTIVE APPROACH FOR RHUBARB	219
IN VITRO SYSTEMS FOR THE PRODUCTION OF SECONDARY METABOLITES IN RHUBARB	221
PRECURSOR ADDITION AS A MEANS TO IMPROVE THE PRODUCTION OF BIO-ACTIVE CONSTITUENTS IN RHUBARB	223
HAIRY ROOTS: A POTENTIAL SOURCE FOR SECONDARY CHEMICAL CONSTITUENTS	225
CONCLUSION AND FUTURE PROSPECTS	227
REFERENCES	229

CHAPTER 9 *IN VITRO* PLANT REGENERATION FROM NODAL SEGMENTS AND BIOCHEMICAL FIDELITY ANALYSIS OF OPERCULINA TURPETHUM, A THREATENED MEDICINAL PLANT OF ODISHA 245
Kumari Monalisa, Shashikanta Behera, Shasmita, Debasish Moha-patra, Anil K. Biswal and *Soumendra K. Naik*

INTRODUCTION	246
MATERIALS AND METHODS	247
Collection and Surface Sterilization of the Explants	247
Culture Medium and Culture Conditions	247
Acclimatization of Plantlets of *O. turpethum*	248
Phytochemical Analysis	248
Preparation of the Sample	248
Data Recording and Statistical Analysis	248
RESULTS	249
Evaluation of Growth Regulators for Axillary Shoot Proliferation	249
Rooting of *in vitro* Regenerated Shoots	250
Acclimatization of *in vitro* Regenerated Plantlets	252
Phytochemical Analysis	253
DISCUSSION	254
CONCLUSION	255
ACKNOWLEDGEMENT	255
REFERENCES	255

CHAPTER 10 TISSUE AND CELL CULTURE OF TEA (*CAMELLIA SP.*) 259
Abhishek Mazumder, Urvashi Lama, Meghali Borkotoky, Sangeeta Borchetia, Shabana Begam and *Tapan Kumar Mondal*

INTRODUCTION	260
MICROPROPAGATION OF TEA	260
Stage I: Selection of Explants and Establishment of Culture Through Explants	260
Stage II: Initiation, Multiplication, and Elongation of Shoots	261

Stage III: Rhizogenesis	262
Stage IV: Hardening & Acclimatization: Transfer from *in vitro* to *ex vitro* condition	262
Studies of Micropropagated Raised Plants on Field	263
Problems of Micropropagation in Tea: Explant Oxidative Browning and Microbial Contamination	263
SOMATIC EMBRYOGENESIS	264
Different Stages of Somatic Embryogenesis	264
Induction	264
Multiplication or Maturation of Callus	265
Development of Embryo	265
Maturation of Embryo	267
Germination	267
Molecular Mechanisms of Somatic Embryogenesis (SE) in Tea	269
CELL CULTURE AND SECONDARY METABOLITE BIOSYNTHESIS	271
Secondary Metabolite Production in Tea Through Cell Culture Method (*Camellia sp.*)	271
CONCLUSION	273
REFERENCES	273

CHAPTER 11 *IN VITRO* STRATEGIES FOR ISOLATION AND ELICITATION OF PSORALEN, DAIDZEIN AND GENISTEIN IN COTYLEDON CALLUS OF *CULLEN CORYLIFOLIUM* (L.) MEDIK 282

Tikkam Singh, Renuka Yadav and *Veena Agrawal*

INTRODUCTION	283
BIOSYNTHETIC PATHWAYS OF PSORALEN, DAIDZEIN AND GENISTEIN	284
SCREENING FOR HIGH PSORALEN, DAIDZEIN AND GENISTEIN YIELDING PLANT PART OF *C. CORYLIFOLIUM*	285
STRATEGIES FOR EXTRACTION AND ISOLATION OF BIOACTIVE COMPOUNDS IN *C. CORYLIFOLIUM*	286
SCALING UP OF BIOACTIVE COMPOUNDS IN *C. CORYLIFOLIUM* THROUGH VARIOUS STRATEGIES	287
Precursor Feeding	289
Hairy Root Cultures	290
In Vitro Elicitation Using Biotic & Abiotic Elicitors	291
Cell Suspension Cultures/Bioreactors	292
Gene Cloning of Key Enzymes and Overexpression	292
CONCLUSION	295
REFERENCES	295

CHAPTER 12 GENETIC IMPROVEMENT OF PELARGONIUM, AN IMPORTANT AROMATIC PLANT, THROUGH BIOTECHNOLOGICAL APPROACHES 302

Pooja Singh, Syed Saema and *Laiq ur Rahman*

INTRODUCTION	302
TISSUE CULTURE STUDIES IN PELARGONIUM	305
AGROBACTERIUM MEDIATED GENETIC TRANSFORMATION IN PELARAGONIUM	310
Terpene Biosynthesis Pathway	311
Importance of Pelargonium	312
Concerns with Pelargonium species	314
CONCLUSION	314
REFERENCES	315
SUBJECT INDEX	321

PREFACE

The extinction of plant species is progressively taking place due to their being trapped in the vicious circle of ever-increasing industrialization, deforestation, global warming, climate change and also unscrupulous human activities. This has led to many species being listed in the Red Data Book or /in the various threat categories of IUCN. Of the total 3000 medicinal plants reported from India, over 1700 species of medicinal value are found in the Indian Himalayan region; nearly 47% are endemic to this region and about 62 species fall under different categories of threat. The situation warrants the acceleration of efforts to develop methods for germplasm preservation. The importance and applications of plant cell and tissue culture in plant science are vast and varied. The last few years of our research investigations have led to the emergence of this technique. Utilizing the biotechnological tools, many tissue culture protocols have been developed for rapid and mass multiplication of valuable medicinal plants to increase planting stock so as to meet the market demand. The rapid increase in knowledge of nutrition, medicine, agriculture, and plant biotechnologies has effectively changed the concept of food and health causing an overwhelming revolution.

India is known for its diverse climatic zones which are habituated of diverse flora having medicinal value, thus there is a wide scope for India to lead global herbal market. The National Medicinal Plant Board of India has recognized more than 7000 medicinal plants, which are currently used in different systems of medicines. The Ayurveda market in India has been valued at INR 300 billion in 2018 and is expected to reach INR 710 billion by 2024. Plants are active biochemical factories of a vast group of secondary metabolites which are indeed the basic source of various commercial pharmaceutical drugs. There are possibilities for year round production of biomass with reduced cost and time. Elicitation and precursor feeding are two important strategies of the *in-vitro* techniques to enhance metabolite production to meet the demands of mankind. Utilization of the existing genetics resources and understanding the biosynthesis, transport, accumulation and modulation of important secondary metabolites are critical issues linked to its improvement.

Overall, the rapid propagation of elite plants will provide high dividends to farmers and the associated herbal industry.

This book provides comprehensive coverage of the fundamental principles, current practices and trends in the field of pharmaceutical industry and provides baseline data for further research in the field. We are grateful to all the contributors and hope the book will be beneficial to students, researchers, scientists and other concerned stake holders who are working in the respective fields. MA acknowledges the much needed moral support of his wife, Humera Anis. We would also like to place on record our sincere thanks to Mr. Mohammad Zohaib Siddiqui for preparing the layout of the contents.

The cooperation and help received from Bentham Science Publishers is duly appreciated.

Mohammad Anis
Plant Biotechnology Laboratory
Department of Botany
Aligarh Muslim University
Aligarh-202002, India

&

Mehrun Nisha Khanam
Plant Biotechnology Laboratory
Department of Botany
Aligarh Muslim University
Aligarh-202002, India

University Centre for Research & Development
Chandigarh University
Mohali-140413
Punjab, India

List of Contributors

Archana Pamulaparthi	Department of Biotechnology, Kakatiya University, Warangal, India
Akshatha Banadka	Department of Life Sciences, CHRIST (Deemed to be University), Bangalore-560029, Karnataka, India
Azra N. Kamili	Department of Botany, Central University of Kashmir, Ganderbal Jammu and Kashmir, India
Awad Y. Shala	Medicinal and Aromatic Plants Research Department, Horticulture Research Institute, Agricultural Research Center, Giza-12619, Egypt
Anil K. Biswal	Department of Botany, Maharaja Sriram Chandra Bhanja Deo University, Takatpur, Baripada -757001, Odisha, India
Abhishek Mazumder	ICAR-National Institute for Plant Biotechnology, (ICAR-NIPB), New Delhi, Pusa, India
Debasish Mohapatra	Department of Botany, Ravenshaw University, Cuttack-753003, Odisha, India
Eman Abd El-Hakim Eisa	Department of Floriculture and Dendrology, Hungarian University of Agriculture and Life Science (MATE), 1118 Budapest, Hungary Botanical Gardens Research Department, Horticulture Research Institute, Agricultural Research Center (ARC), Giza 12619, Egypt
Hayam M. Elmenofy	Fruit Handling Research Department, Horticulture Research Institute, Agricultural Research Center, Giza-12619, Egypt
Ibraq Khurshid	Department of Zoology, Central University of Kashmir, Ganderbal, Jammu and Kashmir, India
Jameel M. Al-Khayri	Department of Agricultural Biotechnology, College of Agriculture and Food Sciences, King Faisal University, Al-Ahsa 31982, Saudi Arabia
Kumari Monalisa	Department of Botany, Ravenshaw University, Cuttack-753003, Odisha, India Center of Excellence in Environment and Public Health, Ravenshaw University, Cuttack-753003, Odisha, India
Laiq ur Rahman	Plant Biotechnology Division, Central Institute of Medicinal and Aromatic Plants, P.O. CIMAP, Picnic Spot Road, Lucknow, India
Madhukar Garg	Chitkara College of Pharmacy, Chitkara University, Rajpura, Patiala, Punjab, India
Mohd. Ishfaq Khan	Plant Biotechnology Section, Department of Botany, Aligarh Muslim University, Aligarh, Uttar Pradesh, India
Meghali Borkotoky	Tocklai Tea Research Institute, Jorhat, Assam, India
Praveen Nagella	Department of Life Sciences, CHRIST (Deemed to be University), Bangalore-560029, Karnataka, India
Priyanka Chaudhary	Department of Botany, DPG Degree College, Gurgaon, India
Pooja Singh	Plant Biotechnology Division, Central Institute of Medicinal and Aromatic Plants, P.O. CIMAP, Picnic Spot Road, Lucknow, India

Rama Swamy Nanna	Department of Botany, Kakatiya University, Warangal, Telangana State, India
Rafiq Lone	Department of Botany, Central University of Kashmir, Ganderbal Jammu and Kashmir, India
Renuka Yadav	Department of Botany, University of Delhi, Delhi-110007, India
Soumi Datta	Dabur Research and Development Center, Dabur India limited, Sahibabad, Ghaziabad-201010, India
Sayeed Ahmad	Hamdard School of Pharmacy, Jamia Hamdard, Hamdard University, Hamdard Nagar, New Delhi, India
Shakir Ahmad Mochi	Department of Botany, Central University of Kashmir, Ganderbal Jammu and Kashmir, India
Samatha Talari	Department of Botany, Kakatiya Universty, Warangal, Telangana State, India
Shivika Sharma	Biochemical Conversion Unit, SSSNIBE, Kapurthala, India
Shahzad A. Pandith	Department of Botany, University of Kashmir, Srinagar, Jammu and Kashmir, India
Shashikanta Behera	Department of Botany, Ravenshaw University, Cuttack-753003, Odisha, India
Shasmita	Department of Botany, Ravenshaw University, Cuttack-753003, Odisha, India
Soumendra K. Naik	Department of Botany, Ravenshaw University, Cuttack-753003, Odisha, India Center of Excellence in Environment and Public Health, Ravenshaw University, Cuttack-753003, Odisha, India
Sangeeta Borchetia	Tocklai Tea Research Institute, Jorhat, Assam, India
Shabana Begam	ICAR-National Institute for Plant Biotechnology, (ICAR-NIPB), New Delhi, Pusa, India
Syed Saema	Environmental Science Department, Integral University, Lucknow, India
Tapan Kumar Mondal	ICAR-National Institute for Plant Biotechnology, (ICAR-NIPB), New Delhi, Pusa, India
Tikkam Singh	Department of Botany, University of Delhi, Delhi-110007, India
Urvashi Lama	Department of Botany, Sovarani Memorial College, Jagatballavpur, Howrah, West Bengal, India
Vamshi Ramana Prathap	Department of Pharmaceutical Sciences, Jawaharlal Nehru Technological University, Hyderabad, India
Vikas Sharma	Biochemical Conversion Unit, SSSNIBE, Kapurthala, India
Veena Agrawal	Department of Botany, University of Delhi, Delhi-110007, India
Wudali Narashima Sudheer	Department of Life Sciences, CHRIST (Deemed to be University), Bangalore-560029, Karnataka, India
Younis Ahmad Hajam	Department of Life Sciences and Allied Health Sciences, Sant Baba Bhag Singh University, Khiala Jalandhar, Punjab, India

CHAPTER 1

Bioactive Components in *Senna Alata* L. Roxb

Archana Pamulaparthi[1], Vamshi Ramana Prathap[2] and Ramaswamy Nanna[1,*]

[1] *Department of Biotechnology, Kakatiya University, Warangal, India*
[2] *Department of Pharmaceutical Sciences, Jawaharlal Nehru Technological University, Hyderabad, India*

Abstract: *Senna alata* is an ethnomedicinal plant. The crude extracts of the plants are said to have a large number of medicinal properties due to their phytochemicals. In the present study, we made an attempt to isolate and screen the phytochemical constituents present in the species. In order to determine the bioactive constituents present in *S. alata*, and the effect of drying on the loss of bioactive constituents, studies on a set of pharmacognostical parameters were conducted on seeds, shade and sun-dried leaves of *S. alata* as per US pharmacopeia and WHO guidelines. The results of the present studies showed the presence of various important bioactive molecules that are responsible for the medicinal properties of the species. The phytochemical analysis of seed extracts revealed the presence of alkaloids, flavonoids, tannins, saponins, anthraquinones, resins and glycosides in all the extracts, while coumarins, phenols, terpenoids, phlobatannins and quinines are completely absent in all the seed extracts. Preliminary phytochemical investigations from shade and sun-dried leaf extracts showed alkaloids, flavonoids, anthraquinones, saponins, glycosides and tannins in high amounts in all the extracts, resins and phenols are present in moderate amounts. Terpenoids and phlobatannins are present only in fresh leaf extracts. Studies were also conducted on the physicochemical and organoleptic properties of leaves of *S. alata* that help in the identification and standardization of the leaf extracts for manufacturing of plant-based drugs of *S. alata*.

Keywords: Bioactive components, Leaf extracts, Preliminary phytochemical screening, *Senna alata*, Sun-dried, Shade dried.

INTRODUCTION

Ever since the origin of the human race, plants have been used as medicine because of their potent therapeutic value. Plants have been a source of therapeutic agents for thousands of years, and the majority of drugs or their derivatives used in the present day have been isolated from plants. Since ancient times, conven-

---

* **Corresponding author Ramaswamy Nanna:** Department of Biotechnology, Kakatiya University, Warangal, India; E-mail: swanynr.dr@gmail.com

Mohammad Anis & Mehrun Nisha Khanam (Eds.)
All rights reserved-© 2024 Bentham Science Publishers

-tional medical systems have been known to play a key role in the primary healthcare needs of the human race [1]. Almost all known civilizations around the world, including the Chinese, Indus Valley, African, and Egyptian, have their own ancient system of medicine that includes various types of naturally occurring compounds derived from medicinal plants.

Indian Vedic literature such as Rig Veda and Atharvana Veda (4500-1600 BC) also mentioned the use of several plants as a source of medicine. Ancient ayurvedic practitioners such as *Charaka* and *Susrutain,* and their respective books *Charaka Samhita* and *Susruta Samhita* referred to the use of more than 700 herbs as medicine and ancient medical systems such as *Ayurveda, Unani, Homeopathic and Siddha* have been surviving over 3000 years, by using plant-based drugs or their preparations and formulations for curing diseases.

World Health Organization [2] defined medicinal plants as follows: "medicinal plant" is any plant in which one or more of its organs possess compounds that can either be used for the therapeutic purposes or act as precursors for the synthesis of practicable drugs. The term "herbal drug" is used for the part/parts of a plant, *viz.* leaves, flowers, fruits, roots, bark, and seeds that are used for the preparation of therapeutic compounds. These definitions distinguish medicinal plants whose bioactive ingredients and therapeutic values have been demonstrated scientifically from plants that are considered medicinal but have not been established scientifically. WHO [3] further defines a medicinal formulation as any medicinal plant preparation obtained by subjecting the crude plant material to physical processes such as extraction, purification, fractionation, concentration, or biological processes that can be used for immediate consumption.

Many people from both developing and developed countries across the world do not have adequate access to basic needs, including food, water, clean environment, and medical and health services. The main concern of public health is still the intense need for basic health care, which is lacking even at the elementary level. According to WHO, more than half of the world's population do not have access to basic healthcare needs as poor people are unable to access the present healthcare services due to their non-affordability. Therefore, the challenge for governments in both developed and developing countries in the near future lies in food and medical security that should necessarily double the production of food and medicine in the next 50 years to meet the needs of the growing population. Medicinal plants not only offer access to medicine to poor people at an affordable price but also help in generating income, employment, and foreign exchange in developing countries, thus contributing significantly to the national economy. It is estimated that plant-derived drugs account for about Rs. 2,00,000 crores in the world market.

During the past century, the formulation and large-scale production of synthetic drugs have brought a revolutionary change in health care across the world. Nevertheless, more than 70-90% of people in both developing as well as developed countries rely on traditional practitioners and herbal medicine as a source of primary medicine [4], which attracted the attention of researchers towards medicinal plants globally. In modern pharmacopoeia, not less than 25% of drugs are derived from plants and many other drugs are synthetic analogues of standard compounds that are already isolated from medicinal plants. Even today, about 121 such active compounds are in use in the pharmaceutical industry [5] and more than 100 herbal-based drugs are under clinical study [6].

Even though modern medicines are effective, they have several disadvantages, including high cost, reducing immunity, causing severe side effects, and physical dependence. On the other hand, plant-based medicines are natural, cost effective, and have minimum or no side effects, that is, leading to an increase in the number of people turning towards herbal medicine, thus being used in achieving the goal of "Health for all" in a cost-effective manner [7]. This interest in phytomedicine can lead to the exploration of about 500 different plant species in the last few decades, and many species are still being studied.

Escalating faith in herbal medicine is one of the several reasons for the increasing need for recognition of medicinal plants [8]. Medicinal plants play a vital role in various traditional, complementary, and alternate systems of medicine as they contain a broad range of secondary metabolites, such as alkaloids, flavonoids, tannins and terpenoids [9, 10], which are found to play a key role in the regulation of diseases in human beings. The presence of these phytochemicals is responsible for the antioxidant, antimicrobial, and antipyretic effects of these medicinal plants [11]. WHO states that medicinal plants are the best source for obtaining a variety of herbal formulations. Hence, plants with such medicinal properties should be studied for a better understanding of their therapeutic properties, efficacy, and safety issues [12].

Plant metabolites can be divided into two groups as primary metabolites that are directly involved in the growth and metabolism of the plant and secondary metabolites are organic compounds which are the byproducts of primary metabolism that are not generally used by plants for metabolic activities. These secondary metabolites serve as interspecific defenses when the plant interacts with their counter biotic and abiotic partners in the environment [13]. These secondary metabolites are structurally and functionally diverse in nature and can be classified as alkaloids, flavonoids, glycopeptides, phenolics, peptides, steroids, terpenoids, and volatile oils [14]. These secondary metabolites act as precursor molecules for pharmaceuticals, agrochemicals, cosmetics, and industrial products and as flavouring and colouring agents in the food processing industry.

Bioactive therapeutic agents produced in plants are the products of natural metabolic processes. Each species has its own genetic makeup that governs the production of these bioactive molecules. In addition to the genetic makeup, other factors such as the effect of environmental factors such as temperature, moisture content, and the difference among cultivars within the species also contribute to the variation in the quality and quantity of the compounds [15].

These bioactive compounds either act on one or different systems of the animal physiologically and/or act by interfering with the metabolism of microbes involved in the infection process, thus regulating the host-microbe interactions in favor of the host. Research is being focused on meeting the challenges of identifying bioactive compounds in plants, and establishing evidences on whether the whole plant or extracted compounds are to be used for therapeutic purposes. Therefore, the identification, isolation, characterization, and purification of the bioactive compound from crude extracts of plants play a key role.

Evaluation of the pharmacological activities of medicinal plants and the subsequent increase in the demand for plant-based drugs is leading to overharvesting, thus creating heavy pressure on high-value medicinal plant populations. Moreover, several of the medicinal plants have low population densities, slow growth rates, and narrow geographic ranges and therefore are more prone to extinction [16]. Furthermore, the knowledge of the use of lesser-known medicinal plants is declining rapidly. Hence, there is a need to spread awareness and conservation of these medicinally important species through various conservation techniques available. Furthermore, the lack of availability of plant material throughout the year and the impracticality of conventional propagation and breeding methods for the production of plants on a large scale act as a barrier to the separation of bioactive molecules. In such cases, alternative and economically feasible approaches for the separation of the desired phytochemicals are to be implied. Biotechnological tools help in solving the problems faced by conventional breeding programmes and act as bioreactors for the production of bioactive metabolites from endangered and medicinally important plants [17]. Thus, these biotechnological tools offer a line of approaches for maintenance, genetic improvement, and efficient use of endangered plant resources and products [18].

In both developing and developed countries, the increase in demand for plant-based crude material in pharma, cosmetic, and herbal industries is leading to the frequent contamination of crude products with extraneous/foreign material or with inferior quality crude drugs that resemble the standard drug. To avoid such problems, systematic approaches towards standardization of crude drugs have been developed in modern pharmacology. These standardization procedures include botanical authentification, microscopic and molecular examination, identification of chemical constituents, and biological activity of the whole plant [19]. Identification of bioactive chemicals and macroscopic microscopic evaluation of plant materials for quality control and standardization have been reported by early researchers [20]. Macroscopic evaluation parameters involve sensory characters such as shape, size, colour, texture, odour, and taste, and microscopic parameters involve comparative microscopic inspection of the powdered herbal drugs. Various modern techniques such as chromatography, spectrophotometry, electrophoresis, polarography, fluorescence analysis alone and in combination are currently employed in the standardization of herbal drugs [21].

Based on the above considerations, plant biotechnology can be regarded as an important tool that enables the conservation of desired species and obtains elite clones of pharmaceutically important medicinal plants, and meets the demands of public healthcare systems and pharmaceutical industries, especially in biodiversity rich areas where there is an urgent need for conservation of germplasm for future generations as important species are becoming extinct due to over-exploitation.

Techniques of plant biotechnology also help in the isolation of therapeutically important compounds from a particular tissue/organ without any loss to the whole plant, thus, helping in the conservation of commercially important medicinal plants. Considering the demand for *Senna alata* L. Roxb. (Syn. *Cassia alata*) due to its ethnic medicinal properties, use as an ornamental shrub, and therapeutic applications, the species needs conservation. Hence, in the present investigations, attempts have been made to evaluate the pharmacological properties of aqueous leaf extracts using various animal models, to multiply and conserve this medicinally important woody legume using various *in vitro* culture techniques. Attempts were also made to study the pharmacognotic properties concentrating mainly on screening for the presence of various phytochemicals from leaf explants dried under different conditions in order to study the effect of drying on the loss of chemical constituents and to study the amount of total anthraquinone and anthraquinone glycosides from various explants and extracts to determine the ideal explants/extracts for commercial production of anthraquinones, and to isolate anthraquinones from five different species of *Cassia/Senna*.

Results and Discussion

About One-fourth of the world's population, accounting for 1.42 billion people generally depends on traditional medicine that particularly consists of plant-based drugs for curing various diseases [8]. Herbal or traditional medicines are gaining importance due to their safety, efficacy, and lack of side effects and are considered a promising choice over modern synthetic drugs [22].

The phytoconstituents present in the plant parts account for various pharmacological activities of these medicinal plants, which form the basis of herbal medicine and herbal industries, which mostly use these fresh or dried plant parts for the manufacturing of herbal drugs. A detailed knowledge of crude drugs is essential in the identification, preparation, safety, and efficacy of herbal products. In order to meet the increasing demand for plant-based crude drugs, phytochemical and pharmaceutical industries in both developing and developed countries are adulterating crude drugs with foreign organic matter or substituting inferior quality crude drugs resembling the standard drugs. Hence, there is a need for the development of a systematic approach to studying crude drugs in modern pharmacognosy.

The process of standardization is a multi-step process and can be achieved by stepwise pharmacognostic studies [23]. Various methods, such as the determination of ash residues, extractive values, and screening of active phytoconstituents, play a significant role in the standardization of indigenous crude drugs [24].

The species *Senna alata* possesses many valuable medicinal properties, but the knowledge of these properties is still confined to tribal areas because of the absence of proper scientific standardization. In order to determine the usefulness of the species in modern medicine, standardization of various parameters, *viz.* morphological, physico-chemical and phytochemical constituents, are essential. Based on these parameters, the plants can be made successfully available for the population and herbal industries across the world. Hence, the present study has been undertaken to evaluate the organoleptic, physicochemical, and phytochemical constituents present in the seed and leaf extracts of *S. alata* in order to elite medicinal properties attributed to the species.

Physicochemical Parameters

Studies on the physicochemical parameters of crude extracts are essential for the identification of the plant material, analyzing the stability of the crude drug, microbial contamination, heavy metal accumulation, avoiding mishandling, and estimating adulteration. In the present study, in order to determine the purity of the drug, various physicochemical parameters have been studied in *S. alata*. The details of the studies are presented in Tables **1 - 3**.

Table 1. Physical characters of various leaf extracts of S. alata.

Extract	Colour	Consistency	% Yield (w/w)
Aqueous	Niger brown	Amorphous	33.9±0.69
Acetone	Green	Amorphous	26.1±0.42
Benzene	Yellowish green	Amorphous	26.9±0.16
Chloroform	Dark green	Amorphous	27.1±0.54
Ether	Dark green	Amorphous	28.6±0.81
Methanol	Greenish brown	Amorphous	28.4±0.39

Various physicochemical properties such as total ash, acid soluble and insoluble ash, and water insoluble ash were determined using powdered leaf material of *S. alata*. The results obtained from the present study are presented in Table **2**. The total ash and acid insoluble ash content of the leaf of *S. alata* is 7.84 (% w/w) and 0.94 (% w/w), respectively. The amount of acid insoluble ash is very less than that of water soluble ash (6.90%). These physicochemical properties related to total ash value and acid insoluble ash values help in determining the evaluation of crude drugs at a large scale.

Table 2. Physicochemical properties of Methanol and Aqueous extracts of S. alata.

Test	Value Obtained (%w/w)
Total Ash content	7.84
Acid insoluble ash	0.94
Water soluble ash	6.90
Moisture content	3.9
Total fiber content	22

The results of the fluorescent studies of powdered leaf material are presented in Table **3**. The extracts exhibited various fluorescent characteristics under normal/UV light. Among the various solvents tested, acetone extract did not show any fluorescence activity. Whereas, all other tests showed characteristic colouration. The colour due to fluorescence was found to be specific to each compound (Table **3**).

Various aspects regarding the physical constituents of five different extracts (Aqueous, Acetone, Benzene, Chloroform and Ether extracts) of *S. alata* obtained by maceration are presented in Table **1**. All the extracts exhibited a characteristic colour ranging from yellowish green to niger brown. The consistency of the extracts was amorphous for all the extracts and the % yield of the extract ranged from 26.1±0.42 (Acetone extract) to 33.9±0.69 (Aqueous extract) (Table **1**).

Table 3. Fluorescent studies of leaf powder of *S. alata*.

Type of Solvent	Appearance Under Visible Light	Appearance Under UV Light
Powder as such	Green	Green
Aqueous extract	Dark brown	Blackish green
Acetone extract	Green	Green
Benzene extract	Dark green	Greenish brown
Chloroform extract	Dark green	Blackish green
Ether extract	Light green	Green
Methanol extract	Light green	Green
Powder + Conc. HCl	Dark green	Greenish Black
Powder + Conc. HNO_3	Reddish brown	Black
Powder + Conc. H_2SO_4	Reddish brown	Greenish Black
Powder + 5% I_2	Brown	Green
Powder + 1N NaOH in H_2O	Greenish brown	Green
Powder + 1N NaOH in Methanol	Light green	Greenish Black
Powder +5% $FeCl_3$	Dark green	Green
Powder + Glacial acetic acid	Yellowish brown	Greenish Brown
Powder + Picric acid	Yellowish brown	Yellowish green

Phytochemical Analysis

Preliminary phytochemical screening of both fresh and shade-dried leaf extracts was carried out to determine the effect of loss of plant material on drying and to determine the appropriate solvent for extraction. Different extracts of seed and fresh/shade-dried leaf material were screened for the presence of phytochemicals. The results of preliminary phytochemical screening of seed fresh/shade-dried leaves of *S. alata* are are shown in Tables **4** and **5**.

The phytochemical analysis of seed extracts revealed the presence of alkaloids, flavonoids, tannins, saponins, anthraquinones, resins, and glycosides in all extracts. Sterols were present only in the chloroform extract and phenols only in the aqueous extracts. Coumarins, phenols, terpenoids, phlobatannins and quinines are completely absent in all seed extracts (Table **4**). Phytochemical screening of fresh and dry leaves showed the presence of alkaloids, flavonoids, anthraquinones, saponins, glycosides, and tannins in high amounts in all extracts except for methanolic extract. Resins and phenols are present in moderate amounts. Sterols are present only in chloroform extracts. Terpenoids and phlobatannins are present only in fresh leaf extracts, and quinones are present only in benzene and aqueous extracts of dried leaves (Table 5).

Table 4. Phytochemical analysis of seed extracts of *S. alata*.

Phytochemical Tests	Methanol Extract	Petroleum Ether Extract	Benzene Extract	Chloroform Extract	Aqueous Extract
ALKALOIDS					
Dragendorff's test	++	++	++	++	++
Mayer's test	++	++	++	++	++
Wagner's test	++	++	++	++	++
Hager's test	++	++	++	++	++
Tannic acid test	++	++	++	++	++
$FeCl_3$ test	++	++	++	++	++
GLYCOSIDES					
Raymond's test	++	++	++	++	++
Legal's test	++	++	++	++	++
Bromine water test	++	++	++	++	++
KellarKiliani test	++	++	++	++	++
Conc. H_2SO_4 test	++	++	++	++	++
Molisch test	++	++	++	++	++
TANNINS					
$FeCl_3$ test	++	++	+	++	++
Gelatin test	++	++	+	++	++
Lead acetate test	++	++	-	++	++
Alkaline reagent test	++	++	+	++	++
Mitchell's test	++	++	+	++	++
FLAVONOIDS					
Lead acetate test	++	++	++	++	++
$FeCl_3$ test	++	++	++	++	++
Shinoda's test	++	++	++	++	++
Alkaline reagent test	++	++	++	++	++
Zn-Hcl	++				
Reduction test	++	++	++	++	++
STEROLS					
LibermannBurchard test	+	-	-	++	-
Salkowski test	+	+	-	++	-
ANTHRAQUINONES	++	++	++	++	++
COUMARINS	-	-	-	-	-
RESINS	-	++	++	++	++

(Table 4) cont.....

Phytochemical Tests	Methanol Extract	Petroleum Ether Extract	Benzene Extract	Chloroform Extract	Aqueous Extract
LIGNINS	-	-	-	-	-
PHENOLS	-	-	-	-	-
TERPENOIDS	-	-	-	-	-
PHLOBATANINS					
QUINONES	-	-	-	-	-
SAPONINS	+	+	+	+	+

++ = Strongly present, + = Present, - = Absent

Table 5. Phytochemical analysis of fresh/dry leaf extracts of *S. alata*.

Phytochemical Tests	Methanol Extract		Petroleum Ether Extract		Benzene Extract		Chloroform Extract		Aqueous Extract	
	F	D	F	D	F	D	F	D	F	D
ALKALOIDS										
Dragendorff's test	-	-	++	++	++	++	++	+	++	++
Mayer's test	-	-	++	-	++	-	++	-	++	++
Wagner's test	-	-	++	++	++	++	++	++	++	++
Hager's test	-	-	++	++	++	++	++	++	++	++
Tannic acid test	-	-	++	++	++	++	++	++	++	++
GLYCOSIDES										
Raymond's test	-	-	++	++	++	++	++	-	++	++
Legal's test	++	-	++	++	++	++	++	++	++	++
Bromine water test	-	-	++	-	++	-	++	-	++	++
KellarKiliani test	-	-	++	++	++	++	++	++	++	++
Conc. H_2SO_4 test	++	-	++	++	++	++	++	++	++	++
Molisch test	-	-	++	-	++	++	++	++	++	++
TANNINS										
$FeCl_3$ test	+	-	++	++	++	+	++	+	++	+
Gelatin test	-	-	++	++	++	+	++	+	++	+
Lead acetate test	-	-	++	++	++	-	++	+	++	+
Mitchell's test	-	-	++	++	++	-	++	-	++	-
FLAVONOIDS										
Lead acetate test	+	-	++	++	++	++	++	++	++	++
$FeCl_3$ test	-	-	++	++	++	++	++	++	++	++
Shinoda's test	+	-	++	++	++	++	++	++	++	++
Alkaline reagent test	-	-	++	++	++	++	++	++	++	++
Zn-Hcl Reduction test	-	-	++	++	++	++	++	++	++	++
STEROLS										
Libermann Burchard test	-	-	-	-	-	-	++	++	-	-
Salkowski test	-	-	-	-	-	-	-	-	-	-

(Table 5) cont.....

Phytochemical Tests	Methanol Extract		Petroleum Ether Extract		Benzene Extract		Chloroform Extract		Aqueous Extract	
ANTHRAQUINONES	++	++	++	++	++	++	++	++	++	++
COUMARINS	-	-	-	-	-	-	-	-	-	-
RESINS	+	-	++	-	++	++	++	-	++	++
LIGNINS	+	+	+	-	-	-	+	+	-	-
PHENOLS	+	-	++	-	++	++	++	++	++	++
TERPENOIDS	-	-	++	-	++	-	++	-	++	-
PHLOBATANINS	+	-	+	+	+	-	+	-	+	+
QUINONES	-	-	-	-	-	-	-	-	-	-
SAPONINS	++	-	++	++	++	++	++	++	++	++

F = Fresh leaf extract; D = Dry leaf extract;
++=Strongly present, + = Feebly present, - = Absent

DISCUSSION

According to the results of our study, the total ash content of leaves of *S. alata* leaves was found to be 7.84%, of which very low quantity (0.94%) of acid insoluble and high amount (6.90%) of water soluble ash, which indicate that the crude extract of *S. alata* leaves contains more amounts of physiological ash than the non-physiological content. Low moisture content (3.9%) discourages the growth of bacteria, yeast, or fungi during storage of the crude drug, and the high fiber content of the leaves is an indication of the possible microbial contamination due to unfavorable moisture content and rich dietary fiber source of the leaves.

Studies on phytochemical screening revealed the presence of alkaloids, glycosides, resins, tannins, and anthraquinones in all seed extracts except in the methanol extract [25] and the presence of alkaloids, glycosides, saponins, tannins, flavonoids, terpenoids, anthraquinones, resins and steroids in the fresh leaf extracts and glycosides, alkaloids, saponins, tannins, flavonoids and anthraquinones in dry leaf extracts (Table 5). The presence of same phytochemicals was also reported in *S. alata* [25, 26]. The presence of these phytoconstituents may be responsible for various pharmacological activities associated with the species as different compounds are associated with different pharmacological activities.

Due to the presence of these phytochemicals with pharmacological activities, the demand for crude drugs is increasing day by day, leading to the adulteration of crude drugs. Hence, there is a need for standardization of crude drugs in order to identify the presence of adulterating substances [27]. WHO stated that morphological characters like epidermal cell features, stomatal index, vein islets, *etc.*, are to be studied for the proper identification of crude drugs [28, 29].

The preparation of herbal medicines involves the use of fresh or dried plant parts and their extracts. Such preparations usually require a sound knowledge of the crude drugs in order to obtain drugs with high efficacy and safety. A detailed, stepwise pharmacognostic evaluation is therefore required in order to standardize the purity of crude drugs [30]. Pahrmacognostic and organoleptic properties such as extractive value, acid value, fluorescence analysis, and microscopic studies play a major role in the standardization of the native crude drugs and help in the assessment of any adulterating substances in the crude drugs [24].

CONCLUSION

In the present study, leaf extracts of *S. alata* were tested for various phytochemical and organoleptic characteristics for their successful standardization according to WHO guidelines. The results of the study revealed the presence of various phytochemicals that are responsible for the pharmacological activities of the species. These studies assist in the identification and standardization of the leaf extracts and in carrying out further research on the pharmacological activities based on the phytochemicals present.

REFERENCES

[1] Owolabi J, Omogbai EKI, Obasuyi O. Antifungal and antibacterial activities of the ethanolic and aqueous extract of *Kigeliaafricana*(Bignoniaceae) stem bark. Afr J Biotechnol 2007; 6(14): 882-5.

[2] Resolution – Promotion and Development of Training and Research in Traditional Medicine. WHO documentno 1977; 30-49.

[3] World Health Organization (WHO). General guidelines for methodologies on research and evaluation of traditional medicines 2001.

[4] General guidelines for methodologies on research and evaluation of traditional medicines. World Health Organization (WHO) 2005.

[5] Sahoo N, Manchikanti P, Dey S. Herbal drugs: Standards and regulation. Fitoterapia 2010; 81(6): 462-71.
[http://dx.doi.org/10.1016/j.fitote.2010.02.001] [PMID: 20156530]

[6] Li JWH, Vederas JC. Drug discovery and natural products: End of an era or an endless frontier? Science 2009; 325(5937): 161-5.
[http://dx.doi.org/10.1126/science.1168243] [PMID: 19589993]

[7] Anonymous . In: report of the task force on conservation & sustainable use of medicinal plants. New Delhi: Tewari DN. Government of India Planning Commission 2000; pp. 9-24.

[8] Kala CP. Health traditions of Buddhist community and role of *Amchis*in trans- Himalayan region of India. Curr Sci 2005; 89(8): 1331-8.

[9] Hartmann T. From waste products to ecochemicals: Fifty years research of plant secondary metabolism. Phytochemistry 2007; 68(22-24): 2831-46.
[http://dx.doi.org/10.1016/j.phytochem.2007.09.017] [PMID: 17980895]

[10]   Jenke-Kodama H, Müller R, Dittmann E. Evolutionary mechanisms underlying secondary metabolite diversity. Prog Drug Res 2008; 65: 119-140, 121-140.
[http://dx.doi.org/10.1007/978-3-7643-8117-2_3] [PMID: 18084914]

[11]   Adesokan AA, Yakubu MT, Owoyele BV, Akanji MA, Soladoye A, Lawal OK. Effect of administration of aqueous and ethanolic extracts of *Enantiachlorantha*stem bark on brewer's yeastinducedpyresis in rats. Afr J Biochem Res 2008; 2(7): 165-9.

[12]   Nascimento GGF, Locatelli J, Freitas PC, Silva GL. Antibacterial activity of plant extracts and phytochemicals on antibiotic-resistant bacteria. Braz J Microbiol 2000; 31(4): 886-91.
[http://dx.doi.org/10.1590/S1517-83822000000400003]

[13]   Harborne JB, Turner BL. Plant chemosystematics. London: Academic Press 1984.

[14]   Verpoorte R. Secondary metabolism.Metabolic engineering of plant secondary metabolism. The Netherlands: Klumer Academic Publishers 1999; pp. 1-29.

[15]   Thomas SC. Medicinal plants, culture, utilization and phytopharmacology, (1st.. Pennyslvania 2002; pp. 12-265.

[16]   Nautiyal S, Rao KS, Maikhuri RK, Negi KS, Kala CP. Status of medicinal plants on way to Vashuki Tal in Mandakini Valley, Garhwal, Uttaranchal. J Non-Timber For Prod 2002; 9: 124-31.

[17]   Yaseen khan, Aliabbas M, Kumar V S, Rajkumar S. Recent advances in medicinal plant biotechnology. Indian J biotechnol 2009; 8: 9-12.

[18]   Bapat VA, Yadav SR, Dixit GB. Rescue of endangered plants through biotechnological applications. Natl Acad Sci Lett 2008; 31: 201-10.

[19]   Patel PM, Patel NM, Goyal RK. Quality control of herbal products. Indian Pharmacist 2006; 5(45): 26-30.

[20]   Anonymous . Indian Herbal Pharmacopoeia. Mumbai: Indian Drug Manufacturers' Association 2002.

[21]   Mosihuzzaman M, Choudhary MI. Protocols on safety, efficacy, standardization, and documentation of herbal medicine (IUPAC Technical Report). Pure Appl Chem 2008; 80(10): 2195-230.
[http://dx.doi.org/10.1351/pac200880102195]

[22]   Sanmugarajah V, Thabrew I, Sivapalan SR. Phyto,Physicochemical standardization of medicinal plant Enicostemma littorale, Blume. IOSR J Pharm 2013; 3(2): 52-8.
[http://dx.doi.org/10.9790/3013-32205258]

[23]   Tripathi GS, Tripathi YB. Choleretic action of andrographolide obtained fromAndrographis paniculata in rats. Phytother Res 1991; 5(4): 176-8.
[http://dx.doi.org/10.1002/ptr.2650050408]

[24]   Trivedi N, Rawal UM. Hepatoprotective and toxicological evaluation of *Andrographis paniculata* on severe liver damage. Indian J Pharmacol 2000; 32: 288-93.

[25]   Preliminary phytochemical screening from leaf and seed extracts of *Senna alata* L. Roxb-an Ethnomedicinalplant. International Journal of Biological & Pharmaceutical Research 2012; 3(3): 82-9.

[26]   Christy Jeyaseelan E, Tharmila S, Thavaranjit AC. *In vitro* evaluation of different aqueous extracts of *Senna alata* leaves for antibacterial activity. Srilankan J Ind Med 2011; 1(2): 64-9.

[27]   Mukherjee PK. Quality Control of Herbal Drugs. New Delhi, India 2002.

[28]   Kumar S, Kumar V, Prakash O. Microscopic evaluation and physiochemical analysis of Dillenia indica leaf. Asian Pac J Trop Biomed 2011; 1(5): 337-40.
[http://dx.doi.org/10.1016/S2221-1691(11)60076-2] [PMID: 23569789]

[29]   Nasreen S, Radha R. Assessment of Quality of *Withania somnifera* Dunal (Solanaceae) pharmacognostical and physicochemical profile. Int J Pharm Pharm Sci 2011; 3(2): 152-5.

[30] Tripathi GS, Tripathi YB. Choleretic action of andrographolide obtained fromAndrographis paniculata in rats. Phytother Res 1991; 5(4): 176-8.
[http://dx.doi.org/10.1002/ptr.2650050408]

CHAPTER 2

Plant Tissue Culture: A Potential Tool for the Production of Secondary Metabolites

Madhukar Garg[1], Soumi Datta[2] and Sayeed Ahmad[3],*

[1] Chitkara College of Pharmacy, Chitkara University, Rajpura, Patiala, Punjab, India

[2] Dabur Research and Development Center, Dabur India limited, Sahibabad, Ghaziabad-201010, India

[3] Hamdard School of Pharmacy, Jamia Hamdard, Hamdard University, Hamdard Nagar, New Delhi, India

Abstract: Plants are an immense source of phytochemicals with therapeutic effects and are widely used as life-saving drugs, and other products of varied applications. Plant tissue culture is a unique technique employed under aseptic conditions from different plant parts called explants (leaves, stems, roots, meristems, *etc.*) for *in vitro* regeneration and multiplication of plants and synthesis of secondary metabolites (SMs). Selection of elite germplasm, high-producing cell lines, strain enhancements, and optimization of media and plant growth regulators may lead to increased *in vitro* biosynthesis of SMs. Interventions in plant biotechnology, like the synthesis of natural and recombinant bioactive molecules of commercial importance, have attracted attention over the past few decades; and the rate of SMs biosynthesis has increased manifold than the supply of intact plants, leading to a quick acceleration in its production through novel plant cultures. Over the years, the production of SMs *in vitro* has been enhanced by standardising cultural conditions, selection of high-yielding varieties, application of transformation methods, precursor feeding, and various immobilization techniques; however, most often, SM production is the result of abiotic or biotic stresses, triggered by elicitor molecules like natural polysaccharides (pectin and chitosan) that are used to immobilize and cause permeabilization of plant cells. *In vitro* synthesis of SMs is especially promising in plant species with poor root systems, difficulty in harvesting, unavailability of elite quality planting material, poor seed set and germination, and difficult to propagate species. Thus, the present article reviews various biotechnological interventions to enhance commercially precious SMs production *in vitro*.

Keywords: Biotecnology, Callus secondary metabolites, Phytomolecules, Plant tissue culture, Suspension cultures.

---

* **Corresponding author Sayeed Ahmad:** Hamdard School of Pharmacy, Jamia Hamdard, Hamdard University, Hamdard Nagar, New Delhi, India; Tel: 09891374527; E-mail: sahmad_jh@jamiahamdard.ac.in

Mohammad Anis & Mehrun Nisha Khanam (Eds.)
All rights reserved-© 2024 Bentham Science Publishers

INTRODUCTION

Plants are renewable sources and form an important part of our daily diet, and provide essential primary metabolites (*e.g.*, carbohydrates, lipids and amino acids) [1] and phytochemicals (low molecular weight compounds-SMs) for different industrial applications like pharmaceuticals, nutraceutical, textile, construction and cosmetic sectors [2]. The majority of the world population's health and wellness relies on plant-derived components. Therefore, plants with medicinal properties are considered important to support the transition to a bio-economy that is less dependent on fossil resources. The SMs not only play a pivotal role in plants' adaptation to their environment but also represent an important source of active pharmaceuticals [3] and are synthesised by plants to defend themselves against exogenous stresses, both biotic and abiotic. A study [4] proposed the concept of SMs that were known as opposed to primary ones and an entire volume of "plant biochemistry" series named as "endproduct" [5]. It is known that higher plants are a rich source of phyto-pharmaceuticals and are used in the pharmaceutical industry. Some of the plant-derived products include drugs like morphine, codeine, cocaine, pilocarpine, belladonna alkaloids, colchines, phytostigminine, L-DOPA, berberine, reserpine, capsaicin, podophyllotoxin, shikonin derivatives, ajmalicine, vincristine and vinblastine [6] and steroids like ginsenosides, anti-cancer (taxol), diosgenin, digoxin and digitoxin. Significant synthetic substitutes of these drugs with the same efficacy and pharmacological specificity are yet to be found [7].

Previously, chemical synthesis for the production of SMs was achieved through field cultivation; however, the plants originating from particular biotypes were difficult to grow outside their ecosystems and thus led scientists and biotechnologists to consider plant cell, tissue and organ cultures as an alternative to produce secondary metabolites. The major advantages of *in vitro* synthesis of bioactive secondary metabolites within controlled conditions include: these are climatic and soil stipulations independent, minimal inferences of negative biological parameters affecting the SM production, possible choice of elite germplasm with respect to the presence of SMs, computerization of cell growth control, metabolic processes regulation, and cost price, which can be decreased with increased production. Plants produce alkaloids, flavonoids, lactones, glycosides, quinines, phenylpropanoids, resins, tannins, terpenoids, saponins, sesquiterpene, and steroids [8]. The first large-scale production of commercial plant cells application was carried out in stirred tank reactors to synthesis shikonin by cell cultures of *Lithospermum erythrorhizon* [9, 10].

SECONDARY METABOLITES

Plants are capable of producing different organic molecules called secondary metabolites, having unique carbon skeletons with basic properties. SMs are not necessarily for a cell (organism) to live but also for interaction with its environment. These are organ, tissue and cell-specific with low molecular weights and often differ amid individuals from the same population with respect to their type. SMs protect plants against stress; and are used as drugs, flavors, fragrances, insecticides and dyes and hence are of great economic value. SMs have evolved as molecules imperative for organisms producing them, the majority of these interfere with the pharmacological targets, and thus make them significant for several biotechnological applications.

Primary *vs* Secondary Metabolites

Primary metabolites (PMs) are compounds that are universally present in all plants, but are not species-specific and, thus might be identical in some organisms. These are directly involved in metabolic activities like growth, development, nutrition and reproduction of a plant whereas secondary metabolites are produced in other metabolic pathways that, although important, but are not essential to the functioning of the plant. Whereas, SMs are species specific and, therefore, unique for each species. The major differences between PMs and SMs are listed in Table **1**.

Table 1. Comparison between primary & secondary metabolites in plants.

Basis for Comparison	Primary Metabolites (PMs)	Secondary Metabolites (SMs)
Function	These are directly involved in the metabolic pathways of an organism required for its growth, development, and reproduction.	These are not directly involved in the growth, development, or reproduction of the organism but are essential in ecological and other activities.
Synonym	Also known as central metabolites.	Also known as specialized metabolites.
Phase of growth	Are produced during the growth phase of the organism, called 'trophophase'.	Are produced during the stationary phase of the organism, called 'idiophase'.
Quantity of production	Synthesized in large quantities.	Synthesized in small quantities.
Process of extraction	These are easy to extract.	These are difficult to extract.
Specificity	They are not species-specific and thus may be identical in some organisms.	These are species-specific and thus are different in different organisms.

(Table 1) cont.....

Basis for Comparison	Primary Metabolites (PMs)	Secondary Metabolites (SMs)
Function	Are involved in the growth, development, nutrition and reproduction of organisms.	Are involved in ecological functions and species interactions.
Structural framework	Are mostly formed from the molecular structure in organisms.	Are not a part of the molecular structure of the organism
Significance	Have applications in various industries for different purposes.	Are applied in various biotechnological procedures for the formation of drugs and other compounds.
Defensive mode of action	Are not active in the defense mechanism.	Are active in a defense mechanism against foreign invaders.
Examples	PMs include proteins, enzymes, carbohydrates, lipids, vitamins, ethanol, lactic acid, butanol, *etc.*	SMs include steroids, essential oils, phenolics, alkaloids, pigments, antibiotics, *etc.*

The Biosynthesis of Secondary Metabolites

SMs are synthesized by diverting energy-generating directions in metabolic pathways like photosynthesis, glycolysis, and Krebs cycle to biosynthetic intermediates; and are classified in separate categories depending upon their biosynthesis, structures and functions. SMs are mostly biosynthesized from acetyl coenzyme A, mevalonic acid, shikimic acid, deoxyxylulose 5-phosphate or various combined pathways [11]. Accordingly, they are classified into terpenoids, steroids, alkaloids, saponin, terpenes, lipids and enzyme cofactors [12, 13]. There are three major pathways to produce SMs-Shikimate, isoprenoid and polyketide. Formation of the fundamental skeleton is followed by further modifications, resulting in synthesis of plant specific compounds. The shikimate pathway is mostly found in microorganisms and plants, but not in mammals, and is a major source of aromatic compounds; thus making it an important target for insecticides and antibiotics, which are harmless on mammalian systems. In the glyphosate pathway, enzymes chorismate mutase and anthranilate synthase channelise chorismate into aromatic amino acids. However, the most popular pathway for the synthesis of SMs is the phenylpropanoid pathway as it leads to the synthesis of major SMs like lignin, lignans, flavonoids, and anthocyanins; it is found in all plants, while others may have several co-enzymes. Phenylalanine Ammonia Lyase (PAL) is the key enzyme that converts phenylalanine into trans-cinnamic acid by non-oxidative deamination. Few others like isoprenoid pathways are involved in the synthesis of terpenoids. The C_5 building block when incorporated into other skeletons forms an array of SMs like anthraquinones, cannabinoids, furanocoumarines, indole alkaloids, and napthaquinones; while incorporation into basic skeletons results in the synthesis of hop bitter acids, flavonoids and isoflavonoids [14].

The Physiological Function of Secondary Metabolites

Secondary plant metabolites are abundant chemical compounds produced by the plant cell through different metabolic pathways derived from the primary metabolic pathways. Some of the major roles they play in plants are:

- Act as signalling functions that influence the activities of other cells, control their metabolic activities and coordinate the development of the complete plant.
- SMs like terpenoids, alkaloids and flavonoids have therapeutic applications in the pharmaceutical industry as drugs and dietary supplements.
- Protection against harmful environmental conditions.
- Protection against pathogens and herbivores in the form of volatile monoterpenes or essential oils.
- The volatile terpenoids also play a major function in plant-plant interactions and serve as attractants for pollinators [15].
- SMs are often used as flower colors which serve to communicate with pollinators or protect plants from feeding deterrence by producing specific phytoalexines after fungi infections that inhibit the spreading of the fungi mycelia in plants.
- Plants use SMs in the form of volatile essential oils colored flavonoids or tetraperpenes to attract insects for pollination and seed disposal.
- They constitute important UV absorbing compounds, thus preventing serious leaf damage from the light.

Classification of Secondary Metabolites

The different classes of SMs synthesized *in vitro* (callus and cell suspension) by various culture conditions are:

Alkaloids

Acridines, Betalaines, Furoquinolines, Galanthamine, Harringtonines, Isoquinolines, Lobeline, Quinolizidines, Indole alkaloids, Isoquinoline alkaloids, Piperidine, Thebaine, Trigonelline, and Tropane alkaloids.

Terpenoids

Artemisinin, Cucurbitacins, Diterpenes, Ginsenoside, Meroterpenes, Monoterpenes. Paclitaxel, Sesquiterpenes, Thapsigargin, Triterpenes, Ursane, and Withanolides.

Steroids

Ajugalactone, Asiaticosid, Asiatic acid, β-sitosterol, Brassinolid, Bufadienolides, Catasterone, Conesine, Cyasterone, Cardenolides, Digoxin, Digitoxin, Digitoxigenin, Diosgenin, Ecdysteroids, Ecdysone, Feruginol, Gagaminine, Gentipicroside, Guggulusterones, Harpagoside, Helleborin, Paclitaxel, Physodine, Polypodine B, Phytosterols, Pterosterone, Ponasterone, Madecassic Acid, Madecassoside, Ouabain, Saponins, Saikosaponins, Scillaridine, Sengoterone, Solasodine, Solarmargine, Spirostanol Saponin, Steroidal glycosides, Steroidal lactones, Swertiamarin, Tanshinone, Taxol, Taxoids, Turkesteron, Typhasterol, 29-norcyasterone, and 20 E Cryptotanshinone.

Quinones

Aloe-emodin, Anthraquinones, Benzoquinones, Chrysophanol, Emodin, β-Lapacho, Naphthoquinones, Phenanthrenequinone, Plumbagin, Rhein, Shikonin, Thymoquinone, and Uglone.

Phenylpropanoid

Anthocyanins, Caffeic acid, Coumarins, Eugenol, Ferulic acid, Flavonoids, Hydroxycinnamoyl derivatives, Isoflavonoids, Lignans, Phenalinones, Proanthocyanidins, Stilbenes, and Tannins.

Secondary Metabolites and Plant Tissue Culture

The primary and age-old practice is to grow selected plants in greenhouses or protected areas and to extract the biomolecule from them. In this context, plant tissue culture (especially cell and organ culture) plays an immense role in *in vitro* propagation of desired germplasm with selected traits. For example, genetic engineering has directly enhanced the production of scopolamine in *Atropa belladonna* by transforming a gene that encodes the enzyme converting L-hyoscyamine into L-scopolamine; similarly, the metabolism of other SMs has been genetically altered. Therefore, there is a gamut to isolate the genes of biosynthetic pathways and to express them either in transgenic plants or in microorganisms; thus combinatorial biosynthesis is the cutting-edge technology, where successful recombinant bacteria or yeasts are used to produce desired SMs. Undifferentiated cell cultures have often failed to produce desirable SMs, whereas differentiated organ cultures (transformed root cultures) have been successful, as cell and tissue-specific gene expression appears to control such mechanisms. The key merits of an *in vitro* synthesis over the traditional cultivation of whole plants are as follows:

- Under traditional conditions, desired plant metabolites may take many years to reach the point where they produce SMs commercially. Alternatively, PTC techniques may be used to surpass such situations and effectively produce SM commercially.
- Desired biomolecules are produced under controlled aseptic conditions independent of environmental parameters (climate and soil).
- Obtained culture cells are microbes/ virus-free.
- The desired trait can be used *via* selective cell culture and multiplied to produce their specific metabolites.
- Automation of cell growth and rational regulation of SMs processes would reduce labor costs and improve productivity.
- Desired SMs are easily extractable from callus cultures.

Therefore in the future, plant tissue culture will be instrumental to improve plant cell cultures as biotechnological production systems. The biosynthesis of SMs *in vitro* involves two stage procedure:

i. Aggregation of biomass
ii. Synthesis of SMs [16, 17]

Callus Cultures

Organized cell specific cultures, like shoots and roots, calli, cell suspension, *etc.*, were supposedly engaged for the production of secondary metabolites [18]. Under *in vitro* aseptic culture conditions, induced plant cells grown in auxin rich media form an undifferentiated mass of cells known as callus; and are used to extract desired SMs.

The growth and development of callus culture is mainly divided into three stages (Fig. **1**). Extracts can be harvested during different stages of growth, identified, and quantified further to detect the presence of other compounds using techniques like HPLC, LC-MS, *etc* [19]. The technique is widely used for the extraction of a wide variety of commercially important secondary metabolites with the help of different elicitors and yield is quantitatively and qualitatively assessed. Different types of plant cultures (Fig. **2**) are used to produce SMs *in vitro*.

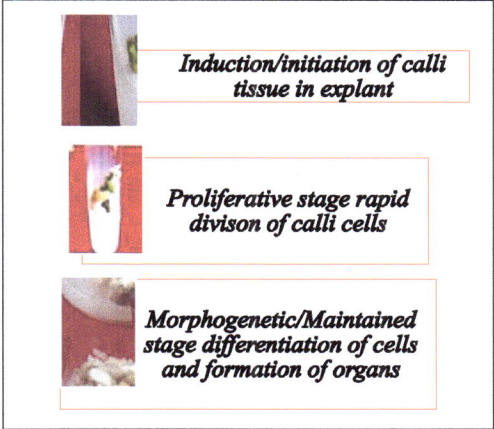

Fig. (1). Stages of development of callus culture.

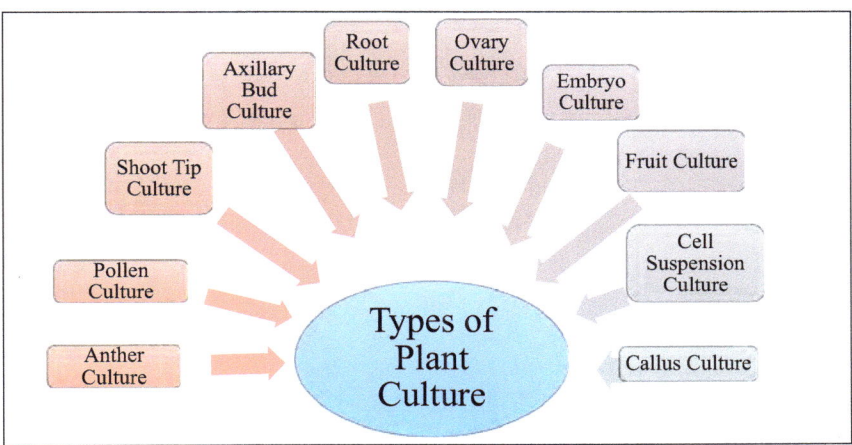

Fig. (2). Different types of plant cell culture.

Callus cultures are obtained from various plant parts, called 'explants' (leaves, stem, *etc.*), cultured on selected media fortified with plant growth regulators in various permutations, combinations and concentrations, and are maintained under aseptic culture conditions (16 h light, 8 h dark) at (25 ± 2°C) with regular sub-culturing over a period of 21-days in general. Morphologically, calluses are differentiated into two types-compact and friable; where the latter is an ideal choice to generate single-cell cultures that are maintained as suspension cultures [20]. The first callus was induced from roots and embryos in the 20th century. Over decades, the callus cultures have obtained commercial potential for the production of therapeutically significant SMs [21, 22] and are considered to be more reliable than those obtained from the wild [23]. The technique of micropropagation may be employed for the generation of multiple copies of plants, single-cell suspension cultures using either batch or continuous fermentation to produce the preferred secondary metabolites [24]. Moreover, callus and suspension cultures have the ability to synthesize secondary metabolites by manipulating the biosynthetic pathway. The production of SMs like tropane alkaloids, a-tocopherol, ajmaline, anthocyanins, flavonoids, paclitaxel, resveratrol, reserpine, serpentine, and scopolamine, stilbene has been achieved from callus cultures [25].

Organ Cultures

Differentiated organ cultures (shoots or roots) are reported for SMs production and have similar metabolomic profile with wild plants [26]; plant hairy root cultures (for example *Agrobacterium rhizogenes*) are the most widely used alternative for the development of SMs in plant roots [27 - 29]; also they have established an effective biological system to study the biosynthesis of several bioactive compounds like nicotine and tropane alkaloids [30], ginsenosides [31], anthraquinones [32] and artemisinin [33] as the hairy roots have advantages over others in terms of high yield, consistency and competence [34 - 36]. Leaf-derived callus induction and establishment of *Picrorhiza kurroa* followed by its subsequent regeneration of micro-shoots on MS medium supplemented with 2,4-D (2 mg l^{-1}) + IBA (0.5 mg l^{-1}) resulted in increased production of picrosides [37], as shown in Fig. (**3**).

Fig. (3). Induction and establishment of callus from leaf explants of *Picrorhiza kurroa* and regeneration of micro-shoots from callus; (**A**)- Curling of leaf; (**B**)- Induction of callus; (**C,D**)- Proliferation of callus after 35 days on MS+ 2,4-D (2 mg/l) + IBA (0.5 mg/l), (**E,F**)- Shooting from callus after 60 days maintained on MS+ 2,4-D (2 mg/l) + IBA (0.5 mg/l).

Cell differentiation and/or the differentiation–dedifferentiation is the underlying reason for the callus formation technique; where stem cell-related genes are crucial as far as dedifferentiation is concerned and this expression is regulated by transcription factors, histone modifications and DNA methylation. These are mostly heterogeneous in nature, but also homogenous too to allow micro propagation to take place to generate similar copies of cells with required features. SMs synthesis depends on many factors like climatic factors (light, temperature, the pH of the medium, and the aeration of cultures); media type (solid/liquid; composition and combination of nutrients). It is reported that usually SMs synthesis is inhibited at high concentrations of ammonium and phosphate ions, while enhanced by low concentrations of both ions [38]. Callus and suspension cultures are totipotent harbored with complete genetic information and have the ability to manipulate the biosynthetic pathway, such as biotransformation with desired characteristics to cater to the market need. Some important SMs produced by significant medicinal plants through callus cultures are listed in Table **2** [39 - 180].

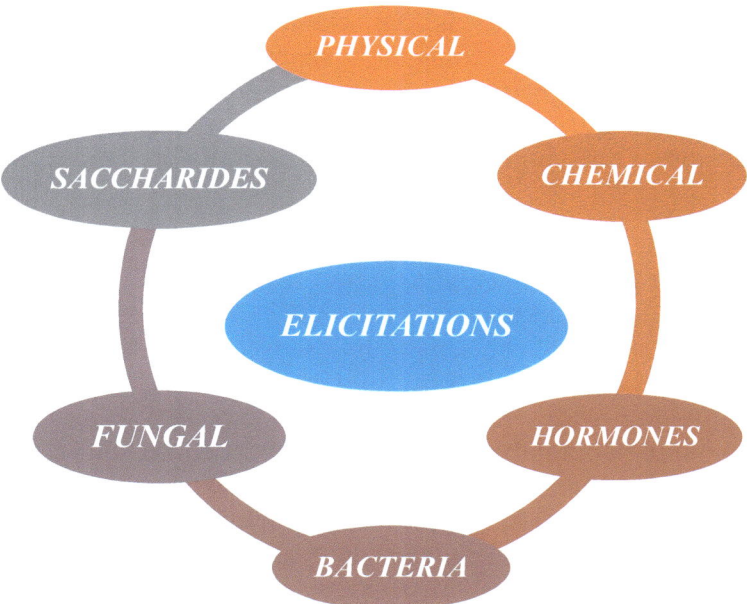

Fig. (4). Different types of elicitors used to modify the SMs production.

Cell Suspension Cultures

A suspension culture comprises cells dispersed in a liquid medium resulting in the formation of a suspension that continues to grow during the incubation period until the maximum yield point is attained [181]. Suspension cultures are induced by using a friable callus of approx. 15 days old, with the active dividing stage (maximum concentrations of meristematic tissues in) on selected media supplemented with various growth hormones in several permutations, combinations and concentrations. The cultures are prepared in 250 ml conical flasks, and are firmly fixed on incubator-cum shaker at 100 rpm, $25 \pm 2°$ C, 70% relative humidity and maintained at a photoperiod of 16/8 h light/dark cycle. The growth of cell suspension culture is studied by determining its increase in the fresh weight and dry weight, number of cells per ml and packed cell volume [182]; and capacity of developed suspension cultures to enhance the therapeutically important SMs production [183]. There are several advantages of suspension culture which include better control for up-scale SMs production, constantly monitored over various operational stages, pocket-friendly, easy to handle in terms of inoculation and harvest, it's a continuous process with shorter duration, enhanced uptake of nutrients in submerged liquid culture, and higher yield of bioactive compounds.

Over the years, several interventions have been reported in suspension culture to increase the amount of desired SMs like adding various agents and chemicals; static magnetic field and Fe_2O_3 magnetic nanoparticles (MNP) enhanced the contents of rosmarinic acid, naringin, apigenin, thymol, carvacrol, quercetin and rutin in *Dracocephlum polychaetum* Bornm [184]. The ultrasonic waves effected the gene expression of biosynthesis of crocin and safranal in *Crocus sativus* [185].

Fungal Elicitation

Fungal elicitation by fungus *Aspergillus flavus* resulted in decreased viability and growth of elicited cultures but enhanced the anthocyanin content, and reported the increase in the content of phenols, flavonoids, trans-resvetrol and tocopherols, when the suspension cultures of *Vitis vinifera* were elicitated with cadmium chloride; elicitation with jasmonic acid increased the synthesis of phenylpropanoid and naphthadianthrones and decreased the anthocyanin content in *Hypericum perforatum*; and coronatine and sorbitol together enhanced artemisinin content in *Artemisia annua,* as coronatine pre-treatment increased antioxidant enzyme activity and reduced hydrogen peroxide concentration and oxidative stress whereas the sorbitol treatment increased the concentration of malondialdehyde (MDA), hydrogen peroxide and oxidative stress [186]. Some important SMs produced by significant medicinal plants through callus cultures [187 - 204].

Micropropagation

It is the technique of plant tissue culture, where a large number of plants are produced from a small part called "explant" under aseptic culture conditions. The technique is especially beneficial to plant species which have difficulty in vegetative propagation. Micropropagation is advantageous as it allows the reproduction of a large number of genetically similar disease-free plants with desirable traits within a short period of time; it is time and climate independent in a sustained way [205]; vegetatively reproduced stock is maintained for longer periods in comparatively lesser space with low energy requirements; it is the only method that retains enough vitality to regenerate genetically manipulated cells or cells after protoplast fusion; the plantlets obtained per square meter are higher in number as compared to traditional methods of cultivation with less requirement of the space for storage of propagates for longer periods [206]; and somatic embryogenesis induced production of artificial seeds and thus increased the of production of secondary metabolites [207].

Fig. (5). (**A**) Field growing plants of *Adhatoda vasica,* (**B**) 120 days old callus on Murashige and Skoog (MS) medium supplemented with 2,4-dichlorophenoxyacetic acid (2,4-D) + benzyladenine (BA) + indole acetic acid (IAA) (1.0 ppm each), (**C**) 120 days old callus on MS medium supplemented with BA + indole butyric acid (IBA) (1.0 ppm each), (**D**) 120 days old callus on MS medium supplemented with BA (0.5 ppm) + IBA (1.0 ppm), (**E**) 120 days old callus on MS medium supplemented with 2,4-D + BA + IAA (1.0 ppm each) treated with 28 mM KNO_3, (**F**) 120 days old callus on MS medium supplemented with 2,4-D + BA + IAA (1.0 ppm each) treated with 28 mM KNO_3 + 100 mM NaCl.

Micropropagation automated by different technologies would render feasible commercial output in various plant species like micro tubers of the potato, bulblets of lily, and corms of gladiolus, *etc.*; meristems or grafting-induced micropropagation will produce disease-resistant plants; modified induction of embryos in somatic cells followed by desiccated embryos and their encapsulation led to the production of 'synthetic seeds' and are employed for the mass cloning of vegetative and reproductive plantlets with simultaneous transformations of the same [208]. In the 1970s, SMs were traced in plant cell cultures, and various bioactive molecules like shikonin, diosgenin, caffeine, glutathione and anthraquinone were extracted. Plantlets were successfully regenerated from tissues stored at -196°C in liquid nitrogen for several months to years; recombinant DNA technology, gene transfer and cell and tissue culture techniques it possible for an effective transformation and production of transgenic plants to be utilized as an efficient means to proceed for single-gene breeding or transgenic breeding of crops. The technique is of significant value when high production is to be sustained in higher volumes [209]. Some important SMs produced by significant medicinal plants through callus cultures [210 - 226].

Strategies to Increase *in vitro* Synthesis of Secondary Metabolite

The aim of any pharmaceutical or food industry is to standardize methods to allow the production of SMs from the plant cell culture in a cost-effective manner rather than the traditional extraction method. The gamut to increase the content of SMs in plant cell cultures has resulted in the need for biotechnological intervention on the secondary metabolism of plants [227]. The research in this area could lead to the successful modification of secondary metabolism and result in an increase of bioactive compounds like alkaloids, flavonoids, terpenes, steroids, glycosides, *etc*, using plant-cell-culture technology. The step-by-step strategy for obtaining increased SMs from the plant cell cultures can be represented as a multi-stage process, where each linkage may be optimized separately or in combination with other processes or treatments:

- **Step 1:** Select the desired PLANT according to the type of SMs required.
- **Step 2:** Establish a rapidly proliferated and high yielding *in vitro* CELL LINE culture.
- **Step 3:** Target the specific METABOLISM pathway either by modification of culture environment chemical (minerals, phytohormones, stress, *etc*.) or physical factors (light, pH, temperature, *etc*.) OR by precursor feeding, biotransformation, elicitors, immobilization, *etc*.
- **Step 4:** EXTRACTION by spontaneous release, two-phase system, membrane permeabilization (chemical or physical), adsorption, *etc*.

The first step is to select the parent plant with high contents of the desired SMs based on its molecular and biochemical characteristics. According to the totipotency theory, any plant part can be employed to induce callus tissue;

however, successful production of callus depends upon plant species and their qualities; young and fresh parts with maximum meristematic zone (young nodes) are mostly preferred. Dicotyledons are rather acquiescent for callus tissue induction, compared to monocotyledons, while the calluses of woody plants generally grow slowly.

The next step is to select the cell line with high-yield and fast-growing traits in *in-vitro* cultures. The metabolite content in cell lines can be increased manifold than as naturally found by the production of new genotypes through protoplast fusion or genetic engineering, however, this requires preidentification of the genes encoding key enzymes of secondary metabolic pathways and their expression. A number of factors like environmental conditions (chemical and physical) and special treatments (precursors, elicitors) may alter the expression of many secondary metabolite pathways; and hence targeting metabolism leads to enhanced SMs production. Modification of plant cell culture medium (inorganic, organic components, and phytohormones) influences the extent of SMs production; however high auxin level stimulates cell growth, but often negatively influences secondary metabolite production [228]. Physical conditions (light, temperature, and medium pH); and special treatments (precursor feeding, application of elicitors, bio-transformation, and immobilization) are also known to affect the SMs production [229].

Finally, the most efficient bio-processing way for the SMs production is its spontaneous release into a medium where they can be more easily recovered; several techniques have been introduced for leakage-free synthesis of the metabolite so that it is transported to the vacuole for accumulation, and therefore, there is a need for prevention of vacuolar accumulation, and consequently, enhancement of substances released into the medium.

Production is a crucial technique having enormous commercial application; where steady synthesis and high yields are imperative [230, 231]. Mostly, two approaches are implemented to achieve enhanced SMs production; conventional and biotechnological intervention like metabolic engineering [232].

Conventional Strategies

SMs are the outcome of primary metabolism and hence it is dependent on the rate at which substrates from primary metabolic pathways are re-routed to secondary bio-synthetic pathways. Both biotic and abiotic factors, like growth and physiology, temperature, humidity, light intensity, *etc.*, are responsible for their synthesis. Under aseptic growth conditions, the metabolite productivity is dependant on culture media composition, pH, inoculum density, and culture environment like temperature, light density, agitation, aeration, *etc.*; therefore these factors may be modified to enhance the production of SMs. Similarly, macro and micro-nutrients, vitamins, carbohydrates (sugars), amino acids, plant growth hormones (cytokinins, auxins, gibberellins, *etc*.), jasmonates and salicylates present in culture media influence the metabolite production [233].

Biotic and abiotic stresses help the plants to adapt according to climatic conditions and to survive best in the prevailing conditions; and chemical compounds generated from secondary metabolism are induced by the external signals (foreign pathogens, oxidative stress, elicitors, and injury to plant) that are mediated internally by chemical substances and their derivatives [234].

Use of Elicitors for Secondary Metabolite(s) Enhancement

Elicitors which trigger secondary metabolic pathways to stimulate plant defence mechanisms to protect the plant cell [235] are the most effective biotechnological interventions in modern times to enhance the synthesis of SMs. Elicitors are source based and classified as abiotic and biotic; where the former are non-biological sources, mostly like inorganic compounds, heavy metals, metal ions, metal oxides, *etc* [236, 237]; and physical stresses (cold shock, UV, osmotic, water stress, *etc*.) known to induce enzymatic activity and secondary metabolism [238]. While the latter are based on biological sources (exogenous/endogenous). The exogenous comprises microbial cell walls (chitosan, chitin); while the endogenous polysaccharides that are released by the deterioration of the plant cell wall due to pathogen invasion, intracellular molecules or proteins like salicylic acid or methyl jasmonate produced by the plant in reaction to various stress or pathogenic invasion and these are reported to premeditated signaling molecules for elicitation routes as it induces plant defence responses similar to pathogen invasion. The various different types of elicitors used to modify the production of SMs are shown in Fig. (**4**).

Phytoalexins are developed when biotic elicitors are perceived by specific cell membrane receptors, and subsequently transfer the stimulus to the cell by signal transduction [239]. Elicitation with salicylic acid and methyl jasmonate (produced in plants with response to stress or pathogen invasion), reported as signalling compounds, results in enhanced production of SMs like flavonoids, alkaloids, terpenoids and phenylpropanoids [240 - 244]. The signal transduction pathways are initiated by targeting secondary signals in the nucleus of the cell, which leads to transcriptional stimulation of various genes, thereby inducing the synthesis of an array of proteins involved in defence or resistance and secondary metabolites [245, 246].

Fungal extracts from *Fusarium, Pythium, Yeast, Aspergillus, Penicillium, Trichoderma* species, *etc.* are often used to induce secondary metabolites in medical plants [247 - 249]; as they induce the expression of specific plant genes, activating secondary metabolic pathways thus increasing secondary metabolites [250] and plant phytoalexins in the course of physiological progressions of plant disease resistance [251]. Extracts of *Aspergillus niger* and *Saccharomyces cerevisiae* improved the gymnemic acid content in *Gymnema sylvestre* [252], and *Aspergills flavus* increased the synthesis of terpenoid indole alkaloids in *Catharanthus roseus*. Similarly, bacterial elicitors like Gram-positive and Gram-negative strains were used to increase the biosynthesis of tropane alkaloids in adventitious hairy root cultures of *Scopolia parviflora* [253]; *Staphylococcus aureus* extracts enhanced the synthesis of ginkgolide and bilobalide biosynthesis in *Ginkgo biloba* [254]; slow expansion of hypericin and pseudohypericin in plantlets of *Hypericum perforatum* [255] by Rhizobacterium; phytotoxin coronatine content enhanced by *Pseudomonas syringae*; taxane content increased by accumulation of coronatine in Taxus cell cultures [256]; coronatine triggered the synthesis of viniferins in the *Vitis vinifera* suspension culture [257], and the content of glycyrrhizic acid was enhanced in *Taverniera cuneifolia* root cultures [258] by *Rhizobium leguminosarum* and *Agrobacterium tumefaciens*.

Seedling-derived callus cultures of *Artocarpus lakoocha* Roxb. resulted in the production of 4 prenyl-flavones, prenylated stilbenes, and nine polyphenolic compounds [259] when elicitated with hormones (2,4-D and BAP) in medium; inorganic elicitors (28 mM KNO_3 + 100 mM NaCl) in MS medium fortified with 1 mg l^{-1} of each PGR (2,4-D + BAP + IAA) improved the vasicine content in leaf explants of *Adhatoda vasica* [260] (Figs. **5** and **6**); and leaf-derived callus cultures of *Stevia rebaudiana* Bert showed that the cultures grown under blue light enhanced the total (phenolic, flavonoid, and antioxidant) content and antioxidant capacity whereas green and red lights improved reducing power assay and DPPH-radical scavenging activity [261]. studied the effect of various concentrations of 2,4-D, NAA, IAA, IBA, BAP and KIN on the synthesis of SMs using leaf-derived callus, followed by HPLC analysis of callus biomass extracts to identify and quantify the concentration of metabolites. NaCl and chitosan increased the thymol and p-cymene concentration in calli and seedlings of *Carum copticum* L [262]. demonstrated that Erlenmeyer flasks and extracts of CNB callus reported significant antiradical activity and increased phenols and flavonoids content. Some important SMs produced by significant medicinal plants through callus cultures [263 - 309].

Fig. (6). (**A**) Field growing *Adhatoda vasica* (**B**) 120 days old callus on Murashige and Skoog + 2,4-D + 6benzyladenine + indole acetic acid (1.0 ppm each) (**C**) 120 days old callus on Murashige and Skoog + 6-benzyladenine + indole butyric acid (1.0 ppm each) (**D**) 120 days old callus on Murashige and Skoog + 6-benzyladenine (0.5 ppm) + indole butyric acid (1.0 ppm each) (**E**) 120 days old callus on Murashige and Skoog + 2,4-D + 6-benzyladenine + indole acetic acid (1.0 ppm each) treated with 28 mM KNO_3 (**F**) 120 days old callus on Murashige and Skoog + 2,4-D + 6-benzyladenine + indole acetic acid (1.0 ppm each) treated with 28 mM KNO_3 + 100 mM NaCl.

Table 2. Efforts made for *in-vitro* regeneration and secondary metabolites production in some selected medicinal plants.

Plant Name	Technique	Secondary Metabolites	References
Verbasceae thapus L.mullein	Cultivation under abiotic elicitors (sugar nutrition, exposure to different light conditions)	Coumarin, eugenol and thymol	[263]
Stevia rebaudiana	ZnO and CuO Nanoparticles, artificial culturing	Rebaudioside-A and Stevioside	[264]
Centella asiatica (L.)	Bioreactor platform (elicitation)	Centellosides, phenolic acids, p-coumaric acid and flavonoids	[265]
Catharanthus roseus	Hormone free MS medium supplemented with0,50,100,150 mgL-1 of MWCNT	Carotenoids, phenolic compounds, alkaloids	[266]
Coriandrum sativum	Methyl jasmonate (MeJA) was amended in a medium and then compared in different cultivating tissues	Superoxide dismutase, catalase, ascorbate peroxidase	[267]
Garcinia mangostana	Biotechnological methods, addition of plant growth regulators such as	Mangostinone, Tannins, xanthons, flavanoids	[268]
	Combination of 2ppm 2,4-D and 15% coconut water		
Calligonum polygonoides	Elicitation of cell suspension using high performance liquid chromatography technique	Phenolic compounds	[269]

(Table 2) cont.....

Plant Name	Technique	Secondary Metabolites	References
Hypericum perforatum L.	Elicitation technique	Naphthodianthrones, phloroglucinols, xanthones and flavonoids	[270]
Stevia rebaudiana	Salinity stress elicitation	Steviol glycosides (SGs), phenol and flavonoid	[271]
Hyoscyamus reticulatus L.	Sodium nitroprusside (SNP) elicitation	Hyoscyamine and scopolamine (Tropane alkaloids)	[272]
Melissa officinalis	Elicitation by sodium nitroprusside (SNP)	Rosmarinic acid, linalool, neral, thymol, monoterpene and sesquiterpene	[273]
Coleus aromaticus Benth (L)	Cytokinin combined elicitors' supplementation in tissue cultures.	Alkaloids, flavonoids, saponins, terpenoids, phenols and tannins	[274]
Linum usitatissimum	Elicitation by nanoparticles	Mucilage	[275]
Lavandula angustifolia	Varying concentrations of Jasmonic acid added to the culture of the plant to enhance essential oil concentration	σ-cadinene, borneol, caryophyllene oxide, τ-cadinol, beta-caryophyllene, 1.8-cineole, β-pinene, geranyl acetate, myrtenal.	[276]
Artemisia alba Turra	Supplementation of shoot cultures with plant growth regulators like benzyl adenine and indole-3-butyric acid	Polyphenolic compounds	[277]
Catharanthus roseus	Plant tissue culture elicited with yeast extract	Vinblastine and vincristine	[278]
Maytenus ilicifolia	Cell culturing with elicitation with Methyl jasmonate	Sesquiterpene pyridine, alkaloids, flavonoids, primarily quinonemethide, triterpenes	[279]
Rauvolfia serpentine	Cell culture elicited by aluminium chloride	Reserpine	[280]
Dracocephalum moldavica and *Dracocephalum kotschyi*	Callus and suspension culture aided by methyl jasmonate and plant growth regulators	Flavonoids	[281]
Betula platyphylla	BpSS and BpSE stimulation with ethephon and MeJA	Birch triterpenoids (TBP)	[282]
Acaciella angustissima	Tissue culture system [Murashige and Skoog (MS) medium supplemented with 2% sucrose, 0.3% of phytagel, and 2,4-dichlorophenoxyacetic acid (1.5 mg/l) and Benzylaminopurine (5 mg/l)]	Gallic acid	[283]
		Rutin	
		Catechin	

(Table 2) cont.....

Plant Name	Technique	Secondary Metabolites	References
Caralluma tuberculata	Shoot morphogenesis Murashige and Skoog (MS) medium containing 30 g l−1 sucrose and a combination of 2, 4-D (2.0 mg l−1) and BA (1.0 mg l−1)	DPPH	[284]
		Flavonoids	
Curculigo orchioides Gaertn (Kali musli) Family: Hypoxidaceae)	Various concentrations of phenylalanine (Phe), tyrosine, (20, 40, 60, and 80 mg/100 ml), chromium (Cr) and nickel (Ni) (1, 2, 3, 4, and 5 ppm) into Zenk media	Curculigoside	[285]
Ferula gummosa Boiss	Murashige and Skoog medium	Phenols	[286]
Lithospermum, Arnebia, Alkanna, Anchusa, Echium and Onosma	Optimization of culture conditions, elicitation, in situ product removal, genetic transformation and metabolic engineering	Shikonins	[287]
Stevia rebaudiana Bertoni	Varying concentrations of plant growth regulators (PGR) and pH levels of media	Phenols and Flavonoids	[288]
Glycyrrhiza glabrais	Murashige and Skoog's (MS) medium supplemented with NAA (1mg/l), BAP (0.5 mg/l) and various concentrations of cinnamic acid	Licochalcone A, Liquirtigenin and Licoisoflavone B	[289]
Rhaponticum carthamoides	Direct and indirect organogenesis	Chlorogenic acid	[290]
	MS agar medium supplemented with BA, IBA or NAA at concentrations of 0.2 and 0.5 mg L−1	20-Hydroxyecdysone (20-HE)	
Sauropus androgynus (sweet shoot)	Use of elicitor [methyl jasmonate (MJ), Salicylic acid (SA)]	Naringin, TBHQ, kaempferol, papaverine and quercetin	[291]
Bacopa monnieri	Different abiotic elicitors (Jasmonic acid, copper sulphate (CuSO4) and salicylic acid) in shoot culture	Bacoside	[292]
Origanum vulgare	Bioelicitors (thiamine and coconut water)	Phenols	[293]
Tylophora indica	Using precursors like salicylic acid, ornithine, cinnamic acid, tyrosine and phenylalanine	Kaempferol	[294]
Glycrrhiza glabra	Elicitation with biotic or abiotic stimuli	Licochalcone, Liquirtigenin and Licoisoflavone B	[295]
Papaver somniferum L.	Abiotic nano elicitors (nano TiO2 and nano-Silver), shoot meristem and root suspension cell cultures using Real-time PCR.	Sanguinarine	[296]

(Table 2) cont.....

Plant Name	Technique	Secondary Metabolites	References
Taxus brevifolia nutt	Plant tissue and cell culture, plant growth hormones 2, 4 dichlorophenoxyacetic acid and Kinetin	Paclitaxel	[297]
Taverniera cuneifolia	Elicitors used *in vitro* production of GA from root culture	Glycyrrhizic acid (GA)	[298]
Ficus religiosa	Shoot organogenesis, on mediums- FCd1, FCt7, FCt13 and FCt16	Phenols, flavonoids, quercetin and myricetin	[299]
Bacopa monniera (L.)	Aseptical culture on Murashige and Skoog (MS) medium with auxins and cytokinins like 2,4-D, IAA, NAA, 6-BA and Kinetin // elicitation treatment// extraction	Saponins- Bacosides A1, bacosides A3, bacosaponins A, B, C, D, E & F Alkaloids, Carbohydrates, Phenolic compounds and Tannins, Flavonoids, Proteins and free amino acids,Steroids and Terpenoids	[300]
Exacum affine Balf. f. ex Regel	Shoot culture medium supplementation with precursors, elicitors and changing the amounts of the medium components.	Phenolic acids and cinnamic acid	[301]
Mitracarpus hirtus L	Transgenic hairy root of *M. hirtus* and the application of CPPU as an elicitor to induce variations in plant secondary metabolites that show their potential to apply to the bio-reactor system	Chrysophanol and 2-methoxy-4-vinylphenol, Eseroline, 7-bromo-, methylcarbamate	[302]
Scutellaria andrachnoides	Grown in the hormone-free liquid Gamborg nutrient medium	Wogonoside, a wogonin glucuronide, balcalin, a baicalein glucuronide	[303]
Azadirachta indica	Calli in agitated woody Plant Medium (WPM) liquid medium, supplemented with glucose (Gl), hydrolyzed casein (HC) and methyl jasmonate (MeJ) as an elicitor agent	Azadirachtin (AZA)	[304]
Tinospora cordifolia (Willd.) Menispermaceae	Shoot induction on Murashige and Skoog basal medium with kinetin or kinetin and BAP in combination, Rhizogenesis	-	[305]
Andrographis paniculata (Burm. f.) Nees	Cell suspension culture was treated with *Aspergillus niger* and *Penicillium expansum* elicitors to enhance the synthesis of andrographolide	Andrographolide	[306]

(Table 2) cont.....

Plant Name	Technique	Secondary Metabolites	References
Stemona sp.	Stemona plantlets were cultured on various concentrations of sodium acetate and sucrose and the alkaloids extract was determined	1′,2′-didehydrostemofoline and stemofoline	[307]
Solanum lycopersicum	Nitrogen-rich medium	Phenolic molecules (chlorogenic acid and rutin)	[308]
		polyphenols	
Phlomis armeniaca Wild.	Murashige and Skoog medium (MS) supplemented with 75 plant growth regulator (PGR) combinations	Apigenin, caffeic acid, p-coumaric acid, luteolin, rutin hydrate, vanillic acid, ferulic acid, salicylic acid, sinapic acid, and chlorogenic acid	[309]

Biotechnological Strategies (Metabolic Engineering, etc.)

The alteration of various metabolic pathways in an organism using contemporary biotechnological interventions like genomics, proteomics, metabolomics, *etc.*, to produce metabolites of commercial importance, is called metabolic engineering [310]; and it allows manipulation of endogenous biochemical pathways by overexpression or down regulation of metabolic pathways or by diverting common precursors, enzymes, regulatory proteins with the help of recombinant DNA technology.

Since the plants contain several metabolic pathways required for the biosynthesis of complex metabolites which can be reconstituted in heterologous hosts for higher yield and isolation of economically important SMs. In plants, the major metabolic pathways for *in vitro* synthesis of SMs are *via* shikimate, terpenoid, and polyketide routes. The key source of phenylpropanoids and aromatic compounds is the shikimate pathway which is highly conserved [311] and cogitated as an important metabolic pathway in plants in relation to carbon flux, and therefore is expected that more than 30% of carbon is fixed*via* this pathway [312]. A few examples of SMs derived from the shikimate pathway are phenols (from phenylalanine); and coumarins, flavonoids, lignans, stilbenoids, catechins, vanillin, gallic acid, betalains (from p-coumaroyl-CoA as origin).

The second type of pathway is the terpenoid pathway or the isoprenoid pathway; and in plants, it uses two autonomous pathways, *i.e.*, mevalonic acid pathway (MVA) and methyl-D-erythritol (MEP) pathway. The former serves as the precursors for the synthesis of brassinosteroids, sesquiterpenoids, phytosterols, triterpenoids, and polyprenols; while the latter is used for the biosynthesis of monoterpenoids, diterpenoids, hemiterpenoids, tocopherols, plastoquinones, plant growth regulators like cytokinins, gibberellins, *etc*. About more than one-third of the plant SMs are terpenoids and these processes have several functions in plants [313], like as antimicrobial agents, antifeedants and protection against abiotic stress (phytoalexins: sesquiterpeniods, diterpenoids and triterpenes); pollinator attractants [volatile terpenoids (monoterpenoids) and carotenoids]; resistance against several predators (killers) [monoterpenes and sesquiterpenes]. Terpenoids are also a source of anthraquinones, cannabinoids, furanocoumarines, naphthoquinones, and terpenoid indole alkaloids.

Finally, the polyketide pathway executes a crucial role in fatty acid biosynthesis, and is derived from either acetyl CoA or malonyl CoA. SMs that derived this pathway are acetogenins, jasmonates 6-methylsalicylic acid, plumbagin, coniine and anthraquinones [314]. The various new-edge interventions in the field of biotechnology for an increase of SMs are listed below:

Metabolic Pathways: Up-regulation

The transcription factors (TFs)or sequence-specific DNA-binding factor is a protein that controls the rate of transcription of genetic information from DNA to messenger RNA, by binding to a specific DNA sequence at promoter or enhancer region; and decode the genome information and modulate the rate of transcription [315, 316]. Many TFs are known for the production of SMs; like MYB TPs control several biosynthetic pathways in various plant species; while TFs like R2R3-MYB, WD repeat (beta-transducin repeat), a elementary helix-loop-helix genes (bHLH genes), adenine nucleotide translocator (or adenine nucleotide translocase or Ant) type 1, APETALA2/ethylene response factor (AP2/ERF), WRKY, NAC, SQUAMOSA Promoter Binding Protein-Like (SPL) have been reported to control anthocyanin accumulation as well as flavonoids in plant species like *Arabidopsis, Petunia hybrida, Zea mays, Solanum lycopersicum, Ipomoea batatas, etc* [317 - 319]; TF GaWRKY1 bonds to the promoter of (+)-d-cadinene synthase genes and regulate gossypol biosynthesis [320]; and biosynthesis of artemisinin in *Artemisia annua* is enhanced by binding TF AaWRKY1to activate the sesquiter-pene synthase gene promoter [321]. However, the perplexing aspect in *up-regulating pathways* is to find suitable TFs that will perform on precise pathway genes [322]; one may produce artificial TFs, which target more than a single significant gene [323].

Alteration of Common Precursors

SMs synthesis *in vitro* may be enhanced by modifying the targeted biosynthetic metabolite precursor by deterring the competitive pathway or by inducing overexpression of genes in the precursor pathway. Taxadiene biosynthesis in tomato mutants may be increased by enhancing the gene expression that decodes precursor in the carotenoid pathway, *i.e.* taxadiene synthase; overexpression of precursor pathway genes results in enhanced monoterpene content in essential oils like peppermint and lavender [324, 325]; over-expression of 1-deoxy-d-xylulose-5-phosphate synthase (GrDXS) resulted to increased terpenoidal SMs accumulation in *Pelargonium spp.* (Essential oil) and *Withania somnifera* (withanolides) [326]; and the alkaloid content was increased in *Catharanthus roseus* by direct overexpression of alkaloid pathway related genes [327].

Targeting Metabolites to Specific Cell Compartments

The enzymes and metabolites of SMs are compartmentalized in cytoplasm and organelles like peroxisomes, vesicles, vacuoles, *etc.*; enzymes for SMs pathways synchronized with multifunctional enzyme complexes associated with substrate channeling effects where metabolic output of one enzyme act as a substrate for the subsequent enzyme [328, 329]; and specific amino acid sequences target and retain proteins to particular organelles [330]. For example, the enzymes responsible for the biosynthesis of monoterpenoid indole alkaloids are located in roots, leaves, flowers, buds, *etc.*, while the metabolites collected in special cells are called idioblasts and laticifers in *C. roseus*; therefore, the enzymes and metabolites synthesised need to be transported not to a specific cell but between cells belonging to different types of tissue [331]. Similarly, in *Nicotiana tabacum*, alkaloid berberine syntheisized by *Coptis japonica* and nicotine are sequestered in vacuoles [332, 333]; flavonoid biosynthesis is cumulatively achieved by three enzymes (chalcone synthase, chalcone isomerase, and dihydroflavonol 4-reductase) [334] and multifunctional dimeric enzyme glutathione S-transferases and a vacuolar flavonoid/H+-antiporter channelizes the vacuolar transport of anthocyanin and proanthocyanidin [335] in *Arabidopsis thaliana*; and the transport of anthocyanins from petals to the central vacuole by intracellular vesicles or pre-vacuolar compartments in *Eustoma grandiflorum* is also reported [336].

Metabolic Pathways: Down-regulation (Silencing)

Yet another technique to increase the desirable SMs content *in vitro* is to block the concealing genes, upregulate the metabolic pathway or increase their catabolism and carbon flux into competitive pathways, hence, synthesis of unwanted metabolites are diminished by restricting the level of enzymes [337]. For this purpose, modern molecular biology techniques like antisense RNA, RNA interference (RNAi), and co-supression are used; and increased production of phenylpropanoids, alkaloids and terpenoids has been successful by using the RNA interference method; however, this process has resulted in abundant coloration in flowers, low seed coat pigment; enhanced the codeine and morphine content in the *Papaver somniferum*; reduced cyanogenic glucoside content in *Manihot esculenta*; increased flavonoids and carotenoids in *Solanum lycopersicum*; and regulated cologne in Petunia flowers, *etc* [338].

Use of Nanoparticles for Secondary Metabolite (s) Enhancement

Nanoscience and nanotechnology are the new cutting edge technologies; where there is the application of extremely small things (size range of 1–100 nm; called nanoparticles-NPs) and can be used across all the other science fields, such as chemistry, biology, physics, materials science, and engineering; and this have been widely used in plant tissue culture to support germination efficiency, boost plant growth, and to improve bioactive metabolites, *etc* [339, 340]. Several reports have suggested the effect of vital metal oxide nanoparticles like titanium oxide, zinc oxide, iron oxide and copper oxide in the enhancement of SMs [341, 342]. A few examples are, increased content of gallic acid, chlorogenic, acid, o-coumaric acid, tannic acid and cinnamic acid in embryonic-derived calli of *Cicer arietinum* by titanium oxide NPs; increased ferulic acid and isovitexin in *Hordeum vulgare* by Cadmium oxide NPs [343]; silver NPs treatment increased the concentration of artemisinin content in hairy root cultures of *Artemisia annua* L.; and diosgenin concentration in *Trigonella foenum-graecum* [344].

Large-scale Production of SMs by Bioreactors

The large-scale production of the desired SMs is done using bioreactor systems; where plant cells in liquid suspension are a unique combination of both physical and chemical environments that must be provided in a large-scale bioreactor process. The advanced technique to scale up the SMs is the immobilization technique that successfully produces the desired metabolites to greater yield. Immobilization in calcium alginate (10 mM $CaCl_2$) increased the production of plumbagin by three-, two-, and one-folds compared with that of control, un-cross linked alginate and $CaCl_2$-treated cells, respectively in *Plumbago rosea* [345].

CONCLUSION AND PERSPECTIVES

In the modern era, people prefer to use more natural products, therefore, the demand for plants has increased manifold. By now, it is a well-established fact that tissue-cultured plant cells are a potent source of bioactive molecules; though only a handful of them are been exploited commercially as stable and sustainable sources of secondary metabolites. A huge number of plants with therapeutic value and their bioactive compounds have been produced by several *in vitro* techniques in a short time as compared to conventional approaches. To fill this demand-supply gap of the secondary metabolites, various biotechnological inventions and strategies like media manoeuvring, phytohormone regulation, precursor feeding, plant cell immobilization, biotransformation, and bioconversion, *etc.*, have been employed over the years to standardise the *in vitro* synthesis of desired traits from plant cell line cultures in a substantial quantity in a pocket-freindly manner. The enhanced use of genetic engineering, various biotechnological interventions like precursor feeding, application of elicitors, *etc.*, and a better understanding of the regulation of pathways for a specific biomolecule are necessary to develop the elite type of plant cell-derived products and, thus will lead to produce commercially important secondary products. Innovative techniques in molecular biology and the production of transgenic cultures may influence the expression and regulation of biosynthetic pathways. Since the nature of plant cells in *in vitro* cultures is very complicated and the mechanism behind is yet to be understood; specific examples have been explicated for the synthesis of secondary metabolites.

However, the *in vitro* synthesis of secondary metabolites can only be achieved at a commercial scale, only if it is more economical with the existing traditional mode of production, like extraction from the field-grown plants, substitute chemical synthesis, microbial fermentation, or improvement of the plant itself through somaclonal variation, and genetic engineering. The *in vitro* synthesis of bioactive compounds can be escalated by overcoming the two current impediments-firstly, the lack of adequate monitoring process/control system for plant cells and secondly, the heterogeneity and instability of the cells. The cumulative efforts of plant scientists from various disciplines may exploit the potential of these plant cells for the production of plant secondary metabolites *in vitro*, developing at a commercial scale.

ABBREVIATIONS

2,4-D	2,4-dichlorophenoxyacetic acid
B_5	Gamborg's medium
BAP	6-benzyloaminopurine Bioactive compounds
$CaCl_2$	Calcium Chloride
HMG-CoA	β-Hydroxy β-methylglutaryl-CoA
IAA	indole-3-acetic acid
IBA	indole-3-butyric acid
Kin	Kinetin
L-DOPA	levodopa and l-3,4-dihydroxyphenylalanine
MeJA	Methyl Jasmonate
MS	Murashige and Skoog medium
MVA	Mevalonic Acid
MVL	Mevalonic Lacton

NAA	α-Naphthalene Acetic Acid
Nanoparticle	Secondary metabolites
SH	Schenk and Hildebrandt medium
WPM	McCown Woody Plant Medium

REFERENCES

[1] Namdeo A. Plant cell elicitation for production of secondary metabolites: A review. Pharmacogn Rev 2007; 1: 69-79.

[2] Guerriero G, Berni R, Muñoz-Sanchez J, *et al.* Production of plant secondary metabolites: Examples, tips and suggestions for biotechnologists. Genes 2018; 9(6): 309.
[http://dx.doi.org/10.3390/genes9060309] [PMID: 29925808]

[3] Bourgaud F, Gravot A, Milesi S, Gontier E. Production of plant secondary metabolites: A historical perspective. Plant Sci 2001; 161(5): 839-51.
[http://dx.doi.org/10.1016/S0168-9452(01)00490-3]

[4] Kossel A. Uber die ChemischeZusammensetzung der Zelle,Archiv fü¨r. Physiologie 1891; 181-6.

[5] Czapek F. SpezielleBiochemie, Biochemie der Pflanzen, *G.* Fischer Jena 1921; 3: 369.

[6] DiCosmo F, Misawa M. Plant cell and tissue culture: Alternatives for metabolite production. Biotechnol Adv 1995; 13(3): 425-53.
[http://dx.doi.org/10.1016/0734-9750(95)02005-N] [PMID: 14536096]

[7] Balandrin MF, Klocke JA. Medicinal, Aromatic and Industrials Products from Plants: Biotechn In Agric and Forest, 4 (Medicinal and Aromatic plants I). Berlin, New York: Springier-Verlag 1998.

[8] Patel SS, Savjani JK. Systematic review of plant steroids as potential antiinflammatory agents: Current status and future perspectives. Journal of Phytopharmacology 2015; 4(2): 121-5.
[http://dx.doi.org/10.31254/phyto.2015.4212]

[9] Fujita Y, Hara Y, Ogino T, Suga C. Production of shikonin derivatives by cell suspension cultures of Lithospermum erythrorhizon. Plant Cell Rep 1981; 1(2): 59-60.
[http://dx.doi.org/10.1007/BF00269272] [PMID: 24258859]

[10] Bajaj YPS. Biotechnology in agriculture and forestry. Med Aromat Plants 1988; I.

[11] Sandra GR. Production of plant secondary metabolites by usingbiotechnological tools, secondary metabolites Sources and applications.Secondary Metabolites Sources and Applications. IntechOpen 2018.
[http://dx.doi.org/10.5772/intechopen.79766]

[12] Hussein R. Plants secondary metabolites: The key drivers ofthepharmacological actions of medicinal plants.HerbalMedicine. Intech Open 2018; pp. 11-30.
[http://dx.doi.org/10.5772/intechopen.76139]

[13] McMurry JE. Organic chemistry with biological applications, Secondary Metabolites.An Introduction to Natural Products Chemistry. Stanford,USA 2015; pp. 1016-46.

[14] Pagare S, Bhatia M, Tripathi N, Pagare S, Bansal YK. Secondary metabolites of plants and their role: Overview. Curr Trends Biotechnol Pharm 2015; 9(3): 293-304.

[15] Tholl D. Terpene synthases and the regulation, diversity and biological roles of terpene metabolism. Curr Opin Plant Biol 2006; 9(3): 297-304.
[http://dx.doi.org/10.1016/j.pbi.2006.03.014] [PMID: 16600670]

[16] Yu KW, Murthy HN, Hahn E-J, Paek K-Y. Ginsenoside production by hairy root cultures of Panax ginseng: Influence of temperature and light quality. Biochem Eng J 2005; 23(1): 53-6.

[http://dx.doi.org/10.1016/j.bej.2004.07.001]

[17] Murthy HN, Lee EJ, Paek KY. Production of secondary metabolites from cell and organ cultures: Strategies and approaches for biomass improvement and metabolite accumulation. Plant Cell Tissue Organ Cult 2014; 118(1): 1-16. [M.Ochoa.].
[http://dx.doi.org/10.1007/s11240-014-0467-7]

[18] Hussain MS, Fareed S, Ansari S, Rahman MA, Ahmad IZ, Saeed M. Current approaches toward production of secondary plant metabolites. J Pharm Bioallied Sci 2012; 4(1): 10-20.
[http://dx.doi.org/10.4103/0975-7406.92725] [PMID: 22368394]

[19] Srivastava P, Singh M, Devi G, Chaturvedi R. Herbal medicine and biotechnology for the benefit of human health.Animal Biotechnology. Academic Press 2014; pp. 563-75.
[http://dx.doi.org/10.1016/B978-0-12-416002-6.00030-4]

[20] Ramachandra Rao S, Ravishankar GA. Plant cell cultures: Chemical factories of secondary metabolites. Biotechnol Adv 2002; 20(2): 101-53.
[http://dx.doi.org/10.1016/S0734-9750(02)00007-1] [PMID: 14538059]

[21] Ogita S. Plant cell, tissue and organ culture: The most flexible foundations for plant metabolic engineering applications. Nat Prod Commun 2015; 10(5): 1934578X1501000.
[http://dx.doi.org/10.1177/1934578X1501000527] [PMID: 26058164]

[22] Wu CF, Karioti A, Rohr D, Bilia AR, Efferth T. Production of rosmarinic acid and salvianolic acid B from callus culture of *Salvia miltiorrhiza* with cytotoxicity towards acute lymphoblastic leukemia cells. Food Chem 2016; 201: 292-7.
[http://dx.doi.org/10.1016/j.foodchem.2016.01.054] [PMID: 26868579]

[23] Efferth T. Biotechnology applications of plant callus cultures. Engineering 2019; 5(1): 50-9.
[http://dx.doi.org/10.1016/j.eng.2018.11.006]

[24] Fischer R, Emans N, Schuster F, Hellwig S, Drossard J. Towards molecular farming in the future: Using plant-cell-suspension cultures as bioreactors. Biotechnol Appl Biochem 1999; 30(2): 109-12.
[PMID: 10512788]

[25] Xu J, Ge X, Dolan MC. Towards high-yield production of pharmaceutical proteins with plant cell suspension cultures. Biotechnol Adv 2011; 29(3): 278-99.
[http://dx.doi.org/10.1016/j.biotechadv.2011.01.002] [PMID: 21236330]

[26] Karuppusamy S. A review on trends in production of secondary metabolitesfromhigher plants by *in vitro* tissue, organ and cell cultures. J Med Plants Res 2009; 3: 1222-39.

[27] Pence VC. Evaluating costs for the *in vitro* propagation and preservation of endangered plants. In Vitro Cell Dev Biol Plant 2011; 47(1): 176-87.
[http://dx.doi.org/10.1007/s11627-010-9323-6]

[28] Chen L, Cai Y, Liu X, *et al.* Soybean hairy roots produced *in vitro* by *Agrobacterium rhizogenes*-mediated transformation. Crop J 2018; 6(2): 162-71.
[http://dx.doi.org/10.1016/j.cj.2017.08.006]

[29] Palazón J, Piñol MT, Cusido RM, Morales C, Bonfill M. Application oftransformedroot technology to the production of bioactive metabolites. *Recent Res.Dev.* Plant Physiol 1997; 1: 125-43.

[30] Zhang L, Ding R, Chai Y, *et al.* Engineering tropane biosynthetic pathway in *Hyoscyamus niger* hairy root cultures. Proc Natl Acad Sci 2004; 101(17): 6786-91.
[http://dx.doi.org/10.1073/pnas.0401391101] [PMID: 15084741]

[31] Ha LT, Pawlicki-Jullian N, Pillon-Lequart M, Boitel-Conti M, Duong HX, Gontier E. Hairy root cultures of *Panax vietnamensis*, a promising approach for the production of ocotillol-type ginsenosides. Plant Cell Tissue Organ Cult 2016; 126(1): 93-103.
[http://dx.doi.org/10.1007/s11240-016-0980-y]

[32] Perassolo M, Cardillo AB, Mugas ML, Núñez Montoya SC, Giulietti AM, Rodríguez Talou J.

Enhancement of anthraquinone production and release by combination of culture medium selection and methyl jasmonate elicitation in hairy root cultures of *Rubia tinctorum*. Ind Crops Prod 2017; 105: 124-32.
[http://dx.doi.org/10.1016/j.indcrop.2017.05.010]

[33] Patra N, Srivastava AK. Artemisinin production by plant hairy root cultures in gas- and liquid-phase bioreactors. Plant Cell Rep 2016; 35(1): 143-53.
[http://dx.doi.org/10.1007/s00299-015-1875-9] [PMID: 26441056]

[34] Thiruvengadam M, Rekha K, Chung IM. Induction of hairy roots by *Agrobacterium rhizogenes* - mediated transformation of spine gourd (*Momordica dioica* Roxb. ex. willd) for the assessment of phenolic compounds and biological activities. Sci Hortic (Amsterdam) 2016; 198: 132-41.
[http://dx.doi.org/10.1016/j.scienta.2015.11.035] [PMID: 32287883]

[35] Thakore D, Srivastava AK, Sinha AK. Mass production of Ajmalicine by bioreactor cultivation of hairy roots of *Catharanthus roseus*. Biochem Eng J 2017; 119: 84-91.
[http://dx.doi.org/10.1016/j.bej.2016.12.010]

[36] Cardoso JC, Oliveira MEBS, Cardoso FCI. Advances and challenges on the *in vitro* production of secondary metabolites from medicinal plants. Hortic Bras 2019; 37(2): 124-32.
[http://dx.doi.org/10.1590/s0102-053620190201]

[37] Rehman H, Jameel U, Chester A, Ennus K, Tajuddin T, Ahmad S. Chemo profiling and tissue culture studies on *Picrorhizakurroa*Royle ex Benth. for production of picroside II. Ann Phytomed 2014; 3(2): 70-7.

[38] Zhao J, Davis LC, Verpoorte R. Elicitor signal transduction leading to production of plant secondary metabolites. Biotechnol Adv 2005; 23(4): 283-333.
[http://dx.doi.org/10.1016/j.biotechadv.2005.01.003] [PMID: 15848039]

[39] Chen Y, Zhang M, Jin X, *et al*. Transcriptional reprogramming strategies and miRNA-mediated regulation networks of Taxus media induced into callus cells from tissues. BMC Genomics 2020; 21(1): 168.
[http://dx.doi.org/10.1186/s12864-020-6576-2] [PMID: 32070278]

[40] Khan MA, Wallace WT, Sambi J, *et al*. Nanoharvesting of bioactive materials from living plant cultures using engineered silica nanoparticles. Mater Sci Eng C 2020; 106: 110190.
[http://dx.doi.org/10.1016/j.msec.2019.110190] [PMID: 31753369]

[41] Srivastava S, Singh VP. Characterization of biocontrol microorganisms from the rhizoplane of *Decalepisarayalpathra* and screening of secondary metabolites. Vegetos 2020.

[42] Razavizadeh R, Adabavazeh F, Komatsu S. Chitosan effects on the elevation of essential oils and antioxidant activity of *Carum copticum* L. seedlings and callus cultures under *in vitro* salt stress. J Plant Biochem Biotechnol 2020; 29(3): 473-83.
[http://dx.doi.org/10.1007/s13562-020-00560-1]

[43] Felipe D, Brambilla L, Porto C, Pilau E, Cortez D. Phytochemical analysis of *Pfaffia glomerata* inflorescences by LC-ESI-MS/MS. Molecules 2014; 19(10): 15720-34.
[http://dx.doi.org/10.3390/molecules191015720] [PMID: 25268723]

[44] Silva TD, Batista DS, Fortini EA, *et al*. Influence of growth regulators on the development, quality, and physiological state of *in vitro* propagated*Lamprocapnos spectabilis* (L.) Fukuhara. *In Vitro*. Cell Dev Biol Plant 2020.

[45] Ali A, Mohammad S, Khan MA, *et al*. Silver nanoparticles elicited *in vitro* callus cultures for accumulation of biomass and secondary metabolites in *Caralluma tuberculata*. Artif Cells Nanomed Biotechnol 2019; 47(1): 715-24.
[http://dx.doi.org/10.1080/21691401.2019.1577884] [PMID: 30856344]

[46] Adil M, Haider Abbasi B, ul Haq I. Red light controlled callus morphogenetic patterns and secondary metabolites production in *Withania somnifera* L. Biotechnol Rep 2019; 24: e00380.

[http://dx.doi.org/10.1016/j.btre.2019.e00380] [PMID: 31641624]

[47] Wang Y, Yang B, Zhang M, Jia S, Yu F. Application of transport engineering to promote catharanthine production in *Catharanthus roseus* hairy roots. Plant Cell Tissue Organ Cult 2019; 139(3): 523-30.
[http://dx.doi.org/10.1007/s11240-019-01696-2]

[48] Muchate NS, Rajurkar NS, Suprasanna P, Nikam TD. NaCl induced salt adaptive changes and enhanced accumulation of 20-hydroxyecdysone in the *in vitro* shoot cultures of Spinacia oleracea (L.). Sci Rep 2019; 9(1): 12522.
[http://dx.doi.org/10.1038/s41598-019-48737-6] [PMID: 31467324]

[49] Hadian F, Koohi-Dehkordi M, Golkar P. Evaluation of *in vitro* mucilage and lepidine biosynthesis in different genotypes of *Lepidium sativum* Linn originated from Iran. S Afr J Bot 2019; 127: 91-5.
[http://dx.doi.org/10.1016/j.sajb.2019.08.028]

[50] Munim Twaij B, Jazar ZH, Hasan MN. The effects of elicitors and precursor on *in-vitro* cultures of *Trifolium resupinatum* for sustainable metabolite accumulation and antioxidant activity. Biocatal Agric Biotechnol 2019; 22: 101337.
[http://dx.doi.org/10.1016/j.bcab.2019.101337]

[51] Felipe SHS, Batista DS, Chagas K, *et al.* Accessions of Brazilian ginseng (*Pfaffia glomerata*) with contrasting anthocyanin content behave differently in growth, antioxidative defense, and 20-hydroxyecdysone levels under UV-B radiation. Protoplasma 2019; 256(6): 1557-71.
[http://dx.doi.org/10.1007/s00709-019-01400-3] [PMID: 31209575]

[52] Quintanilha-Peixoto G, Torres RO, Reis IMA, Oliveira TAS, Bortolini DE, Duarte EAA, *et al.* Calm before the storm: A glimpse into the secondary metabolism of *Aspergillus welwitschiae*, the etiologic agent of the sisal bole rot. Toxins 2019; 11(11): 631.
[http://dx.doi.org/10.3390/toxins11110631] [PMID: 31671681]

[53] Reyes-Martínez A, Valle-Aguilera JR, Antunes-Ricardo M, Gutiérrez-Uribe J, Gonzalez C, Santos-Díaz MS. Callus from pyrostegia venusta (ker gawl.) miers: A source of phenylethanoid glycosides with vasorelaxant activities. Plant Cell Tissue Organ Cult 2019; 139(1): 119-29.
[http://dx.doi.org/10.1007/s11240-019-01669-5]

[54] Boonsnongcheep P, Sae-foo W, Banpakoat K, *et al.* Artificial color light sources and precursor feeding enhance plumbagin production of the carnivorous plants *Drosera burmannii* and *Drosera indica*. J Photochem Photobiol B 2019; 199: 111628.
[http://dx.doi.org/10.1016/j.jphotobiol.2019.111628] [PMID: 31610432]

[55] Pal M, Kumar V, Yadav R, Gulati D, Yadav RC. Potential and prospects of shikonin production enhancement in medicinal plants. Proc Natl Acad Sci, India, Sect B Biol Sci 2019; 89(3): 775-84.
[http://dx.doi.org/10.1007/s40011-017-0931-3]

[56] Gutiérrez-Rebolledo GA, Estrada-Zúñiga ME, Garduño-Siciliano L, *et al. In vivo* anti-arthritic effect and repeated dose toxicity of standardized methanolic extracts of *Buddleja cordata* Kunth (Scrophulariaceae) wild plant leaves and cell culture. J Ethnopharmacol 2019; 240: 111875.
[http://dx.doi.org/10.1016/j.jep.2019.111875] [PMID: 31034952]

[57] Ghanadian M, Abbaspour J, Ehsanpour AA, Aghaei M. Sesquiterpene lactones from shoot culture of *Artemisia aucheri* with cytotoxicity against prostate and breast cancer cells. Res Pharm Sci 2019; 14(4): 329-34.
[http://dx.doi.org/10.4103/1735-5362.263557] [PMID: 31516509]

[58] Adil M, Ren X, Jeong BR. Light elicited growth, antioxidant enzymes activities and production of medicinal compounds in callus culture of *Cnidium officinale* Makino. J Photochem Photobiol B 2019; 196: 111509.
[http://dx.doi.org/10.1016/j.jphotobiol.2019.05.006] [PMID: 31128431]

[59] Bonello M, Gašić U, Tešić Ž, Attard E. Production of stilbenes in callus cultures of the Maltese indigenous grapevine variety. Molecules 2019; 24(11): 2112.

[http://dx.doi.org/10.3390/molecules24112112] [PMID: 31167390]

[60] Sarmadi M, Karimi N, Palazón J, Ghassempour A, Mirjalili MH. Improved effects of polyethylene glycol on the growth, antioxidative enzymes activity and taxanes production in a *Taxus baccata* L. callus culture. Plant Cell Tissue Organ Cult 2019; 137(2): 319-28.
[http://dx.doi.org/10.1007/s11240-019-01573-y]

[61] Abd El-Kader EM, Serag A, Aref MS, Ewais EEA, Farag MA. Metabolomics reveals ionones upregulation in MeJA elicited *Cinnamomum camphora* (camphor tree) cell culture. Plant Cell Tissue Organ Cult 2019; 137(2): 309-18.
[http://dx.doi.org/10.1007/s11240-019-01572-z]

[62] Ruiz-Molina N, Ortega-Bedoya I, Arias-Zabala M. Protonema suspension cultures of *PolytrichumJuniperinum*as a potential production platform for bioactive compounds. J Herbs Spices Med Plants 2019; 25(2): 114-27.
[http://dx.doi.org/10.1080/10496475.2019.1577321]

[63] Bose B, Tripathy D, Chatterjee A, Tandon P, Kumaria S. Secondary metabolite profiling, cytotoxicity, anti-inflammatory potential and *in vitro* inhibitory activities of *Nardostachys jatamansi* on key enzymes linked to hyperglycemia, hypertension and cognitive disorders. Phytomedicine 2019; 55: 58-69.
[http://dx.doi.org/10.1016/j.phymed.2018.08.010] [PMID: 30668444]

[64] Mohammad S, Khan MA, Ali A, Khan L, Khan MS, Mashwani ZR. Feasible production of biomass and natural antioxidants through callus cultures in response to varying light intensities in olive (*Olea europaea*. L) cult. Arbosana. J Photochem Photobiol B 2019; 193: 140-7.
[http://dx.doi.org/10.1016/j.jphotobiol.2019.03.001] [PMID: 30852387]

[65] Inyai C, Boonsnongcheep P, Komaikul J, Sritularak B, Tanaka H, Putalun W. Alginate immobilization of *Morus alba* L. cell suspension cultures improved the accumulation and secretion of stilbenoids. Bioprocess Biosyst Eng 2019; 42(1): 131-41.
[http://dx.doi.org/10.1007/s00449-018-2021-1] [PMID: 30284036]

[66] Tyunin AP, Suprun AR, Nityagovsky NN, *et al.* The effect of explant origin and collection season on stilbene biosynthesis in cell cultures of *Vitis amurensis* Rupr. Plant Cell Tissue Organ Cult 2019; 136(1): 189-96.
[http://dx.doi.org/10.1007/s11240-018-1490-x]

[67] Abbasi B, Siddiquah A, Tungmunnithum D, *et al. Isodonrugosus* (Wall. ex Benth.) codd *in vitro* cultures: Establishment, phytochemical characterization and *in vitro* antioxidant and anti-aging activities. Int J Mol Sci 2019; 20(2): 452.
[http://dx.doi.org/10.3390/ijms20020452] [PMID: 30669669]

[68] Maurya B, Rai KK, Pandey N, Sharma L, Goswami NK, Rai SP. Influence of salicylic acid elicitation on secondary metabolites and biomass production in *in vitro* cultured *Withaniacoagulans* (l.) dunal. Plant Arch 2019; 19: 1308-16.

[69] Al Khateeb W, Alu'datt M, Al Zghoul H, Kanaan R, El-Oqlah A, Lahham J. Enhancement of phenolic compounds production in *in vitro* grown *Rumex cyprius* Murb. Acta Physiol Plant 2017; 39(1): 14.
[http://dx.doi.org/10.1007/s11738-016-2312-6]

[70] Siatka T. Effects of growth regulators on production of anthocyanins in callus cultures of *Angelica archangelica.* Nat Prod Commun 2019; 14(6): 1934578X1985734.
[http://dx.doi.org/10.1177/1934578X19857344]

[71] Parmar VR, Jasrai YT. Studies on the flavonoid synthesis in *Urariapicta* callus culture. Med Plant 2019; 11(2): 195-9.

[72] Rady MR, Saker MM, Matter MA. *In vitro* culture, transformation and genetic fidelity of Milk Thistle. J Genet Eng Biotechnol 2018; 16(2): 563-72.
[http://dx.doi.org/10.1016/j.jgeb.2018.02.007] [PMID: 30733774]

[73] Martínez ME, Poirrier P, Prüfer D, *et al.* Kinetics and modeling of cell growth for potential anthocyanin induction in cultures of Taraxacum officinale G.H. Weber ex Wiggers (Dandelion) *in vitro*. Electron J Biotechnol 2018; 36: 15-23.
[http://dx.doi.org/10.1016/j.ejbt.2018.08.006]

[74] Sarmadi M, Karimi N, Palazón J, Ghassempour A, Mirjalili MH. The effects of salicylic acid and glucose on biochemical traits and taxane production in a *Taxus baccata* callus culture. Plant Physiol Biochem 2018; 132: 271-80.
[http://dx.doi.org/10.1016/j.plaphy.2018.09.013] [PMID: 30240989]

[75] Pech-Kú R, Muñoz-Sánchez JA, Monforte-González M, Vázquez-Flota F, Rodas-Junco BA, Hernández-Sotomayor SMT. Caffeine extraction, enzymatic activity and gene expression of *caffeine synthase* from plant cell suspensions. J Vis Exp 2018; (140): 140.
[PMID: 30346406]

[76] Bibi A, Khan MA, Adil M, Mashwani ZUR. Production of callus biomass and antioxidant secondary metabolites in black cumin. J Anim Plant Sci 2018; 28(5): 1321-8.

[77] Reis A, Scopel M, Zuanazzi J, A,S. Trifolium pratense: Friable calli, cell culture protocol and isoflavones content in wild plants, *in vitro* and cell cultures analyzed by UPLCBrazilian. J Pharmacog 2018; 28(5): 542-50.

[78] De Sousa-Machado IB, Felippe T, Garcia R, Pacheco G, Moreira D, Mansur E. Total phenolics, resveratrol content and antioxidant activity of seeds and calluses of pinto peanut (*Arachis pintoi* Krapov. & W.C. Greg.). Plant Cell Tissue Organ Cult 2018; 134(3): 491-502.
[http://dx.doi.org/10.1007/s11240-018-1438-1]

[79] Srivastava S, Sanchita , Singh R, Srivastava G, Sharma A. Comparative study of withanolide biosynthesis-related miRNAs in root and leaf tissues of *withania somnifera*. Appl Biochem Biotechnol 2018; 185(4): 1145-59.
[http://dx.doi.org/10.1007/s12010-018-2702-x] [PMID: 29476318]

[80] Marthe F. Tissue culture approaches in relation to medicinal plant improvement. BiotechnolCrop Improv 2018; 1: 487-97.
[http://dx.doi.org/10.1007/978-3-319-78283-6_15]

[81] Al-Mahdawe MM, Al-Mallah MK, Ahmad TA. Isolation and identification of rutin from tissues cultures of *Ruta graveolens* l. J Pharmac Sci Res 2018; 10(6): 1517-20.

[82] Lattanzio V, Caretto S, Linsalata V, Colella G, Mita G. Signal transduction in artichoke [*Cynara cardunculus* L. subsp. scolymus (L.) Hayek] callus and cell suspension cultures under nutritional stress. Plant Physiol Biochem 2018; 127: 97-103.
[http://dx.doi.org/10.1016/j.plaphy.2018.03.017] [PMID: 29571004]

[83] Fidemann T, De Araujo Pereira GA, Bossard Nascimento L, *et al.* Holistic protocol for callus culture optimization using statistical modelling. Nat Prod Res 2018; 32(9): 1109-17.
[http://dx.doi.org/10.1080/14786419.2017.1380026] [PMID: 28956460]

[84] Kayani WK, Kiani BH, Dilshad E, Mirza B. Biotechnological approaches for artemisinin production in Artemisia. World J Microbiol Biotechnol 2018; 34(4): 54.
[http://dx.doi.org/10.1007/s11274-018-2432-9] [PMID: 29589124]

[85] Marques HP, Barbosa S, Nogueira DA, Santos MH, Santos BR, Santos-Filho PR. Proteic and phenolics compounds contents in Bacupari callus cultured with glutamine and nitrogen sources. Braz J Biol 2017; 78(1): 41-6.
[http://dx.doi.org/10.1590/1519-6984.03416] [PMID: 28562777]

[86] Pant B, Shah S, Shrestha R, Pandey S, Joshi PR. An overview on orchid endophytes Mycorrhiza - Nutrient Uptake, Biocontrol. 4th. Ecorestoration 2018; pp. 503-24.

[87] Çölgeçen H, Atar H, Toker G, Akgül G. Callus production and analysis of some secondary metabolites in *Globularia trichosantha* subsp. trichosantha. Turk J Bot 2018; 42(5): 559-67.

[http://dx.doi.org/10.3906/bot-1712-13]

[88] Kundu D, Talukder P, Sen Raychaudhuri S. *In Vitro* Biosynthesis of Polyphenols in the Presence of Elicitors and Upregulation of Genes of the Phenylpropanoid Pathway in Plantago ovata. Studies in Natural Products Chemistry 2019; 60: 299-344.
[http://dx.doi.org/10.1016/B978-0-444-64181-6.00008-5]

[89] Fazeli-Nasab B. The effect of explant, BAP and 2,4-D on callus induction of *Trachyspermum ammi*. Potravinárstvo 2018; 12(1): 578-86.
[http://dx.doi.org/10.5219/953]

[90] Rubin E, Aziz ZA, Surugau N. Changes in shoot proliferation and chemical components of *in vitro* cultured Dendrobium officinale due to organic additives. J Appl Horticul 2018; 20(1): 24-8.

[91] Tuan TT, Thien NS, Nguyen HC, *et al*. Changes in shoot proliferation and chemical components of *in vitro* cultured *Dendrobium officinale* due to organic additives. J Appl Hortic 2018; 20(1): 24-8.
[http://dx.doi.org/10.37855/jah.2018.v20i01.04]

[92] Ganesh KY, Ramarao A, Veeresham C. Picroside I and picroside II from tissue cultures of *Picrorhizakurroa*. Pharmacognosy Res 2017; 9(5): 553-6.

[93] Li RZ, Lin J, Wang XX, Yu XM, Chen CL, Guan YF. Nontargeted metabolomic analysis of *Anoectochilus roxburghii* at different cultivation stages. Zhongguo Zhongyao Zazhi 2017; 42(23): 4624-30.
[PMID: 29376262]

[94] Kapare V, Satdive R, Fulzele DP, Malpathak N. Impact of gamma irradiation induced variation in cell growth and phytoecdysteroid production in *Sesuviumportulacastrum*. J Plant Growth Regul 2017; 36(4): 919-30.
[http://dx.doi.org/10.1007/s00344-017-9697-3]

[95] Ismail Z, Ahmad A, Muhammad TST. Phytochemical screening of *in vitroAglaonema simplex* plantlet extracts as inducers of SR-B1 ligand expression. J Sustain Sci Manag 2017; 12(2): 34-44.

[96] Dusek J, Carazo A, Trejtnar F, *et al*. Steviol, an aglycone of steviol glycoside sweeteners, interacts with the pregnane X (PXR) and aryl hydrocarbon (AHR) receptors in detoxification regulation. Food Chem Toxicol 2017; 109(Pt 1): 130-42.
[http://dx.doi.org/10.1016/j.fct.2017.09.007] [PMID: 28887089]

[97] Javed R, Yucesan B, Zia M, Gurel E. Differential effects of plant growth regulators on physiology, steviol glycosides content, and antioxidant capacity in micropropagated tissues of *Stevia rebaudiana*. Biologia 2017; 72(10): 1156-65.
[http://dx.doi.org/10.1515/biolog-2017-0133]

[98] Gurel E, Karvar S, Yucesan B, Eker I, Sameeullah M. An overview of cardenolides in digitalis : More than a cardiotonic compound. Curr Pharm Des 2017; 23(34): 5104-14.
[PMID: 28847302]

[99] Soni D, Rana P, Goyal S, Singh A, Jamal S, Grover A. Plant tissue culture based strategies for production of withanolides. Science of Ashwagandha: Preventive and Therapeutic Potentials 2017; 479-93.
[http://dx.doi.org/10.1007/978-3-319-59192-6_23]

[100] Kaur K, Singh P, Guleri R, *et al*. Biotechnological approaches in propagation and improvement of Withaniasomnifera (L.) dunal. Science of Ashwagandha: Preventive and Therapeutic Potential 2017; 459-78.

[101] Thakore D, Srivastava AK. Production of biopesticide azadirachtin using plant cell and hairy root cultures. Eng Life Sci 2017; 17(9): 997-1005.
[http://dx.doi.org/10.1002/elsc.201700012] [PMID: 32624850]

[102] Chen C, Luo X, Jin G, *et al*. Shading effect on survival, growth, and contents of secondary metabolites in micropropagated *Anoectochilus* plantlets. Rev Bras Bot 2017; 40(3): 599-607.

[http://dx.doi.org/10.1007/s40415-017-0365-4]

[103] Raju DC, Varghese RJ, Biji N. Research. J Pharm Technol 2017; 10(7): 2095-100.

[104] Deep A, Rana P, Soni G. Iron chelation and iron reducing activity of tissue cultured and tissue culture derived Mentha spp. J Appl Pharm Sci 2017; 7(5): 78-83.

[105] Mascarello C, Sacco E, Pamato M, *et al. Rosmarinus officinalis* L.: Micropropagation and callus induction for cell biomass development. Acta Hortic 2017; (1155): 631-6.
[http://dx.doi.org/10.17660/ActaHortic.2017.1155.92]

[106] Bankole AE, Uchendu EE, Adekunle AA. *In vitro* germination of *Markhamia tomentosa* Benth K. Schum ex. Engl. and preliminary phytochemical screening for medicinal compounds. Indian J Plant Physiol 2017; 22(1): 85-93.
[http://dx.doi.org/10.1007/s40502-016-0279-3]

[107] Eid HH, Metwally GF. Phytochemical and biological study of callus cultures of *Tulbaghia violacea* Harv. Cultivated in Egypt. Nat Prod Res 2017; 31(15): 1717-24.
[http://dx.doi.org/10.1080/14786419.2017.1289206] [PMID: 28278648]

[108] El-Sharabasy S, El-Dawayati M. Bioreactor steroid production and analysis of date palm embryogenic callus. Methods Mol Biol 2017; 1637: 309-18.
[http://dx.doi.org/10.1007/978-1-4939-7156-5_25] [PMID: 28755355]

[109] Farvardin A, Ebrahimi A, Hosseinpour B, Khosrowshahli M. Effects of growth regulators on callus induction and secondary metabolite production in *Cuminum cyminum*. Nat Prod Res 2017; 31(17): 1963-70.
[http://dx.doi.org/10.1080/14786419.2016.1272105] [PMID: 28044460]

[110] Ghezelbash S, Qaderi A, Bodaghi H. The effect of antioxidant compounds and media on biosynthesis of limiter phenolic compounds during *in vitro* culture of Mentha arvensis L. JMed Plants 2017; 16: 1-9.

[111] Rumiyati , Sismindari , Semiarti E, *et al.* Callus induction from various organs of dragon fruit, apple and tomato on some mediums. Pak J Biol Sci 2017; 20(5): 244-52.
[http://dx.doi.org/10.3923/pjbs.2017.244.252] [PMID: 29023036]

[112] Ibrahim IR, Ameen SKM. Influence of stress on secondary metabolites production from callus of *Moringa oleiferain vitro*. Iraqi J Agricul Sci 2017; 48(4): 1011-107.

[113] Teixeira da Silva JA, Jha S. Micropropagation and genetic transformation of *Tylophora indica* (Burm. f.) Merr.: A review. Plant Cell Rep 2016; 35(11): 2207-25.
[http://dx.doi.org/10.1007/s00299-016-2041-8] [PMID: 27553812]

[114] Ng TLM, Karim R, Tan YS, *et al.* Amino acid and secondary metabolite production in embryogenic and non-embryogenic callus of fingerroot ginger (Boesenbergia rotunda). PLoS ONE 2016; 11(6): e0156714.

[115] Patil RH, Patil MP, Maheshwari VL. Bioactive secondary metabolites from endophytic fungi: A review of biotechnological production and their potential applications. StudNat Prod Chem 2016; 49: 189-205.
[http://dx.doi.org/10.1016/B978-0-444-63601-0.00005-3]

[116] Gerolino E. Evaluation of limonoid production in suspension cell culture of *Citrus sinensis* (2015) Brazilian. J Pharmacognosy 2015; 25(5): 455-61.

[117] Daud NH, Mohamed R. Flavonoid production by calli and cell suspension cultures of *Aquilaria malaccensis*: A threatened tropical tree. Med Plant 2015; 7(3): 208-18.

[118] Ghaderi N, Jafari M. Efficient plant regeneration, genetic fidelity and high-level accumulation of two pharmaceutical compounds in regenerated plants of *Valeriana officinalis* L. S Afr J Bot 2014; 92: 19-27.
[http://dx.doi.org/10.1016/j.sajb.2014.01.010]

[119] AbouZid S. Yield improvement strategies for the production of secondary metabolites in plant tissue culture: silymarin from *Silybum marianum* tissue culture. Nat Prod Res 2014; 28(23): 2102-10.
[http://dx.doi.org/10.1080/14786419.2014.927465] [PMID: 24947979]

[120] Cetin E, Zehra B. The effects of cadmium chloride on secondary metabolite production in Vitisvinifera cv. cell suspension cultures. Biological Research 2014; 47: 1-47.

[121] Yadnya-Putra AAGR, Chahyadi A, Elfahmi E. Production of panduratin A, cardamomin and sitosterol using cell cultures of fingerroot (*Boesenbergiapandurata*Roxb. schlechter). Biosci Biotechnol Res Asia 2014; 11(1): 43-52.
[http://dx.doi.org/10.13005/bbra/1231]

[122] Gonda S, Kiss-Sziksai A, Szűcs Z, Máthé C, Vasas G. Effects of N source concentration and NH_4^+/NO_3^- ratio on phenylethanoid glycoside pattern in tissue cultures of Plantago lanceolata L.: A metabolomics driven full-factorial experiment with LC–ESI–MS 3. Phytochemistry 2014; 106: 44-54.
[http://dx.doi.org/10.1016/j.phytochem.2014.07.002] [PMID: 25081104]

[123] D'Angiolillo F, Pistelli L, Noccioli C, *et al*. In vitro cultures of *Bituminaria bituminosa*: Pterocarpan, furanocoumarin and isoflavone production and cytotoxic activity evaluation. Nat Prod Commun 2014; 9(4): 1934578X1400900.
[http://dx.doi.org/10.1177/1934578X1400900411] [PMID: 24868860]

[124] Maliński MP, Michalska AD, Tomczykowa M, Tomczyk M, Thiem B. Ragged Robin (*Lychnis flos-cuculi*) : A plant with potential medicinal value. Rev Bras Farmacogn 2014; 24(6): 722-30.
[http://dx.doi.org/10.1016/j.bjp.2014.11.004]

[125] Piovan A, Filippini R, Innocenti G. Coumarin compounds in coronillascorpioides callus cultures. Nat Prod Commun 2014; 9(4): 489-92.

[126] Amali P, Kingsley SJ, Ignacimuthu S. Frequency callus induction and plant regeneration from shoot tip explants of *Sorghum bicolor* L. moench. Int J Pharm Pharm Sci 2014; 6(6): 213-6.

[127] Deshpande J, Labade D, Shankar K, *et al*. In vitro callus induction and estimation of plumbagin content from *Plumbago auriculata* Lam. Indian J Exp Biol 2014; 52(11): 1122-7.
[PMID: 25434108]

[128] Lodha D, Patel AK, Rai MK, Shekhawat NS. *In vitro* plantlet regeneration and assessment of alkaloid contents from callus cultures of *Ephedra foliata* (Unth phog), a source of anti-asthmatic drugs. Acta Physiol Plant 2014; 36(11): 3071-9.
[http://dx.doi.org/10.1007/s11738-014-1677-7]

[129] Ncube B, Finnie JF, Van Staden J. Carbon–nitrogen ratio and *in vitro* assimilate partitioning patterns in *Cyrtanthus guthrieae* L. Plant Physiol Biochem 2014; 74: 246-54.
[http://dx.doi.org/10.1016/j.plaphy.2013.11.007] [PMID: 24321874]

[130] Siddiqui ZH, Mujib A. Mahmooduzzafar, Aslam, J., Rehman Hakeem, K., Parween, T. In vitro production of secondary metabolites using elicitor in *Catharanthus roseus*: A case study (2013) Crop Improvement: New Approaches and Modern Techniques, pp. 401-419.

[131] Ayob Z, Wagiran A, Abd Samad A. Potential of tissue cultured medicinal plants in Malaysia. J Teknol 2013; 62(1): 111-7. [Sciences and Engineering].
[http://dx.doi.org/10.11113/jt.v62.1775]

[132] Tavares S, Vesentini D, Fernandes JC, *et al*. Vitis vinifera secondary metabolism as affected by sulfate depletion: Diagnosis through phenylpropanoid pathway genes and metabolites. Plant Physiol Biochem 2013; 66: 118-26.
[http://dx.doi.org/10.1016/j.plaphy.2013.01.022] [PMID: 23500714]

[133] Grech-Baran M, Pietrosiuk A. Artemisia species *in vitro* cultures for production of biologically active secondary metabolites. Biotechnologia 2013; 93(4): 371-80.

[134] Gonçalves S, Romano A. *In vitro* culture of lavenders (*Lavandula* spp.) and the production of

secondary metabolites. Biotechnol Adv 2013; 31(2): 166-74.
[http://dx.doi.org/10.1016/j.biotechadv.2012.09.006] [PMID: 23022737]

[135] Constantin D, Coste A, Mircea T. Epilobium Sp. (willow herb): Micropropagation and production of secondary metabolites. Biotechnology for Medicinal Plants: Micropropagation and Improvement 2013; 149-70.

[136] Downey PJ, Levine LH, Musgrave ME, McKeon-Bennett M, Moane S. Effect of hypergravity and phytohormones on isoflavonoid accumulation in soybean (*Glycine max* L.) callus. Microgravity Sci Technol 2013; 25(1): 9-15.
[http://dx.doi.org/10.1007/s12217-012-9322-9]

[137] Wojakowska A, Muth D, Narożna D, Mądrzak C, Stobiecki M, Kachlicki P. Changes of phenolic secondary metabolite profiles in the reaction of narrow leaf lupin (*Lupinus angustifolius*) plants to infections with *Colletotrichum lupini* fungus or treatment with its toxin. Metabolomics 2013; 9(3): 575-89.
[http://dx.doi.org/10.1007/s11306-012-0475-8] [PMID: 23678343]

[138] Nguyen TKO, Dauwe R, Bourgaud F, Gontier E. From bioreactor to entire plants: Development of production systems for secondary metabolites. Adv Bot Res 2013; 68: 205-32.
[http://dx.doi.org/10.1016/B978-0-12-408061-4.00008-0]

[139] Rawat B, Rawat JM, Mishra S, Mishra SN. Picrorhiza kurrooa: Current status and tissue culture mediated biotechnological interventions. Acta Physiol Plant 2013; 35(1): 1-12.
[http://dx.doi.org/10.1007/s11738-012-1069-9]

[140] Ozarowski M, Mikołajczak PL, Thiem B. Medicinal plants in the phytotherapy of alcohol or nicotine addiction. Implication for plants *in vitro* cultures. Przegl Lek 2013; 70(10): 869-74.
[PMID: 24501814]

[141] Tusevski O, Petreska Stanoeva J, Stefova M, Simic SG. Phenolic profile of dark-grown and photoperiod-exposed Hypericum perforatum L. Hairy root cultures. ScientWorldJ 2013; 2013: 1-9.
[http://dx.doi.org/10.1155/2013/602752] [PMID: 24453880]

[142] Kumar V, Kumar CS, Hari G, Venugopal NK, Vijendra PD, B GB. Homology modeling and docking studies on oxidosqualene cyclases associated with primary and secondary metabolism of *Centella asiatica*. Springerplus 2013; 2(1): 189.
[http://dx.doi.org/10.1186/2193-1801-2-189] [PMID: 25247142]

[143] Silva da Rocha A, Rocha EK, Alves LM, *et al.* Production and optimization through elicitation of carotenoid pigments in the *in vitro* cultures of Cleome rosea Vahl (Cleomaceae). J Plant Biochem Biotechnol 2015; 24(1): 105-13.
[http://dx.doi.org/10.1007/s13562-013-0241-7]

[144] Hunaefi D, Smetanska I. The effect of tea fermentation on rosmarinic acid and antioxidant properties using selected *in vitro* sprout culture of *Orthosiphon aristatus* as a model study. Springerplus 2013; 2(1): 167.
[http://dx.doi.org/10.1186/2193-1801-2-167] [PMID: 23667816]

[145] C Jain S, Pancholi B, Jain R. *In-vitro* callus propagation and secondary metabolite quantification in *Sericostomapauciforum*. Iran J Pharm Res 2012; 11(4): 1103-9.
[PMID: 24250543]

[146] Gharehmatrossian S, Popov YU, Ghorbanli M, Safaeian S. Antioxidant activities and cytotoxic effects of whole plant and isolated culture of *Artemisia aucheri*Boiss. Asian J Pharm Clin Res 2012; 5(4): 95-8.

[147] Aremu AO, Bairu MW, Novák O, *et al.* Physiological responses and endogenous cytokinin profiles of tissue-cultured 'Williams' bananas in relation to roscovitine and an inhibitor of cytokinin oxidase/dehydrogenase (INCYDE) treatments. Planta 2012; 236(6): 1775-90.
[http://dx.doi.org/10.1007/s00425-012-1721-z] [PMID: 22886380]

[148] Mangal M, Sheoryan A, Mangal A, Kajla K, Choudhury S, Dhawan A. A biotechnological advances in *Tinospora cordifolia* (Willd.) Miers ex Hook. F. &Thoms: Overview of present status and future prospects. Vegetos 2012; 25(2): 182-91.

[149] Bernard F, Moghbel N, Hassannejad S. Treatment of licorice seeds with colchicine: changes in seedling DNA levels and anthocyanin and glycyrrhizic acid contents of derived callus cultures. Nat Prod Commun 2012; 7(11): 1934578X1200701.
[http://dx.doi.org/10.1177/1934578X1200701112] [PMID: 23285806]

[150] Moglia A, Menin B, Comino C, Lanteri S, Beekwilder J. Globe artichoke callus as an alternative system for the production of dicaffeoylquinic acids. Acta Hortic 2012; (961): 261-5.
[http://dx.doi.org/10.17660/ActaHortic.2012.961.33]

[151] Tanveer H, Ali S, Rafique Asi M. Appraisal of an important flavonoid, quercetin, in callus cultures of *Citrullus colocynthis*. Int J Agric Biol 2012; 14(4): 528-32.

[152] Petrova M, Zayova E, Vassilevska-Ivanova R, Vlahova M. Biotechnological approaches for cultivation and enhancement of secondary metabolites in *Arnica montana* L. Acta Physiol Plant 2012; 34(5): 1597-606.
[http://dx.doi.org/10.1007/s11738-012-0987-x]

[153] Mirmazloum I, György Z. Review of the molecular genetics in higher plants towards salidrosid and cinnamyl alcohol glycosides biosynthesis in *Rhodiola rosea* L. Acta Aliment 2012; 41(1) (1): 133-46.
[http://dx.doi.org/10.1556/AAlim.41.2012.Suppl.13]

[154] Kuo CI, Chao CH, Lu MK. Effects of auxins on the production of steroidal alkaloids in rapidly proliferating tissue and cell cultures of *Solanum lyratum*. Phytochem Anal 2012; 23(4): 400-4.
[http://dx.doi.org/10.1002/pca.1371] [PMID: 22009634]

[155] Mathew L, Rashmi PA. Development of salinity resistant somaclones of *Justicia adhatoda*-An important medicinal plant with bronchodilatory effect. Res J Pharm Biol Chem Sci 2012; 3(3): 132-7.

[156] Gathungu RM, Oldham JT, Bird SS, Lee-Parsons CWT, Vouros P, Kautz R. Application of an integrated LC-UV-MS-NMR platform to the identification of secondary metabolites from cell cultures: Benzophenanthridine alkaloids from elicited Eschscholzia californica (california poppy) cell cultures. Anal Methods 2012; 4(5): 1315-25.
[http://dx.doi.org/10.1039/c2ay05803k] [PMID: 22707983]

[157] El-Dawayati MM, Zaid ZE, Elsharabasy SF. Effect of conservation on steroids contents of callus explants of date palm cv. Sakkoti. Aust J Basic Appl Sci 2012; 6(5): 305-10.

[158] Butiuc-Keul AL, Vlase L, Crăciunaş C. Clonal propagation and production of cichoric acid in three species of Echinaceae. In Vitro Cell Dev Biol Plant 2012; 48(2): 249-58.
[http://dx.doi.org/10.1007/s11627-012-9435-2]

[159] Gerszberg A, Hnatuszko-Konka K, Kowalczyk T. *In vitro* regeneration of eight cultivars of *Brassica oleracea* var. *capitata*. In Vitro Cell Dev Biol Plant 2015; 51(1): 80-7.
[http://dx.doi.org/10.1007/s11627-014-9648-7] [PMID: 25774081]

[160] Gupta SK, Kuo CL, Chang HC, *et al*. In vitropropagation and approaches for metabolites production in medicinal plants. Adv Bot Res 2012; 62: 35-55.
[http://dx.doi.org/10.1016/B978-0-12-394591-4.00002-7]

[161] Patil KS, Powar PV. Antibacterial and antifungal activities in endophytic fungi isolated from *Celastruspaniculatus*. INDIAN DRUGS 2012; 49(8): 21-6.
[http://dx.doi.org/10.53879/id.49.08.p0021]

[162] Jain SC, Pancholi B, Jain R. Assessment of secondary metabolites and bio efficacies of *Peltophorumpterocarpum* (DC.) Baker ex K. Heyne cell cultures. Med Plant 2011; 3(4): 265-71.

[163] Piątczak E, Królicka A, Wysokińska H. Morphology, secoiridoid content and RAPD analysis of plants regenerated from callus of centaurium erythraea RAFN. Acta Biol Cracov Ser; Bot 2011; 53(2): 79-86.

[http://dx.doi.org/10.2478/v10182-011-0030-3]

[164]   Negi RS. Fast *in vitro* callus induction in *Catharanthus roseus* -A medicinally important plant used in cancer therapy. Res J Pharm Biol Chem Sci 2011; 2(4): 597-603.

[165]   Bekheet SA. *In vitro* biomass production of liver-protective compounds from globe artichoke (*Cynara scolymus* L.) and milk thistle (*Silybum marianum*) plants. Emir J Food Agric 2011; 23(5): 473-81.

[166]   El-Beltagi HS, Ahmed OK, El-Desouky W. Effect of low doses γ-irradiation on oxidative stress and secondary metabolites production of rosemary (*Rosmarinus officinalis* L.) callus culture. Radiat Phys Chem 2011; 80(9): 968-76.
[http://dx.doi.org/10.1016/j.radphyschem.2011.05.002]

[167]   Zare K, Khosrowshahli M, Nazemiyeh H, Movafeghi A, Azar AM, Omidi Y. Callus culture of *Echium italicum* L. towards production of a shikonin derivative. Nat Prod Res 2011; 25(16): 1480-7.
[http://dx.doi.org/10.1080/14786410902804857] [PMID: 20635302]

[168]   Reis RV, Borges APPL, Chierrito TPC, *et al.* Establishment of adventitious root culture of *Stevia rebaudiana* Bertoni in a roller bottle system. Plant Cell Tissue Organ Cult 2011; 106(2): 329-35.
[http://dx.doi.org/10.1007/s11240-011-9925-7]

[169]   Fariz A, Arvind B, Chan LK, Gunawan I, Shaida FS. Effect of yeast extract and chitosan on shoot proliferation, morphology and antioxidant activity of *Curcuma mangga in vitro* plantlets. Afr J Biotechnol 2011; 10(40): 7787-95.
[http://dx.doi.org/10.5897/AJB10.1261]

[170]   Shekhawat MS, Shekhawat NS. Micropropagation of Arnebia hispidissima (Lehm). DC. and production of alkannin from callus and cell suspension culture. Acta Physiol Plant 2011; 33(4): 1445-50.
[http://dx.doi.org/10.1007/s11738-010-0680-x]

[171]   Gautam YK, Rani R, Vimala Y. *In vitro* accumulation of active metabolites in *Physalis peruviana* (Linn.) callus. Vegetos 2011; 24(1): 58-60.

[172]   Lee Y, Lee DE, Lee HS, *et al.* Influence of auxins, cytokinins, and nitrogen on production of rutin from callus and adventitious roots of the white mulberry tree (Morus alba L.). Plant Cell Tissue Organ Cult 2011; 105(1): 9-19.
[http://dx.doi.org/10.1007/s11240-010-9832-3]

[173]   Pise M, Rudra J, Bundale S, Begde D, Nashikkar N, Upadhyay A. Shatavarin production from *in vitro* cultures of *Asparagus racemosus* Wild. J Med Plants Res 2011; 5(4): 507-13.

[174]   Krzyzanowska J, Janda B, Pecio L, *et al.* Determination of polyphenols in *Mentha longifolia* and *M. piperita* field-grown and *in vitro* plant samples using UPLC-TQ-MS. J AOAC Int 2011; 94(1): 43-50.
[http://dx.doi.org/10.1093/jaoac/94.1.43] [PMID: 21391480]

[175]   Crockett SL, Poller B, Tabanca N, *et al.* Bioactive xanthones from the roots of *Hypericum perforatum* (common St John's wort). J Sci Food Agric 2011; 91(3): 428-34.
[http://dx.doi.org/10.1002/jsfa.4202] [PMID: 21218475]

[176]   Basavaraju R. Plant tissue culture-Agriculture and health of man. Indian J Sci Technol 2011; 4(3): 333-5.
[http://dx.doi.org/10.17485/ijst/2011/v4i3.34]

[177]   De Dios-López A, Montalvo-González E, Andrade-González I, Gómez-Leyva JF. Induction of anthocyanins and phenolic compounds in cell cultures of roselle (*hibiscus sabdariffa* L.) *in vitro*. Rev Chapingo Ser Hortic 2011; XVII(2): 77-87.
[http://dx.doi.org/10.5154/r.rchsh.2011.17.011]

[178]   Rahiman FA, Taha RM. Plant regeneration and induction of coloured callus from henna (*Lawsoniainermis* syn. *Lawsonia alba*). J Food Agric Environ 2011; 9(2): 397-9.

[179]   Ćalić-Dragosavac D, Stević S, Zdravković-Korać S, Milojević J, Cingel A, Vinterhalter B.

Secondary metabolite of horse chestnut *in vitro* culture. Adv Environ Biol 2011; 5(2 SPEC. ISSUE): 267-70.

[180] Singh O, Khanam Z, Misra N, Srivastava M. Chamomile (Matricaria chamomilla L.): An overview. Pharmacogn Rev 2011; 5(9): 82-95.
[http://dx.doi.org/10.4103/0973-7847.79103] [PMID: 22096322]

[181] Torres KC. Overview of cell suspension culture.Tissue Culture Techniques for Horticultural Crops. Boston, MA: Springer 1989; pp. 151-60.
[http://dx.doi.org/10.1007/978-1-4615-9756-8_18]

[182] Dodds J, H, Roberts L,W. Experiments in plant tissue culture. London: Cambridge University press 1986.

[183] Ahmad N, Rab A, Ahmad N. Light-induced biochemical variations in secondary metabolite production and antioxidant activity in callus cultures of Stevia rebaudiana (Bert). J Photochem Photobiol B 2016; 154: 51-6.
[http://dx.doi.org/10.1016/j.jphotobiol.2015.11.015] [PMID: 26688290]

[184] Taghizadeh M, Nasibi F, Kalantari KM, Ghanati F. Evaluation of secondary metabolites and antioxidant activity in *Dracocephalum polychaetum* Bornm. cell suspension culture under magnetite nanoparticles and static magnetic field elicitation. Plant Cell Tissue Organ Cult 2019; 136(3): 489-98.
[http://dx.doi.org/10.1007/s11240-018-01530-1]

[185] Taherkhani T, Asghari Zakaria R, Omidi M, Zare N. Effect of ultrasonic waves on crocin and safranal content and expression of their controlling genes in suspension culture of saffron (*Crocus sativus* L.). Nat Prod Res 2019; 33(4): 486-93.
[http://dx.doi.org/10.1080/14786419.2017.1396598] [PMID: 29124962]

[186] Salehi M, Karimzadeh G, Naghavi MR. Synergistic effect of coronatine and sorbitol on artemisinin production in cell suspension culture of *Artemisia annua* L. cv. Anamed. Plant Cell Tissue Organ Cult 2019; 137(3): 587-97.
[http://dx.doi.org/10.1007/s11240-019-01593-8]

[187] Amos S, R, Premnath D, Paul D, Raj R, S. Strategy for early callus induction and identification of anti-snake venom triterpenoids from plant extracts and suspension culture of Euphorbia hirta L. 3 Biotech 2019; 9(7): 266.

[188] Shasmita R, Rai MK, Naik SK. Exploring plant tissue culture in *Withania somnifera* (L.) Dunal: *in vitro* propagation and secondary metabolite production. Crit Rev Biotechnol 2018; 38(6): 836-50.
[http://dx.doi.org/10.1080/07388551.2017.1416453] [PMID: 29278928]

[189] Kašparová M, Martin J, Tůmová L, Spilková J. Production of podophyllotoxin by plant tissue cultures of *Juniperusvirginiana*. Nat Prod Commun 2017; 12(1): 1934578X1701200.
[http://dx.doi.org/10.1177/1934578X1701200129] [PMID: 30549838]

[190] Andre CM, Hausman JF, Guerriero G. *Cannabis sativa*: The plant of the thousand and one molecules. Front Plant Sci 2016; 7: 19.
[http://dx.doi.org/10.3389/fpls.2016.00019] [PMID: 26870049]

[191] Billore V, Khatediya L, Jain M. Source system of *in vitro* suspension culture of *Celastruspaniculatus* under regulation of monochromatic lights. Plant Tissue Cult Biotechnol 2016; 26(2): 175-85.
[http://dx.doi.org/10.3329/ptcb.v26i2.30567]

[192] Mahjouri S, Movafeghi A, Zare K, Kosari-Nasab M, Nazemiyeh H. Production of naphthoquinone derivatives using two-liquid-phase suspension cultures of *Alkanna orientalis*. Plant Cell Tissue Organ Cult 2016; 124(1): 201-7.
[http://dx.doi.org/10.1007/s11240-015-0877-1]

[193] Dwivedi S, Alam A, Shekhawat GS. Relative production and quantification of stevioside from *In-vitro* generated shoots, callus, suspension culture and synseeds of *Stevia rebaudiana* (Bertoni). Plant Cell Biotechnol Mol Biol 2016; 17(3-4): 155-66.

[194] Farag MA, El Sayed AM, El Banna A, Ruehmann S. Metabolomics reveals distinct methylation reaction in MeJA elicited *Nigella sativa* callus *via* UPLC–MS and chemometrics. Plant Cell Tissue Organ Cult 2015; 122(2): 453-63.
[http://dx.doi.org/10.1007/s11240-015-0782-7]

[195] Post J, Eisenreich W, Huber C, Twyman RM, Prüfer D, Schulze Gronover C. Establishment of an *ex vivo* laticifer cell suspension culture from *Taraxacum brevicorniculatum* as a production system for cis-isoprene. J Mol Catal, B Enzym 2014; 103: 85-93.
[http://dx.doi.org/10.1016/j.molcatb.2013.07.013]

[196] Jevremović S, Jeknić Z, Subotić A. Micropropagation of *Iris* sp. Methods Mol Biol 2013; 11013: 291-303.
[PMID: 23179708]

[197] Siatka T, Kašparová M, Spilková J. Effects of zinc and cadmium ions on cell growth and production of coumarins in cell suspension cultures of *Angelica archangelica* L. Ceska Slov Farm 2012; 61(6): 261-6.
[PMID: 23387854]

[198] Rahman MA, Bari MA. Callus induction and cell culture of castor (Ricinus communis L. CV. Shabje). J Biosci 2014; 20: 161-9.
[http://dx.doi.org/10.3329/jbs.v20i0.17738]

[199] Collakova E, Yen JY, Senger RS. Are we ready for genome-scale modeling in plants? Plant Sci 2012; 191-192: 53-70.
[http://dx.doi.org/10.1016/j.plantsci.2012.04.010] [PMID: 22682565]

[200] Xu Y, Xu TF, Zhao XC, *et al.* Co-expression of VpROMT gene from Chinese wild *Vitis pseudoreticulata* with VpSTS in tobacco plants and its effects on the accumulation of pterostilbene. Protoplasma 2012; 249(3): 819-33.
[http://dx.doi.org/10.1007/s00709-011-0335-9] [PMID: 22038118]

[201] Zhu JH, Wen W, Hu YS, Tang Y, Yu RM. Hydroxyl octadecenoic acids biosynthesized by crown galls of *Panax quinquefolium* induced by artermisinic acid. Zhong Yao Cai 2012; 35(6): 869-72.
[PMID: 23236817]

[202] Staszków A, Swarcewicz B, Banasiak J, Muth D, Jasiński M, Stobiecki M. LC/MS profiling of flavonoid glycoconjugates isolated from hairy roots, suspension root cell cultures and seedling roots of Medicago truncatula. Metabolomics 2011; 7(4): 604-13.
[http://dx.doi.org/10.1007/s11306-011-0287-2] [PMID: 22039365]

[203] Vijendra PD, Jayanna SG, Kumar V, Sannabommaji T, J R, Gajula H. Product enhancement of triterpenoid saponins in cell suspension cultures of Leucas aspera Spreng. Ind Crops Prod 2020; 156(15): 112857.
[http://dx.doi.org/10.1016/j.indcrop.2020.112857]

[204] Zhang X, Deng M. Hyper-production of 13C-labeled trans-resveratrol in Vitis vinifera suspension cell culture by elicitation and *in situ* adsorption. Biochem Eng J 2011; 53(3): 292-6.

[205] Adams AN. An improved medium for strawberry meristem culture. J Hortic Sci 1972; 47(2): 263-4.
[http://dx.doi.org/10.1080/00221589.1972.11514466]

[206] Razdan MK. Introduction to Plant Tissue Culture. Enfield, Plymouth: Science Publishers, Inc. 2003.

[207] Dodds JH, Roberts LW. Experiments in Plant Tissue Culture. Cambridge, UK: Cambridge University Press 1995.

[208] Gosal SS, Wani SH. Cell and tissue culture approaches in relation to crop improvement. Biotechnol Crop Improv 2018; 1: 1-55.
[http://dx.doi.org/10.1007/978-3-319-78283-6_1]

[209] Jain S, Jenks M, Rout MA, Radejovigl GR. Micropropagation of ornamental potted plants. Propag

Ornam Plants 2006; 6: 67-82.

[210] Saha PS, Sarkar S, Jeyasri R, Muthuramalingam P, Ramesh M, Jha S. *In Vitro* Propagation, phytochemical and neuropharmacological profiles of *Bacopa monnieri* (L.) Wettst.: A Review. Plants 2020; 9(4): 411.
[http://dx.doi.org/10.3390/plants9040411] [PMID: 32224997]

[211] Trujillo-Chacón LM, Pastene-Navarrete ER, Bustamante L, Baeza M, Alarcón-Enos JE, Cespedes-Acuña CL. *In vitro* micropropagation and alkaloids analysis by GC–MS of Chilean Amaryllidaceae plants: *Rhodophiala pratensis*. Phytochem Anal 2020; 31(1): 46-56.
[http://dx.doi.org/10.1002/pca.2865] [PMID: 31304645]

[212] Bakhshipour M, Mafakheri M, Kordrostami M, *et al. In vitro* multiplication, genetic fidelity and phytochemical potentials of *Vaccinium arctostaphylos* L.: An endangered medicinal plant. Ind Crops Prod 2019; 141: 111812.
[http://dx.doi.org/10.1016/j.indcrop.2019.111812]

[213] Singh N, Kumaria S. combinational phytomolecular-mediated assessment in micropropagated plantlets of CoelogyneovalisLindl.: A horticultural and medicinal orchid. Proceedings of the National Academy of Sciences India Section B - Biological Sciences 2019.

[214] Yarra R, Mushke R, Velmala M. *In vitro* approaches for conservation and sustainable Utilization of Butea monosperma (Lam.) Taub. Var. Lutea (Witt.) Maheshwari: A Highly Valuable Medicinal Plant Biotechnological Approaches for Medicinal and Aromatic Plants. Conservation, Genetic Improvement and Utilization 2018; 345-52.

[215] Tikhomirova LI, Bazarnova NG, Ilicheva TN, Martirosian IuTs, Afanasenkova IV. Obtaining plant materials *Siberian iris (Irissibirica* L.) by methods of biotechnology. KhimiyaRastitel'nogoSyr'ya 2018; 21(4): 235-45.

[216] Trettel JR, Gazim ZC, Gonçalves JE, Stracieri J, Magalhães HM. Volatile essential oil chemical composition of basil (*Ocimum basilicum* L. 'Green') cultivated in a greenhouse and micropropagated on a culture medium containing copper sulfate. In Vitro Cell Dev Biol Plant 2017; 53(6): 631-40.
[http://dx.doi.org/10.1007/s11627-017-9868-8]

[217] Hayta S, Bayraktar M. Direct plant regeneration from different explants through micro-propagation and determination of secondary metabolites in the critically endangered endemic *Rhaponticoidesmykalea.* Plant Biosyst 2017; 151(1): 20-8.

[218] Sharma M, Kumari A, Mahant E. Micropropogation and phytochemical profile analysis of tissue culture grown *Plantago ovata*Forsk. Asian J Pharm Clin Res 2017; 10(4): 202-6.
[http://dx.doi.org/10.22159/ajpcr.2017.v10i4.16532]

[219] Wesołowska A, Grzeszczuk M, Wilas J, Kulpa D. Gas Chromatography-Mass Spectrometry (GC-MS) analysis of indole alkaloids isolated from *Catharanthusroseus*(L.) G. don cultivated conventionally and derived from *in vitro* cultures. Not Bot Horti Agrobot Cluj-Napoca 2016; 44(1): 100-6.
[http://dx.doi.org/10.15835/nbha44110127]

[220] Goyali JC, Igamberdiev AU, Debnath SC. Micro-propagation affects not only the fruit morphology of Lowbush blueberry (*Vacciniumangustifolium*Ait.) but also its medicinal properties. Acta Hortic 2015; (1098): 137-42.
[http://dx.doi.org/10.17660/ActaHortic.2015.1098.14]

[221] Rybczyński JJ, Davey MR, Mikuła A. The gentianaceae volume 2: Biotechnology and applications.The Gentianaceae - Volume 2: Biotechnology and Applications. FisicalBook 2015; pp. 1-452.

[222] Chen CC, Chang HC, Kuo CL, Agrawal DC, Wu CR, Tsay HS. *In vitro* propagation and analysis of secondary metabolites in *Glossogyne tenuifolia* (Hsiang-Ju) - a medicinal plant native to Taiwan. Bot Stud 2014; 55(1): 45.
[http://dx.doi.org/10.1186/s40529-014-0045-7] [PMID: 28510942]

[223] Da Silva JAT. Orchids: Advances in tissue culture, genetics, phytochemistry and transgenic biotechnology. Floric Ornam Biotechnol 2013; 7(1): 1-52.

[224] Koca U, Çölgeçen H, Reheman N. Progress in biotechnological applications of diverse species in Boraginaceaejuss. Biotechnological Production of Plant Secondary Metabolites 2012; pp. 200-14.

[225] Wijaya BK, Hardjo PH, Emantoko S. Menthol from the stem and leaf *in-vitro Mentha piperita* Linn. IOP Conf Ser Earth Environ Sci 2019; 293(1): 012009.
[http://dx.doi.org/10.1088/1755-1315/293/1/012009]

[226] Tasheva K, Kosturkova G. Establishment of callus cultures of *Rhodiolarosea* Bulgarian ecotype. Acta Hortic 2012; (955): 129-35.
[http://dx.doi.org/10.17660/ActaHortic.2012.955.17]

[227] Dixon RA. Engineering of plant natural product pathways. Curr Opin Plant Biol 2005; 8(3): 329-36.
[http://dx.doi.org/10.1016/j.pbi.2005.03.008] [PMID: 15860431]

[228] Zenk MH. The impact of plant cell cultures on industry.Frontiers of plant tissue culture. Calgary: The International Association of Plant Tissue Culture 1978; pp. 1-14.

[229] Johnson TS, Ravishankar GA. Precursor biotransformation in immobilizedplacental tissues of *Capsicum frutescens* Mill: I. Influence of feeding intermediatemetabolites of the capsaicinoid pathway on capsaicin and dihydrocapsaicin accumulation. J Plant Physiol 1996; 147: 481-5.

[230] Vijaya SN, Udayasri PVV, Kumar YA, Babu BR, Kumar YP, Varma VM. Advancements in the production of secondary metabolites. J Nat Prod 2010; 3: 112-23.

[231] Gonçalves S, Romano A. Production of plant secondary metabolites by usingbiotechnologicaltools.SecondaryMetabolites-Sources and Applications. Intech Open 2018.
[http://dx.doi.org/10.5772/intechopen.76414]

[232] Isah T, Umar S, Mujib A, *et al.* Secondary metabolism of pharmaceuticals in the plant *in vitro* cultures: strategies, approaches, and limitations to achieving higher yield. Plant Cell Tissue Organ Cult 2018; 132(2): 239-65.
[http://dx.doi.org/10.1007/s11240-017-1332-2]

[233] Ochoa-Villarreal M, Howat S, Hong S, *et al.* Plant cell culture strategies for the production of natural products. BMB Rep 2016; 49(3): 149-58.
[http://dx.doi.org/10.5483/BMBRep.2016.49.3.264] [PMID: 26698871]

[234] Radman R, Bucke C, Keshavarz T. Elicitor effects on *Penicillium chrysogenum* morphology in submerged cultures. Biotechnol Appl Biochem 2004; 40(Pt 3): 229-33.
[PMID: 15134576]

[235] Baenas N, García-Viguera C, Moreno D. Elicitation: A tool for enriching the bioactive composition of foods. Molecules 2014; 19(9): 13541-63.
[http://dx.doi.org/10.3390/molecules190913541] [PMID: 25255755]

[236] Gorelick J, Bernstein N. Elicitation: An underutilized tool in the developmentofmedicinal plants as a source of therapeutic secondary metabolites.Advances in Agronomy. Amsterdam. TheNetherlands: Elsevier 2014; 124: pp. 201-30.

[237] Ramirez-Estrada K, Vidal-Limon H, Hidalgo D, *et al.* Elicitation, an effective strategy for the biotechnologicalproduction of bioactive high-added value compounds inplant cell factories. Molecules 2016; 21(2): 182.
[http://dx.doi.org/10.3390/molecules21020182] [PMID: 26848649]

[238] Akula R, Ravishankar GA. Influence of abiotic stress signals on secondary metabolites in plants. Plant Signal Behav 2011; 6(11): 1720-31.
[http://dx.doi.org/10.4161/psb.6.11.17613] [PMID: 22041989]

[239] Nascimento NC, Fett-Neto AG. Plant secondary metabolism and challenges in modifying its operation: an overview. Methods Mol Biol 2010; 643: 1-13.

[http://dx.doi.org/10.1007/978-1-60761-723-5_1] [PMID: 20552440]

[240] Van Der Fits L, Memelink J. ORCA3, a jasmonate-responsive transcriptional regulator of plant primary and secondary metabolism. Science 2000; 289(5477): 295-7.
[http://dx.doi.org/10.1126/science.289.5477.295] [PMID: 10894776]

[241] Walker T, Pal Bais H, Vivanco JM. Jasmonic acid-induced hypericin production in cell suspension cultures of Hypericum perforatum L. (St. John's wort). Phytochemistry 2002; 60(3): 289-93.
[http://dx.doi.org/10.1016/S0031-9422(02)00074-2] [PMID: 12031448]

[242] Kang SM, Min JY, Kim YD, et al. Effects of methyl jasmonate and salicylic acid on the production of bilobalide and ginkgolides in cell cultures of *Ginkgo biloba*. *In Vitro*. Cell Dev Biol Plant 2006; 42(1): 44-9.
[http://dx.doi.org/10.1079/IVP2005719]

[243] Siddiqui ZH, Mujib A, Aslam MJ, Hakeem KR, Parween T. *In vitro* production of secondary metabolites using elicitor in *Catharanthus roseus*:A case study. In: Ahmad P, Ozturk M, Eds. Hakeem,K, R. Berlin, Germany: CropImprovement,New Approaches and Modern Techniques., Springer ScienceandBusiness Media 2013; pp. 401-41.

[244] Giri CC, Zaheer M. Chemical elicitors versus secondary metabolite production *in vitro* using plant cell, tissue and organ cultures: Recent trends and a sky eye view appraisal. Plant Cell Tissue Organ Cult 2016; 126(1): 1-18.
[http://dx.doi.org/10.1007/s11240-016-0985-6]

[245] Siddiqui MS, Thodey K, Trenchard I, Smolke CD. Advancing secondary metabolite biosynthesis in yeast with synthetic biology tools. FEMS Yeast Res 2012; 12(2): 144-70.
[http://dx.doi.org/10.1111/j.1567-1364.2011.00774.x] [PMID: 22136110]

[246] Singh A, Dwivedi P. Methyl-jasmonate and salicylic acid as potent elicitorsforsecondary metabolite production in medicinal plants: A review. J Pharmacogn Phytochem 2018; 7: 750-7.

[247] Takeuchi C, Nagatani K, Sato Y. Chitosan and a fungal elicitor inhibit tracheary element differentiation and promote accumulation of stress lignin-like substance in *Zinnia elegans* xylogenic culture. J Plant Res 2013; 126(6): 811-21.
[http://dx.doi.org/10.1007/s10265-013-0568-0] [PMID: 23732634]

[248] Namdeo A, Patil S, Fulzele DP. Influence of fungal elicitors on production of ajmalicine by cell cultures of Catharanthus roseus. Biotechnol Prog 2002; 18(1): 159-62.
[http://dx.doi.org/10.1021/bp0101280] [PMID: 11822914]

[249] Liang C, Chen C, Zhou P, et al. Effect of *Aspergillus flavus*fungal elicitor on the production of terpenoid indole alkaloids in *Catharanthus roseus* cambial meristematic cells. Molecules 2018; 23(12): 3276.
[http://dx.doi.org/10.3390/molecules23123276] [PMID: 30544939]

[250] Han X, Xu X, Cui C, Gu Q. Alkaloidal compounds produced by a marine derived fungus, *Aspergillus fumigatus* H1-04, and their antitumor activities. Zhongguo Yaowu Huaxue Zazhi 2007; 17: 232-7.

[251] Zhang B, Zheng LP, Yi Li W, Wen Wang J. Stimulation of artemisinin productionin*Artemisia annua* hairy roots by Ag-SiOcore-shell nanoparticles. Curr Nanosci 2013; 9(3): 363-70.
[http://dx.doi.org/10.2174/1573413711309030012]

[252] Chodisetti B, Rao K, Gandi S, Giri A. Improved gymnemic acid production in the suspension cultures of Gymnema sylvestre through biotic elicitation. Plant Biotechnol Rep 2013; 7(4): 519-25.
[http://dx.doi.org/10.1007/s11816-013-0290-3]

[253] Jung H, Y, Kang S,M, Kang Y,M, et al. Enhanced production of scopolamine by bacterial elicitors in adventitioushairy root cultures of Scopolia parviflora. Enzyme Microb Technol ;2003 :33 987-90.

[254] Kang SM, Min JY, Kim YD, et al. Effect of biotic elicitors on the accumulation of bilobalide and ginkgolides in *Ginkgo biloba* cell cultures. J Biotechnol 2009; 139(1): 84-8.
[http://dx.doi.org/10.1016/j.jbiotec.2008.09.007] [PMID: 18983879]

[255] Mañero FJG, Algar E, Martín Gómez MS, Saco Sierra MD, Solano BR. Elicitation of secondary metabolism in *Hypericum perforatum* by rhizosphere bacteria and derived elicitors in seedlings and shoot cultures. Pharm Biol 2012; 50(10): 1201-9.
[http://dx.doi.org/10.3109/13880209.2012.664150] [PMID: 22900596]

[256] Onrubia M, Moyano E, Bonfill M, Cusidó RM, Goossens A, Palazón J. Coronatine, a more powerful elicitor for inducing taxane biosynthesis in *Taxus* media cell cultures than methyl jasmonate. J Plant Physiol 2013; 170(2): 211-9.
[http://dx.doi.org/10.1016/j.jplph.2012.09.004] [PMID: 23102875]

[257] Taurino M, Ingrosso I, D'amico L, *et al*. Jasmonates elicit different sets of stilbenes in *Vitis vinifera* cv. *Negramaro* cell cultures. Springerplus 2015; 4(1): 49.
[http://dx.doi.org/10.1186/s40064-015-0831-z] [PMID: 25674504]

[258] Vithal A, Aniket K, Abhay H. Microbial elicitation in root cultures of Tavernieracuneifolia (Roth) Arn. for elevated glycyrrhizic acid production. Ind Crops Prod 2014; 54: 13-6.

[259] Maneechai S, De-Eknamkul W, Umehara K, Noguchi H, Likhitwitayawuid K. Flavonoid and stilbenoid production in callus cultures of *Artocarpus lakoocha*. Phytochemistry 2012; 81: 42-9.
[http://dx.doi.org/10.1016/j.phytochem.2012.05.031] [PMID: 22769436]

[260] Ahmad S, Garg M, Tamboli ET, Kamal YT, Ansari SH. Quantification of vasaka alkaloids in *in vitro* cultures and in natural leaves from Indian subcontinents by reversed phase- high performance liquid chromatography. Drug Development and Therapeutics 2016; 7(1): 51-4.
[http://dx.doi.org/10.4103/2394-6555.180165]

[261] Janarthanam B, Gopalakrishnan M, Sekar T. Secondary metabolite production in callus cultures of *Stevia rebaudiana*Bertoni. Bangladesh J Sci Ind Res 1970; 45(3): 243-8.
[http://dx.doi.org/10.3329/bjsir.v45i3.6532]

[262] Grąbkowska R, Matkowski A, Grzegorczyk-Karolak I, Wysokińska H. Callus cultures of *Harpagophytum procumbens* (Burch.) DC. ex Meisn.; production of secondary metabolites and antioxidant activity. S Afr J Bot 2016; 103: 41-8.
[http://dx.doi.org/10.1016/j.sajb.2015.08.012]

[263] Abed A. *In vitro* influences of sugar nutrition and light condition on accumulation of some phytochemicals in *Verbascum thapsus* L. Culture. Plant Arch 2020; 20(1): 1126-30.

[264] Ahmad M. Engineered ZnO and CuO nanoparticles ameliorate morphological and biochemical response in tissue culture regenerants of candy leaf (Stevia rebaudiana). Molecules 2020; 25(6): 25061356.

[265] Skrzypczak-Pietraszek E, Urbańska A, Żmudzki P, Pietraszek J. Elicitation with methyl jasmonate combined with cultivation in the Plantform™ temporary immersion bioreactor highly increases the accumulation of selected centellosides and phenolics in *Centella asiatica* (L.) Urban shoot culture. Eng Life Sci 2019; 19(12): 931-43.
[http://dx.doi.org/10.1002/elsc.201900051] [PMID: 32624983]

[266] Ghasempour M, Iranbakhsh A, Ebadi M. Multi-walled carbon nanotubes improved growth, anatomy, physiology, secondary metabolism, and callus performance in Catharanthus roseus: An *in vitro* study. 3 Biotech 2019; 9(11): 404.

[267] Ali M, Mujib A, Gulzar B, Zafar N. Essential oil yield estimation by Gas chromatography–mass spectrometry (GC-MS) after Methyl jasmonate (MeJA) elicitation in *in vitro* cultivated tissues of Coriandrum sativum L. 3 Biotech 2019; 9(11): 414.

[268] Nasution NH, Nasution IW. The effect of plant growth regulators on callus induction of Mangosteen (*Garcinia mangostana* L.). IOP Conf Ser Earth Environ Sci 2019; 305(1): 012049.
[http://dx.doi.org/10.1088/1755-1315/305/1/012049]

[269] Owis A, Abdelwahab N, Abul-Soad A. Analysis of phenolics in *Calligonum polygonoides in vitro* cultured roots. JRep Pharmac Sci 2019; 8(2): 124-7.

[http://dx.doi.org/10.4103/jrptps.JRPTPS_62_18]

[270] Shakya P, Marslin G, Siram K, Beerhues L, Franklin G. Elicitation as a tool to improve the profiles of high-value secondary metabolites and pharmacological properties of *Hypericum perforatum*. J Pharm Pharmacol 2018; 71(1): 70-82.
[http://dx.doi.org/10.1111/jphp.12743] [PMID: 28523644]

[271] Javed R, Gürel E. Salt stress by NaCl alters the physiology and biochemistry of tissue culture-grown *Stevia rebaudiana* Bertoni. Turk J Agric For 2019; 43(1): 11-20.
[http://dx.doi.org/10.3906/tar-1711-71]

[272] Khezerluo M, Hosseini B, Amiri J. Sodium nitroprusside stimulated production of tropane alkaloids and antioxidant enzymes activity in hairy root culture of *Hyoscyamus reticulatus* L. Acta Biol Hung 2018; 69(4): 437-48.
[http://dx.doi.org/10.1556/018.69.2018.4.6] [PMID: 30587015]

[273] Esmaeilzadeh BS, Rezaei A. Nitric oxide increased the rosmarinic acid and essential oil production in *in vitro*-cultured *Melissa officinalis*. Faslnamah-i Giyahan-i Daruyi 2018; 17(65): 61-72.

[274] Govindaraju S, Indra Arulselvi P. Effect of cytokinin combined elicitors (l-phenylalanine, salicylic acid and chitosan) on in vitro propagation, secondary metabolites and molecular characterization of medicinal herb – Coleus aromaticus Benth (L). J Saudi Soc Agric Sci 2018; 17(4): 435-44.
[http://dx.doi.org/10.1016/j.jssas.2016.11.001]

[275] Kavianifar S, Ghodrati K, Badi NH, Etminan A. Effects of nano elicitors on callus induction and mucilage production in tissue culture of *Linumusitatissimum* L. Faslnamah-i Giyahan-i Daruyi 2018; 17(67): 45-54.

[276] Andrys D, Adaszyńska-Skwirzyńska M, Kulpa D. Jasmonic acid changes the composition of essential oil isolated from narrow-leaved lavender propagated in *in vitro* cultures. Nat Prod Res 2018; 32(7): 834-9.
[http://dx.doi.org/10.1080/14786419.2017.1309533] [PMID: 28421828]

[277] Petrova N, Koleva P, Velikova V, *et al*. Relations between photosynthetic performance and polyphenolics productivity of *Artemisia alba* Turra in *in vitro* tissue cultures. Int J Bioautomation 2018; 22(1): 73-82.
[http://dx.doi.org/10.7546/ijba.2018.22.1.73-82]

[278] Maqsood M, Abdul M. Yeast extract elicitation increases vinblastine and vincristine yield in protoplast derived tissues and plantlets in *Catharanthus roseus*. Rev Bras Farmacogn 2017; 27(5): 549-56.
[http://dx.doi.org/10.1016/j.bjp.2017.05.008]

[279] Paz TA, Dos Santos VAFFM, Inácio MC, *et al*. Proteome profiling reveals insights into secondary metabolism in *Maytenus ilicifolia* (Celastraceae) cell cultures producing quinonemethide triterpenes. Plant Cell Tissue Organ Cult 2017; 130(2): 405-16.
[http://dx.doi.org/10.1007/s11240-017-1236-1]

[280] Zafar N, Mujib A, Ali M, Tonk D, Gulzar B. Aluminum chloride elicitation (amendment) improves callus biomass growth and reserpine yield in *Rauvolfia serpentina* leaf callus. Plant Cell Tissue Organ Cult 2017; 130(2): 357-68.
[http://dx.doi.org/10.1007/s11240-017-1230-7]

[281] Allahverdi-Mamaghani B, Movafeghi A, Hejazi SMH, Mirza M. The effect of plant growth regulatores and methyl jasmonate on flavonoid production and antioxidant activity in callus and suspension cultures of two *dracocephalum* species. Plant Cell Biotechnol Mol Biol 2017; 18(3-4): 131-44.

[282] Zhang M, Wang S, Yin J, *et al*. Molecular cloning and promoter analysis of squalene synthase and squalene epoxidase genes from Betula platyphylla. Protoplasma 2016; 253(5): 1347-763.

[283] Alonso-Herrada J, Rico-Reséndiz F, Campos-Guillén J, Guevara-González RG, Torres-Pacheco I, Cruz-Hernández A. Establishment of *in vitro* regeneration system for *Acaciella angustissima* (Timbe)

a shrubby plant endemic of México for the production of phenolic compounds. Ind Crops Prod 2016; 86: 49-57.
[http://dx.doi.org/10.1016/j.indcrop.2016.03.040]

[284] Zamir R, Khalil SA, Ahmad N, *et al.* The synergistic effects of sucrose and plant growth regulators on morphogenesis and evaluation of antioxidant activities in regenerated tissues of Caralluma tuberculata. Acta Physiol Plant 2016; 38(8): 200.
[http://dx.doi.org/10.1007/s11738-016-2218-3]

[285] Chaturvedi P, Briganza V. Enhanced synthesis of curculigoside by stress and amino acids in static culture of Curculigo orchioides gaertn (kali musli). Pharmacognosy Res 2016; 8(3): 193-8.
[http://dx.doi.org/10.4103/0974-8490.182915] [PMID: 27365988]

[286] Ashouri SA, Hassanpour H, Jonoubi P, Ghorbani NM, Nadimifar MS. The effect of gamma irradiation on *in vitro* total phenolic content and antioxidant activity of *Ferulagummosa*Bioss. Faslnamah-i Giyahan-i Daruyi 2016; 15(59): 122-31.

[287] Malik S, Bhushan S, Sharma M, Ahuja PS. Biotechnological approaches to the production of shikonins: A critical review with recent updates. Crit Rev Biotechnol 2016; 36(2): 327-40.
[http://dx.doi.org/10.3109/07388551.2014.961003] [PMID: 25319455]

[288] Radić S, Vujčić V, Glogoški M, Radić-Stojković M. Influence of pH and plant growth regulators on secondary metabolite production and antioxidant activity of Stevia rebaudiana (Bert). Period Biol 2016; 118(1): 9-19.
[http://dx.doi.org/10.18054/pb.2016.118.1.3420]

[289] Vijayalakshmi U, Shourie A. Cinnamic acid supplementation regulates the production of licochalcone A, liquirtigenin and licoisoflavone B in *Glycyrrhiza glabra* callus cultures. Int J Phytomed 2016; 8(3): 343-52.
[http://dx.doi.org/10.5138/09750185.1859]

[290] Skała E, Grąbkowska R, Sitarek P, Kuźma Ł, Błauż A, Wysokińska H. Rhaponticum carthamoides regeneration through direct and indirect organogenesis, molecular profiles and secondary metabolite production. Plant Cell Tissue Organ Cult 2015; 123(1): 83-98.
[http://dx.doi.org/10.1007/s11240-015-0816-1]

[291] Wee SL, Yap WSP, Alderson PG, Khoo TJ. Effects of elicitors on *in vitro* cultures of *Sauropus androgynus* (Sweet shoot) for sustainable metabolite production and antioxidant capacity improvement. Acta Hortic 2015; (1083): 145-55.
[http://dx.doi.org/10.17660/ActaHortic.2015.1083.16]

[292] Sharma M, Ahuja A, Gupta R, Mallubhotla S. Enhanced bacoside production in shoot cultures of *Bacopa monnieri* under the influence of abiotic elicitors. Nat Prod Res 2015; 29(8): 745-9.
[http://dx.doi.org/10.1080/14786419.2014.986657] [PMID: 25485652]

[293] Darwesh H. *In vitro* investigation for improving secondary metabolites in *Origanum vulgare* plants using tissue culture technique at Taif Governorate, KSA. Res J Pharm Biol Chem Sci 2015; 6(5): 1117-22.

[294] Srivastava M, Singh G, Misra P. Contribution of biotechnological tools in the enhancement of secondary metabolites in selected medicinal climbers. Biotechnological Strategies for the Conservation of Medicinal and Ornamental Climbers 2015; pp. 465-86.

[295] Vijayalakshmi U, Shourie A. Elicitor induced flavonoid production in callus cultures of *Glycyrrhiza glabra* and regulation of genes encoding enzymes of the phenylpropanoid pathway. Pharm Lett 2015; 7(8): 156-66.

[296] Khodayari M, Omidi M. Gene expression involved in sanguinarine biosynthesis is affected by nano elicitors in *Papaver somniferum*L. Faslnamah-i Giyahan-i Daruyi 2015; 14(54): 41-54.

[297] Karimian R, Lahouti M, Davarpanah SJ. Effects of different concentrations of 2, 4-D and kinetin on callogenesis of *Taxus brevifolia*Nutt. J ApplBiotechnolRep 2014; 1(4): 167-70.

[298] Awad V, Kuvalekar A, Harsulkar A. Microbial elicitation in root cultures of *Taverniera cuneifolia* (Roth) Arn. for elevated glycyrrhizic acid production. Ind Crops Prod 2014; 54: 13-6.
[http://dx.doi.org/10.1016/j.indcrop.2013.12.036]

[299] Gill AR, Siwach P. Production of selected secondary metabolites in callus and shoot cultures of *Ficus religiosa* L. - A valuable medicinal plant. Res J Biotechnol 2014; 9(3): 63-73.

[300] Ahmed A, Rahman M, Tajuddin TE, *et al*. Effect of nutrient medium, phytohormones and elicitation treatment on *in-vitro* callus culture of *Bacopa monniera* and expression of secondary metabolites. Nat Prod J 2014; 4(1): 13-7.
[http://dx.doi.org/10.2174/2210315504011407151426244]

[301] Skrzypczak-Pietraszek E, Słota J, Pietraszek J. The influence of L-phenylalanine, methyl jasmonate and sucrose concentration on the accumulation of phenolic acids in *Exacum affine* Balf. f. ex Regel shoot culture. Acta Biochim Pol 2014; 61(1): 47-53.
[http://dx.doi.org/10.18388/abp.2014_1922] [PMID: 24644557]

[302] Pansuksan K, Mii M, Supaibulwatana K. Phytochemical alteration and new occurring compounds in hairy root cultures of Mitracarpus hirtus L. induced by phenylurea cytokinin (CPPU). Plant Cell Tissue Organ Cult 2014; 119(3): 523-32.
[http://dx.doi.org/10.1007/s11240-014-0552-y]

[303] Kuzovkina IN, Prokof'eva MY, Umralina AR, Chernysheva TP. Morphological and biochemical characteristics of genetically transformed roots of *Scutellaria andrachnoides*. Russ J Plant Physiol 2014; 61(5): 697-706.
[http://dx.doi.org/10.1134/S1021443714040116]

[304] Rodrigues M, Festucci-Buselli RA, Silva LC, Otoni WC. Azadirachtin biosynthesis induction in *Azadirachta indica* A. Juss cotyledonary calli with elicitor agents. Braz Arch Biol Technol 2014; 57(2): 155-62.
[http://dx.doi.org/10.1590/S1516-89132014000200001]

[305] Bhalerao BM, Vishwakarma KS, Maheshwari VL. Tinospora cordifolia (Willd.) Miers ex. Hook. f. &Thoms- plant tissue culture and comparative chemo-profiling study as a function of different supporting trees. Indian J Nat Prod Resour 2013; 4(4): 380-6.

[306] Vakil M,M,A, Mendhulkar V,D. Enhanced synthesis of andrographolide by Aspergillus niger and Penicillium expansum elicitors in cell suspension culture of Andrographis paniculata (Burm. f.) Nees. Bot Stud 2013; 54(1): 49.

[307] Dheeranupa S, Chaichana N. Effects of sodium acetate and sucrose on *in vitro* alkaloid production from *Stemona spp.* culture. Asian J Plant Sci 2013; 12(2): 92-6.
[http://dx.doi.org/10.3923/ajps.2013.92.96]

[308] Bénard C, Bourgaud F, Gautier H. Impact of temporary nitrogen deprivation on tomato leaf phenolics. Int J Mol Sci 2011; 12(11): 7971-81.
[http://dx.doi.org/10.3390/ijms12117971] [PMID: 22174644]

[309] Karakas FP, Turker AU. Improvement of shoot proliferation and comparison of secondary metabolites in shoot and callus cultures of *Phlomis armeniaca* by LC-ESI-MS/MS analysis. *In Vitro*. Cell Dev Biol Plant 2016; 52(6): 608-18.
[http://dx.doi.org/10.1007/s11627-016-9792-3]

[310] DellaPenna D. Plant metabolic engineering. Plant Physiol 2001; 125(1): 160-3.
[http://dx.doi.org/10.1104/pp.125.1.160] [PMID: 11154323]

[311] Herrmann KM, Weaver LM. The shikimate pathway. Annu Rev Plant Physiol Plant Mol Biol 1999; 50(1): 473-503.
[http://dx.doi.org/10.1146/annurev.arplant.50.1.473] [PMID: 15012217]

[312] Maeda H, Dudareva N. The shikimate pathway and aromatic amino Acid biosynthesis in plants. Annu Rev Plant Biol 2012; 63(1): 73-105.

[http://dx.doi.org/10.1146/annurev-arplant-042811-105439] [PMID: 22554242]

[313] Chandran H, Meena M, Barupal T, Sharma K. Plant tissue culture as a perpetual source for production of industrially important bioactive compounds. Biotechnol Rep 2020; 26: e00450.
[http://dx.doi.org/10.1016/j.btre.2020.e00450] [PMID: 32373483]

[314] Verpoorte R. Plant secondary metabolites.Metabolic Engineering of Plant Secondary Metabolism. Dodrecht, Boston, London: Kluwer AcademicPublisher 2000; pp. 1-30.
[http://dx.doi.org/10.1007/978-94-015-9423-3_1]

[315] Latchman DS. Eucaryotic transcription factors. San Diego: Academic Press 2003.

[316] Broun P. Transcription factors as tools for metabolic engineering in plants. Curr Opin Plant Biol 2004; 7(2): 202-9.
[http://dx.doi.org/10.1016/j.pbi.2004.01.013]

[317] Yang CQ, Fang X, Wu XM, Mao YB, Wang LJ, Chen XY. Transcriptional regulation of plant secondary metabolism. J Integr Plant Biol 2012; 54(10): 703-12.
[http://dx.doi.org/10.1111/j.1744-7909.2012.01161.x] [PMID: 22947222]

[318] Albert NW, Davies KM, Lewis DH, *et al.* A conserved network of transcriptional activators and repressors regulates anthocyanin pigmentation in eudicots. Plant Cell 2014; 26(3): 962-80.
[http://dx.doi.org/10.1105/tpc.113.122069] [PMID: 24642943]

[319] Xu W, Dubos C, Lepiniec L. Transcriptional control of flavonoid biosynthesis by MYB–bHLH–WDR complexes. Trends Plant Sci 2015; 20(3): 176-85.
[http://dx.doi.org/10.1016/j.tplants.2014.12.001] [PMID: 25577424]

[320] Xu YH, Wang JW, Wang S, Wang JY, Chen XY. Characterization of GaWRKY1, a cotton transcription factor that regulates the sesquiterpene synthase gene (+)-delta-cadinene synthase-A. Plant Physiol 2004; 135(1): 507-15.
[http://dx.doi.org/10.1104/pp.104.038612] [PMID: 15133151]

[321] Ma D, Pu G, Lei C, *et al.* Isolation and characterization of AaWRKY1, an *Artemisia annua* transcription factor that regulates the amorpha-4,11-diene synthase gene, a key gene of artemisinin biosynthesis. Plant Cell Physiol 2009; 50(12): 2146-61.
[http://dx.doi.org/10.1093/pcp/pcp149] [PMID: 19880398]

[322] Kinney A, J. Metabolic engineering in plants for human health and nutrition. Curr Opin Biotechnol 2006; 17(2): 130-8.
[http://dx.doi.org/10.1016/j.copbio.2006.02.006]

[323] Moore M, Ullman C. Recent developments in the engineering of zinc finger proteins. Brief Funct Genomics 2003; 1(4): 342-55.
[http://dx.doi.org/10.1093/bfgp/1.4.342] [PMID: 15239882]

[324] Mahmoud SS, Croteau RB. Metabolic engineering of essential oil yield and composition in mint by altering expression of deoxyxylulose phosphate reductoisomerase and menthofuran synthase. Proc Natl Acad Sci 2001; 98(15): 8915-20.
[http://dx.doi.org/10.1073/pnas.141237298] [PMID: 11427737]

[325] Muñoz-Bertomeu J, Arrillaga I, Ros R, Segura J. Up-regulation of 1-deoxy-D-xylulose-5-phosphate synthase enhances production of essential oils in transgenic spike lavender. Plant Physiol 2006; 142(3): 890-900.
[http://dx.doi.org/10.1104/pp.106.086355] [PMID: 16980564]

[326] Jadaun JS, Sangwan NS, Narnoliya LK, *et al.* Over-expression of *DXS* gene enhances terpenoidal secondary metabolite accumulation in rose-scented geranium and *Withania somnifera* : active involvement of plastid isoprenogenic pathway in their biosynthesis. Physiol Plant 2017; 159(4): 381-400.
[http://dx.doi.org/10.1111/ppl.12507] [PMID: 27580641]

[327] Canel C, Lopes-Cardoso MI, Whitmer S, *et al.* Effects of over-expression of strictosidine synthase and

tryptophan decarboxylase on alkaloid production by cell cultures of *Catharanthus roseus*. Planta 1998; 205(3): 414-9.
[http://dx.doi.org/10.1007/s004250050338] [PMID: 9640666]

[328] Winkel BSJ. Metabolic channeling in plants. Annu Rev Plant Biol 2004; 55(1): 85-107.
[http://dx.doi.org/10.1146/annurev.arplant.55.031903.141714] [PMID: 15725058]

[329] Jørgensen K, Rasmussen AV, Morant M, *et al.* Metabolon formation and metabolic channeling in the biosynthesis of plant natural products. Curr Opin Plant Biol 2005; 8(3): 280-91.
[http://dx.doi.org/10.1016/j.pbi.2005.03.014] [PMID: 15860425]

[330] Lessard PA, Kulaveerasingam H, York GM, Strong A, Sinskey AJ. Manipulating gene expression for the metabolic engineering of plants. Metab Eng 2002; 4(1): 67-79.
[http://dx.doi.org/10.1006/mben.2001.0210] [PMID: 11800576]

[331] St-Pierre B, Vazquez-Flota FA, De Luca V. Multicellular compartmentation of *catharanthus roseus* alkaloid biosynthesis predicts intercellular translocation of a pathway intermediate. Plant Cell 1999; 11(5): 887-900.
[http://dx.doi.org/10.1105/tpc.11.5.887] [PMID: 10330473]

[332] Saunders JA. Investigations of vacuoles isolated from tobacco: I. Quantitation of nicotine. Plant Physiol 1979; 64(1): 74-8.
[http://dx.doi.org/10.1104/pp.64.1.74] [PMID: 16660918]

[333] Otani M, Shitan N, Sakai K, Martinoia E, Sato F, Yazaki K. Characterization of vacuolar transport of the endogenous alkaloid berberine in *Coptis japonica*. Plant Physiol 2005; 138(4): 1939-46.
[http://dx.doi.org/10.1104/pp.105.064352] [PMID: 16024684]

[334] Burbulis IE, Winkel-Shirley B. Interactions among enzymes of the *Arabidopsis* flavonoid biosynthetic pathway. Proc Natl Acad Sci 1999; 96(22): 12929-34.
[http://dx.doi.org/10.1073/pnas.96.22.12929] [PMID: 10536025]

[335] Marinova K, Pourcel L, Weder B, *et al.* The Arabidopsis MATE transporter TT12 acts as a vacuolar flavonoid/H+ -antiporter active in proanthocyanidin-accumulating cells of the seed coat. Plant Cell 2007; 19(6): 2023-38.
[http://dx.doi.org/10.1105/tpc.106.046029] [PMID: 17601828]

[336] Zhang H, Wang L, Deroles S, Bennett R, Davies K. New insight into the structures and formation of anthocyanic vacuolar inclusions in flower petals. BMC Plant Biol 2006; 6(1): 29.
[http://dx.doi.org/10.1186/1471-2229-6-29] [PMID: 17173704]

[337] Gómez-Galera S, Pelacho AM, Gené A, Capell T, Christou P. The genetic manipulation of medicinal and aromatic plants. Plant Cell Rep 2007; 26(10): 1689-715.
[http://dx.doi.org/10.1007/s00299-007-0384-x] [PMID: 17609957]

[338] Wagner GJ, Kroumova AB. The use of RNAi to elucidate and manipulate secondary metabolite synthesis in plants.Current Perspectives in MicroRNAs (miRNA). Dordrecht: Springer 2008.
[http://dx.doi.org/10.1007/978-1-4020-8533-8_23]

[339] Wang P, Lombi E, Zhao FJ, Kopittke PM. Nanotechnology: A new opportunity in plant sciences. Trends Plant Sci 2016; 21(8): 699-712.
[http://dx.doi.org/10.1016/j.tplants.2016.04.005] [PMID: 27130471]

[340] Ruttkay-Nedecky B, Krystofova O, Nejdl L, Adam V. Nanoparticles based on essential metals and their phytotoxicity. J Nanobiotechnology 2017; 15(1): 33.
[http://dx.doi.org/10.1186/s12951-017-0268-3] [PMID: 28446250]

[341] Monica RC, Cremonini R. Nanoparticles and higher plants. Caryologia 2009; 62(2): 161-5.
[http://dx.doi.org/10.1080/00087114.2004.10589681]

[342] Mohammed AE. Green synthesis, antimicrobial and cytotoxic effects of silver nanoparticles mediated by *Eucalyptus camaldulensis* leaf extract. Asian Pac J Trop Biomed 2015; 5(5): 382-6.
[http://dx.doi.org/10.1016/S2221-1691(15)30373-7]

[343] Večeřová K, Večeřa Z, Dočekal B, *et al.* Changes of primary and secondary metabolites in barley plants exposed to CdO nanoparticles. Environ Pollut 2016; 218: 207-18.
[http://dx.doi.org/10.1016/j.envpol.2016.05.013] [PMID: 27503055]

[344] Jasim B, Thomas R, Mathew J, Radhakrishnan EK. Plant growth and diosgenin enhancement effect of silver nanoparticles in Fenugreek (Trigonella foenum-graecum L.). Saudi Pharm J 2017; 25(3): 443-7.
[http://dx.doi.org/10.1016/j.jsps.2016.09.012] [PMID: 28344500]

[345] Vanisree M, Chen Y, Shu-Fung L, *et al.* Studies on the production of some important secondary metabolites from medicinal plants by plant tissue cultures. Bot Bull Acad Sin 2004; 45: 1-22.

CHAPTER 3

In Vitro Propagation and Secondary Metabolite Production from *Withania Somnifera* (L.) Dunal

Praveen Nagella[1,*]**, Wudali Narashima Sudheer**[1] **and Akshatha Banadka**[1]

[1] *Department of Life Sciences, CHRIST (Deemed to be University), Bangalore-560029, Karnataka, India*

Abstract: *Withania somnifera* (L.) Dunal, commonly known as ashwagandha or Indian ginseng, is an important medicinal plant that belongs to the family Solanaceae. Ashwagandha has been used from time immemorial in different systems of medicine and extensively used in the Indian system of medicine, and there is discussion of this plant in different ayurvedic scripts like Charaka samhita, Ashtanga sangraha, *etc.* The plant is extensively used for anti-aging and general well-being, and also has anti-cancer potential. Ashwagandha is also known for its antioxidant, anti-inflammatory, and other therapeutic activities. In the recent days of Covid-19, the plant has been extensively used as an immunostimulant. The plant has great potential for its raw materials, especially for the extraction of bioactive molecules like withanolide-A, withaferin-A, withasomniferin, withanone, *etc.* The conventional mode of propagation could not meet the required commercial demand for either the pharmaceutical industries or the traditional practitioners. The conventional method of obtaining biomass is influenced by a large number of environmental factors, where biomass quality and quantity of bioactive molecules have shown variation. To overcome this, biotechnological approaches such as plant tissue culture techniques have been established for large-scale cultivation using micropropagation and also other techniques like a callus and cell suspension culture, shoot culture, adventitious root culture, and hairy root culture have been extensively used for *in vitro* production of bioactive molecules from ashwagandha. With the advent of metabolic engineering, biosynthetic pathway editing has made it possible to obtain higher yields of desired metabolites. The present chapter focuses on the *in vitro* propagation, biosynthesis of withanolides, and tissue culture strategies for obtaining high biomass and metabolites. The chapter also focuses on different elicitation strategies, metabolic engineering approaches, and the development of elite germplasms for improved metabolite content. The chapter also identifies research lacunas that need to be addressed for the sustainable production of important bioactive molecules from ashwagandha.

Keywords: Ashwagandha, *Agrobacterium rhizogenes*, Callus culture, *Withania somnifera*, Withanolide-A, Withaferin-A.

---

[*] **Corresponding author Praveen Nagella:** Department of Life Sciences, CHRIST (Deemed to be University), Bangalore - 560029, Karnataka, India; E-mail: praveen.n@christuniversity.in

Mohammad Anis & Mehrun Nisha Khanam (Eds.)
All rights reserved-© 2024 Bentham Science Publishers

INTRODUCTION

Withania somnifera (L.) Dunal (Solanaceae) is one of the important medicinal plants which has been cited in different traditional systems of medicine. It is a woody shrub which is rich in diversified phytochemicals. It is commonly distributed and grown in various parts of Africa, tropical regions of Europe, and Asian countries, especially in India, where it is cultivated for its rich pharmacological potential [1]. It is commonly called ashwagandha in Sanskrit for its characteristic smell of horse ("ashwa" means horse and "gandha" means characteristic smell) [2]. Studies suggest that every part of the plant is rich in various kinds of metabolites (Fig. **1**). Phytochemicals like withanolide-A, stigmasterol, withanine, withananine, vitoindosides, sitoindosides, and ashwagandhanolide were reported from the roots of ashwagandha. In leaf, bioactive compounds like withaferin, withanone, withanolides-B, D, E, Z, 27-deoxywithaferin-A, 2, 24-dienolide, trienolide (steroidal lactones), and withanoside-IV and many varieties of phytochemicals were reported [3]. Some of the pharmacologically significant molecules from ashwagandha are shown in Fig. (**2**).

Ashwagandha is rich in secondary metabolites, which contribute to various pharmacological activities. Studies reveal that withaferin-A has great potential as an anti-cancer agent, and it induces apoptosis in human melanoma cells [4]. And also, a synergistic effect, along with X-ray irradiation, enhances apoptosis in U937 cell lines (human myeloid leukemia cells) [5]. It is observed that ashwagandha has great potential in combating arthritis [6]. In collagen induced arthritis, root extracts of ashwagandha showed efficient anti-oxidant properties and helped in the production of antibodies for arthritis [7], proving the anti-inflammatory activity as well [6]. The cardioprotective efficiency of ashwagandha was also explored. Reports suggest that isoproterenol induced myocardial infarction is efficiently suppressed by the hydro-alcoholic root extracts of Ashwagandha [8]. There are many reports which prove the antimicrobial activity of ashwagandha. Significant antibacterial [9 - 11] and antifungal properties have been demonstrated [12] from ashwagandha extracts. Initial studies on ashwagandha as a potential anti-covid 19 agent were in discussion. The multi pharmacological potential of ashwagandha showed promising results with respect to *in-silico* studies and showed good binding efficiency with target proteins [13, 14].

The market value for ashwagandha was rising and reached a million-dollar market due to the rise in demand for raw materials for the extraction of phytochemicals. To meet the increasing demand for the supply of raw materials, conventional methods employed for the production of ashwagandha will not be sufficient.

Moreover, the seed viability for the cultivation of ashwagandha is very poor and has less germination efficiency when stored for a prolonged time [15]. So, there is a need for alternate propagation methods. For industrial-scale production and the maintenance of regenerated plants, the *in vitro* approach of propagation is the most useful. Several biotechnological techniques, including plant cell culture, hairy root culture, multiple shoot culture, different elicitation strategies, metabolic engineering approaches, and the creation of transgenic variants, can be used to produce plant secondary metabolites.

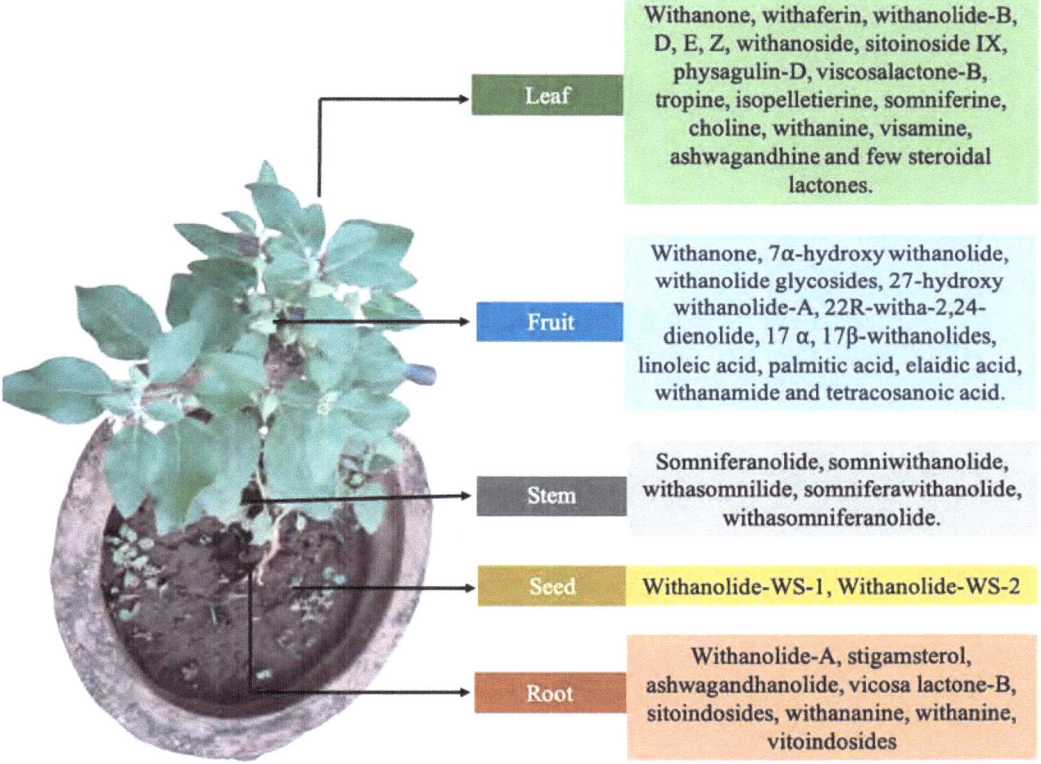

Fig. (1). Phytochemical profile of various parts of *Withania somnifera* (L.) Dunal.

In view of these, the present review focuses on various *in vitro* techniques used for micropropagation, secondary metabolite production, optimization of cultural conditions for withanolides, and withaferin production, and their elicitation strategies using various biotic and abiotic factors. The biosynthesis and metabolic engineering studies related to bioactive molecules production from ashwagandha are also emphasized in this study.

Fig. (2). Chemical structures of some important bioactive molecules from *Withania somnifera* (L.) Dunal. (**A**). Withaferin-A, (**B**). Withanolide - A, (**C**). Withanolide - R, (**D**). Withanone, (**E**). Withasomniferin.

BIOSYNTHESIS OF WITHANOLIDES

Withanolides are a group of C-28 steroidal lactones built on an ergostane skeleton that constitute the major secondary metabolites in *W. somnifera*. The structure of withanolide consists of a steroid backbone attached to a lactone or its derivatives [16]. Withanolides are synthesized from 24-methylene cholesterol, the first branching point, which undergoes a series of processes such as cyclization, chain elongation, desaturation, epoxidation, glycosylation and hydroxylation to ultimately form withanosteroids [17]. Putatively, in plants, withanolides (C-30) are synthesized *via* two independent isoprenoid biosynthetic pathways: mevalonate (MVA) pathway and non-MVA non-mevalonate pathways 1-deoxy-D-xylulose 5-phosphate/2-Cmethyl-D-erythritol 4-phosphate (DOXP)/2-C-methyl-d-erythritol-4-phosphate (MEP) pathways. The mevalonate pathway (MVA) of isoprenoid synthesis occurs in the cytosol and accounts for 75% of carbon contribution in withanolide biosynthesis, whereas the non-mevalonate pathway of isoprenoid synthesis occurs in plastids and accounts for 25% of carbon contribution in withanolide biosynthesis [18].

In the MVA pathway, acetoacetyl-CoA is formed by the activation of acetyl-CoA. The acetoacetyl-CoA formed condenses with acetyl-CoA in the presence of enzyme HMG-CoA synthase (HMGS) to form 3-hydroxy-3-methylglutaryl-CoA (HMG-CoA), which in turn is converted to mevalonate by the enzyme 3-hydroxy-3-methylglutaryl-CoA reductase (HMGR). The mevalonate is phosphorylated to 5-phosphomevalonate by mevalonate kinase and 5-phosphomevalonate is further phosphorylated to 5-pyrophosphomevalonate in the presence of phosphomevalonate kinase. In the presence of enzyme mevalonate-5-pyrophosphate decarboxylase, 5-pyro phosphomevalonate transforms into 3-isopentenyl pyrophosphate (IPP), which isomerises into 3,3-dimethylallyl pyrophosphate (DMAPP) catalysed by IPPI isomerase (Isopentenyl pyrophosphate isomerase).

In the MEP/DOXP pathway, pyruvate condenses with d-glyceraldehyde-3-phosphate (GA-3P) to form 1-deoxy-d-xylulose-5-phosphate (DOXP) catalyzed by DOXP synthase (DOXS). DOXP is then converted by enzyme DOXP reductoisomerase into 2-methyl d-erythritol 4-phosphate (MEP) which is then converted to 4-diphospho-cytidine-2-methyl-d-erythritol (CDP-ME) in a CTP-dependent reaction by 4-(cytidine-5-diphospho)-2-C-methyl-d-erythritol synthase (CMS). CDP-ME is then phosphorylated to 2-C-methyl-d-erythritol-2-phosphate (CDP-MEP) by 4-(cytidine-5-diphospho)-2-C-methyl-d-erythritol kinase (CMK). The latter undergoes further reaction to form 2-C-methyl-d-erythritol-2,4-cyclodiphosphate (ME-cPP) catalyzed by 2-C-methyl-d-erythritol-2,4-cyclodiphosphate synthase (MCS). Subsequently, ME-cPP is converted to hydroxyl methyl butenyl 4-diphosphate (HMBPP) in the presence of hydroxyl methyl butenyl 4-diphosphate synthase (HDS). HMBPP is then converted to IPP and DMAPP by the enzyme hydroxymethyl butenyl 4-diphosphate reductase (HDR). IPP and DMAPP move to the cytosol [18].

The IPP formed condenses with its isomer DMAPP in the presence of geranyl pyrophosphate synthase (GPS) to give geranyl pyrophosphate (GPP), which condenses with IPP to form farnesyl pyrophosphate (FPP) catalyzed by FPPS farnesyl diphosphate synthase (FPPS). Further two molecules of FPP in the presence of squalene synthase (SQS) condense to yield squalene which is then converted to 2,3-oxidosqualene ((S)-squalene-2, 3-epoxide) by squalene epoxidase [18]. 2,3-oxidosqualene catalyzed by cycloartenol synthase (CAS) undergoes a cyclization reaction to form cycloartenol [17]. Sterol methyltransferase 1 (SMT1) catalyzes the transfer of methyl group from S-adenosylmethionine to the double bond in the side chain of cycloartenol, forming 24-methylene cycloartenol which is then converted to cyclocucalenol by sterol-4α-methyl-oxidase [19]. The cyclopropane ring of cyclocucalenol is cleaved by cycloeucalenol-obtusifoliol, yielding obtusifoliol [20]. The obtusifoliol undergoes a series of steps and forms 24- methylene cholesterol. This intermediate is the starting point for the biosynthesis of withanolides [17]. 24-methylene cholesterol is isomerized to 24-methyl desmosterol by sterol Δ24-isomerase (24ISO), which then forms withanolides [21].

Parallelly, 24-methylenecycloartanol is methylated to isofucosterol, which is then isomerized to 24-ethyl desmosterol (Δ24(25)-sitosterol) and further hydrogenated to give sitosterol which further forms stigmasterol and withanolides (directly or *via* stigmasterol). On the other hand, 24 methylene cholesterol is epimerized to 24-methyl desmosterol (Δ24(25)-campesterol) and the epimerized double bond is hydrogenated by Δ7-C-5-desaturase (DW F1) to form campesterol which then forms brassinolide catalyzed by sterol side chain reductase 1(SSR1) [19]. The detailed biosynthetic pathway for withanolide production is represented in Fig. (3).

IN VITRO PROPAGATION STUDIES

W. somnifera (Ashwagandha) is a well-known medicinal plant from the ancient medical system. The phytochemical profile of the plants is very diverse, and their pharmacological activities are well studied. Due to the increased demand for raw materials for the production of pharmaceuticals from Ashwagandha, tissue culture studies were adapted to increase the supply of the essential biomass. Tissue culture strategies also help in deriving good quality and quantity of medicinally significant metabolites from Ashwagandha. Less viability and poor seed germination efficiency of Ashwagandha also make the tissue culture technique an efficient alternative way for growing plants. Various explants like shoot tips, axillary buds, nodal segments, leaves, and cotyledonary regions were used in regeneration studies [22].

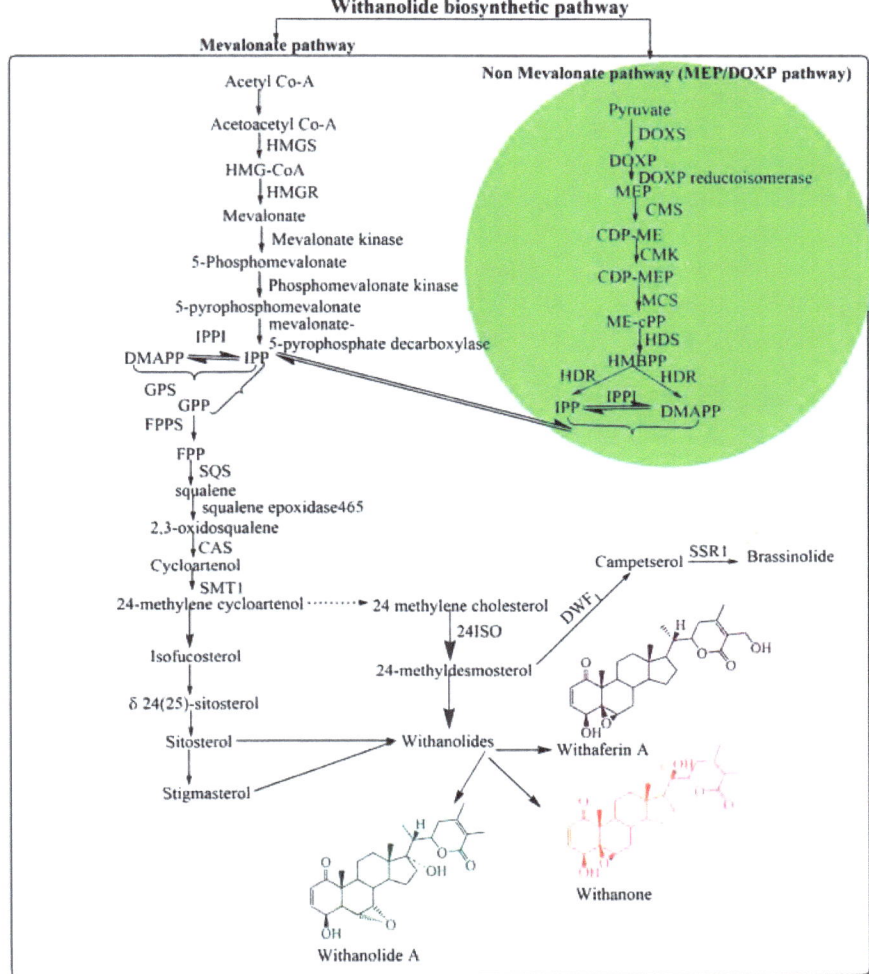

Fig. (3). Biosynthetic pathway of withanolide. 3-hydroxy-3-methylglutaryl-CoA (HMG-CoA); HMG-CoA synthase (HMGS); 3-hydroxy-3-methylglutaryl-CoA reductase (HMGR); IPPI (Isopentenyl pyrophosphate isomerase); DOXP synthase (DOXS); 4-diphospho-cytidyl-2-methyl-d-erythritol (CDP-ME); CDP-ME synthase (CMS). CDP-ME phosphate (CDP-MEP); CDP-ME kinase (CMK); 2-C-methyl-d-erythritol-2,4-cyclodiphosphate (ME-cPP); ME-cPP synthase (MCS); hydroxyl methyl butenyl 4-diphosphate (HMBPP); HMBPP synthase (HDS); hydroxymethyl butenyl 4-diphosphate reductase (HDR); 3-isopentenycal pyrophosphate(IPP); 3,3-dimethylallyl pyrophosphate (DMAPP); IPPI (IPP isomerase); d-glyceraldehyde-3-phosphate (GA-3P); 2-C-methyl-d-erythritol-2,4-cyclodiphosphate (ME-cPP); ME-cPPsynthase (MCS); hydroxyl methyl butenyl 4-diphosphate (HMBPP); HMBPPsynthase (HDS); hydroxymethyl butenyl 4-diphosphate reductase (HDR); geranyl pyrophosphate (GPP); GPP synthase(GPS); farnesyl pyrophosphate (FPP); FPP synthase (FPPS); squalene synthase(SQS); cycloartenol synthase (CAS); Δ7-C-5-desaturase (DWF1); sterol side chain reductase 1(SSR1).

Direct Organogenesis

The phenomenon of direct organogenesis aids in the direct generation of desired organ cultures from cultured explants. In ashwagandha, multiple shoots were induced from leaf explants of *in vitro* grown plantlets when cultured on Murashige and Skoog (MS) medium supplemented with a combination of indole-3-acetic acid (IAA) and N6-benzyl aminopurine (BAP). An increased number of adventitious shoots (13.37) were induced from leaf explants when cultured on 7.99 µM IAA along with 13.2 µM BAP [23]. Leaf explants when cultured on MS medium supplemented with 2.0 mg/L BAP along with 0.1 mg/L naphthalene acetic acid (NAA) induced multiple shoots [24].

Nodal and internodal explants of ashwagandha were used for the induction of multiple shoots. It is seen that when MS medium is supplemented with BAP, there is a positive response for shoot formation. It is observed that hypocotyl explants also show similar responses when cultured on MS medium supplemented with 0.1 to 5.0 mg/L of BAP, which will help in the induction of multiple shoots [25]. MS medium supplemented with 2.5 µM BAP in combination with 0.5 µM NAA showed the highest regeneration efficiency for shooting when nodal explants were inoculated [26]. Nodal explants, when cultured on MS medium supplemented with 1.0 mg/L NAA along with 0.5 mg/L BAP induced callus with shoots; subculturing on MS medium supplemented with 1.0 mg/L BAP along with 0.5 mg/L KN enhanced the shoots [27]. The highest number of (21 ± 0.057) shoots with the highest length was induced from nodal explants when cultured on MS medium supplemented with 2.0 mg/L BAP [28]. The MS medium supplemented with 1.5 mg/L BAP along with 1.5 mg/L IAA helped in the induction of multiple shoots when nodal explants were inoculated, and for shoot proliferation, GA3 at 0.15 mg/L concentration was effective [29]. MS medium supplemented with 3.0 mg/L BAP helped in the induction of multiple shoots from nodal explants and MS medium fortified with a combination of 3.0 mg/L BAP along with 0.5 mg/L NAA helped in the multiplication and proliferation of shoots [30]. Nodal explants when cultured on MS medium supplemented with 2.0 mg/L BAP along with 0.5 mg/L NAA showed the highest number of shoots [31]. Cytokinins like meta-Topolin at a concentration of 2.5 µM was supplemented in the growth medium and showed the highest induction efficiency for shoots when compared to conventional PGR's [32].

In ashwagandha, epicotyl explants were also considered to be potential explants for regeneration. Epicotyl explants, when cultured on MS medium supplemented with 2.0 mg/L BAP along with 0.2 mg/L IAA, showed induction of the maximum number of shoots. Multiplication of induced shoots was done on MS medium supplemented with 1.0 mg/L GA3 [33]. Cotyledonary nodes were cultured on MS medium supplemented with 1.0 mg/L BAP, and the highest shoot initiation and proliferation efficiency (16.93 shoots per explant) was recorded [34].

Shoot tips were also reported to be one of the potential explants for organogenesis. MS medium supplemented with 2.0 mg/L BAP in combination with 2.0 mg/L IAA induced multiple shoots and the elongation of shoots was seen on MS medium supplemented with 0.3 mg/L GA3 [35]. Along with the MS medium, the revised tobacco medium (RT medium) also showed a positive response to the shoot induction from apical bud explants. RT medium supplemented with 1.0 mg/L 2,4-Dichlorophenoxyacetic acid (2,4-D) helped in the induction of shoots from apical bud explants of ashwagandha and sub-sequential rooting was seen in the same medium [36]. Seeds of ashwagandha were also used to induce the multiple shoots when cultured on MS medium supplemented with 1.5 mg/L BAP along with 0.5 mg/L IAA and proliferation is done on MS medium fortified with 0.3 mg/L GA3 along with 3.0 mg/L IBA [37].

Indirect Organogenesis

In indirect organogenesis, through the process of regeneration, a callusing stage is observed, followed by desired organ cultures. In ashwagandha, callus was induced on MS medium supplemented with 2.0 mg/L 2,4-D and 0.6 mg/L KN from epicotyl explants, and further subculturing of callus on 1.0 mg/L BAP enriched media along with 20 mg/L Adenine sulphate induced shoots [33]. MS medium supplemented with 1.0 mg/L BAP along with 0.5 mg/L NAA showed a maximum number of shoots from the callus induced from the nodal segments [31]. When hypocotyl explants were cultured on MS medium supplemented with 2.0 mg/l 2,4-D in combination with 0.2 mg/l KN showed the best callus induction efficiency, and the same media was used for shoot induction from the callus, and maximum proliferation efficiency was recorded on MS medium supplemented with 2.0 mg/L BAP [38].

Leaf explants, when cultured on MS medium supplemented with 2.0 mg/L BAP along with 0.5 mg/L IAA, showed the best callusing efficiency followed by shooting induction [39]. *In vitro* grown leaf explants were cultured on MS medium supplemented with IBA and BAP and callusing was observed. Simultaneous shooting was also seen when the callus was subcultured on 2.0 mg/L BAP supplemented MS media [40]. MS medium supplemented with 0.5 mg/l BAP along with 1.0 mg/L NAA helped in the induction of callus from leaf explants and when subcultured on 0.5 mg/L NAA along with 2 mg/L BAP, helped in the induction and regeneration of shoots [41].

Rooting of the Regenerated Plantlets

In tissue culture techniques, rooting is crucial for survival throughout the acclimatization phase. The *in vitro* generated shoots must be well rooted since roots provide support for the plantlets as they develop and flourish. The MS medium supplemented with BAP medium seems to be suitable for root induction and multiplication in ashwagandha. *In vitro* induced shoots showed the best rooting efficiency when cultured on MS medium supplemented with 0.44 µM BAP [23]. Full strength or half strength MS medium supplemented with 0.01 mg/L BAP showed rooting [25]. Studies suggest that auxin rich media also help in root induction in ashwagandha. *In vitro* grown shoots when cultured on half strength MS medium supplemented with 2.0 mg/L IBA showed root induction and multiplication [30, 35, 38] and, similarly, in another study, rooting was established for the *in vitro* grown shoots when cultured on 1.0 mg/L IBA fortified medium [24, 34]. IBA of 0.8 mg/L, when supplemented in MS medium showed induction of roots for the *in vitro* grown shoots cultured on GA3 [33]. Half strength MS medium supplemented with 1.4 mg/L IBA also showed optimal root inducing efficiency [27]. MS medium of half strength concentration supplemented with 0.5 µM NAA showed 100% rooting efficiency [26]. Higher concentrations of IBA, *i.e.*, 5 mg/L, showed improved rooting and showed better efficiency [29]. In general, various auxins were found to be efficient PGRs for the induction of rooting from *in vitro* regenerated shoots.

Acclimatization

Acclimatization is an important step in the regeneration of plantlets. The process of acclimatization usually happens in a polyhouse or greenhouse, in which plants become accustomed to a new climate or to new conditions with respect to parameters like sunlight intensity, temperature, and relative humidity. In acclimatization, the plantlets were transferred to a specific potting mixture, where they were adjusted and later transferred to the outer environment. Table 1 shows different potting mixtures used for the acclimatization of the *in vitro* raised plantlets.

Table 1. Different potting mixtures used for the acclimatization of *in vitro* grown ashwagandha plantlets.

S.No	Potting Mixture	References
1	Sand: Soil = 1:1	[23, 38]
2	Soil:Perlite = 3:1 along with MS salts in the mixture	[24]
3	Garden soil:Vermicompost:Sand = 1:1:1	[36]

(Table 1) cont.....

S.No	Potting Mixture	References
4	Garden soil:vermicompost = 3:1 along with half strength MS salts without vitamins in the potting mixture	[26]
5	Soil:Sand:Vermiculite = 1:2:1	[33]
6	Vermicompost:Soil:Sand = 1:2:1	[27, 42]
7	Sand:Soil:FYM (Manure) = 1:1:1	[30]
8	Soil:Vermicompost = 1:3	[40]

IN VITRO PRODUCTION OF WITHANOLIDES

Apart from the conventional production of withanolides and other important secondary metabolites, various biotechnological approaches involving tissue culture studies have been established. Cell suspensions, shoot culture, adventitious roots, and transformed hairy roots were the most successful techniques for increasing the production of withanolide. A flow chart of various approaches for withanolide and related metabolites from *W. somnifera* is depicted in Fig. (**4**).

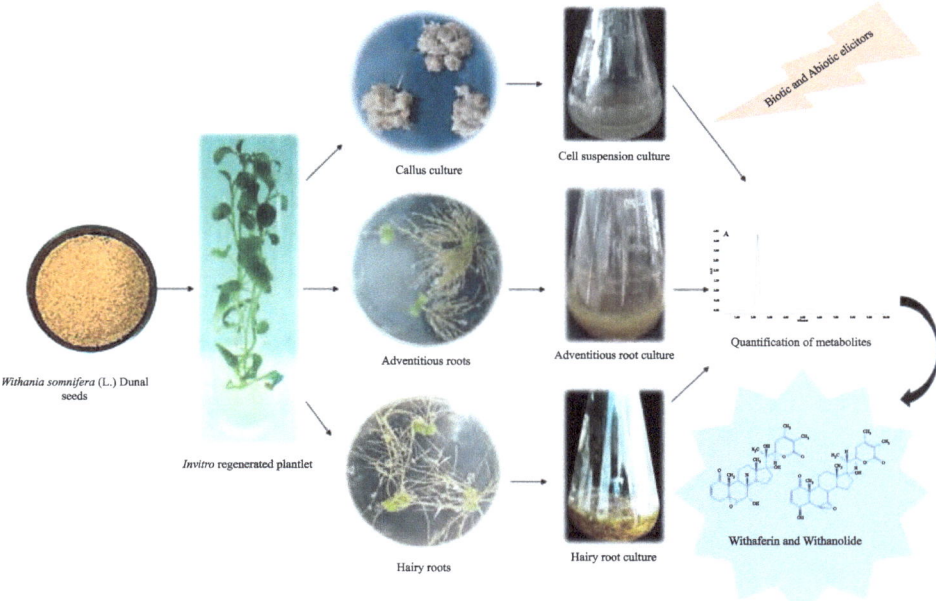

Fig. (4). Flow chart for the *in vitro* production of withanolides and withaferin-A from *Withania somnifera* (L.) Dunal.

Production of Withanolides from Cell/Callus Culture

Callus is an undifferentiated mass of cells that is obtained when the explant is cultured on the media supplemented with either individual PGR or a combination of PGRs, especially from the auxins and or in combination with cytokinins. The induction of callus is dependent on factors like the selection of appropriate explant/s from the source mother plant, the right choice of the nutrient medium, and, very importantly, the PGRs. Studies have shown that the leaf has served as an optimal explant for the induction of callus, and the leaf is also reported to have a higher concentration of withanolide content compared with other explant sources of *W. somnifera*. Studies have also shown that root as an explant has the potential for the production of withanolides (Table 2).

Table 2. Production of withanolides from cell/callus suspension cultures of *W. somnifera*.

S No.	Explant Source	Culture Medium and Optimization	Withanolide Content	References
1	Leaf	MS + 3.0 mg/L 2,4-D + 0.5 mg/L KN	0.46 mg/g DW	[43]
2	Leaf	MS + 2.0 mg/L 2,4-D + 0.5 mg/L KN **Optimized parameters:** 3% sucrose, 10.0 g/L inoculum density, pH- 5.8.	2.26 mg/g DW, Optimized parameters: 2.95 mg/g DW, 2.42 mg/g DW, 2.51 mg/g DW.	[44]
3	Leaf	MS + 2.0 mg/L 2,4-D + 0.5 mg/L KN **Optimized parameters:** 2.0X KNO$_3$, 14.38/37.60 mM of NH$_4^+$/NO$_3^-$ ratio.	2.26 mg/g DW Optimized parameters: 4.36 mg/g DW, 3.96 mg/g DW.	[45]

Studies suggest that MS medium supplemented with 3.0 mg/L 2,4-D along with 0.5 mg/L KN helped in the induction of callus and cell suspension culture was established in the liquid media with the same PGRs [43]. Praveen and Murthy optimized the culture conditions for the production of withanolides. It is observed that leaf explants, when cultured on MS medium supplemented with a combination of 2.0 mg/L 2,4-D and 0.5 mg/L KN with a pH of 5.8, have produced good biomass and produced 2.26 mg/g DW of withanolide content. It is also observed that the optimal inoculum density for withanolide production is 10 g/L [44]. In addition to that, Praveen and Murthy also optimized the macronutrients and nitrogen sources for withanolide production. Studies suggest that MS medium containing 0.5 X of NH$_4$NO$_3$ showed the highest biomass accumulation and medium containing 2.0 X KNO$_3$ produced 4.36 mg/g DW of withanolides. Along with this, the ratio of NH$_4^+$/NO$_3^-$ ions was optimized for optimal withanolide production and it is observed that MS medium containing NH_4^+/NO_3^- in the ratio of 14.38/37.60 mM produced 3.96 mg/g DW of withanolides [45]. 98% of callus induction was recorded when leaf explants were cultured on the MS media supplemented with 0.5 mg/L 2,4-D along with 0.2 mg/L KN. However, HPTLC studies reported that principal phytochemicals like withaferin and withanolides were present only in the callus induced from media supplemented with 1.0 mg/L IBA and (0.5 to 2.0 mg/L) BAP [40].

Production of Withanolides from Shoot Culture

Shoot organogenesis, either through direct or indirect modes, helps in establishing plantlets for sustainable production of bioactive metabolites like withanolides. Shoot tips of Withania if cultured on MS medium supplemented with 1.0 mg/L BAP helped in the induction of a maximum number of shoots. The highest amount of withaferin A (14%) was produced from the *in vitro* induced shoots when 10% of coconut milk was supplemented with the media [46]. Nodal explants, when cultured on MS medium supplemented with 1.0 mg/L BAP along with 0.5 mg/L KN, resulted in the formation of shoot culture and the same when quantified for withanolide- A, which is present at around 0.238g per 100g of dry weight [47]. Shoot tips of *in vitro* grown plants when cultured on MS medium supplemented with 0.4 µM IAA combined with 0.4 µM BAP. The derived *in vitro* shoots were quantified for the presence of various phytochemicals. It was observed that withanolide-A was present at around 2.59 µg/g DW, withanone around 0.44 µg/g DW, withaferin-A around 0.14 µg/g DW and withanolide-B around 0.16 µg/g DW [48]. MS medium supplemented with 1.0 mg/L BAP in combination with 1.0 mg/L KN showed optimal proliferation of multiple shoots from nodal explants. It was observed that 3.647 mg/g DW of withanolides were derived from these shoot cultures [43]. Nodal explants when cultured on MS medium supplemented with 1.5 mg/L BA along with 0.3 mg/L IAA helped in deriving multiple shoots. Respective contents of withanolides, withaferin and withanone were quantified and it was observed that *in-vitro* derived leaves, stems, and roots showed the highest phytochemical yield when compared to *in vivo* plant organs [49]. Nodal explants when cultured on MS medium supplemented with 1.5 mg/L BA along with 0.3 mg/L IAA, 6.0% sucrose, and 20 mg/L of L-glutamine helped in deriving the highest number of shoots and leaves. Bioactive molecules like withanolide-A (0.75 mg/g DW), withanolide-B (1.08 mg/g DW), withanone (1.74 mg/g DW) and withaferin-A (2.05 mg/g DW) were quantified [50].

Production of Withanolides from Adventitious Root Culture

Adventitious roots offer an alternative source for the production of pharmacologically important metabolites. The optimal induction and proliferation of the adventitious roots always aid in the increased production of metabolites when compared to the conventional form of production from field grown plants [51]. Leaf explants when cultured on MS medium supplemented with 0.5 mg/L IBA resulted in the induction of adventitious roots. Praveen and Murthy reported that the withanolide-A content was around 8.8 mg/g DW from the *in vitro* induced adventitious roots [52]. Chief metabolites like withanolides, withaferin, and withanone were produced from adventitious roots, which were induced from callus when subcultured on half strength MS medium supplemented with 0.5 mg/L IBA along with 0.1 mg/L IAA [53]. Murthy and Praveen optimized the medium for the optimal growth and production of withanolides from adventitious roots. Adventitious roots were induced from leaf explants when cultured on MS medium supplemented with 0.5 mg/L IBA. Culturing of adventitious roots in suspension culture having 0.5 X of NH4NO3 showed the highest production of withanolide-A, *i.e.*, 14 mg/g DW and the optimal NH_4^+/NO_3^- ratio for production of withanolide-A was 0.00/18.80mM and yield was recorded as 11.76 mg/g DW [54]. They also reported that MS medium supplemented with 2% sucrose was optimal for the production of withanolides, *i.e.*, 8.93 mg/g DW, and the highest respective biomass was reported in the same conditions. It is also seen that a pH of 5.5 was optimal for the production of withanolide, *i.e.*, 9.09 mg/g DW [55].

Production of Withanolides from Hairy Root Culture

Induction of hairy roots is an event which occurs when *Agrobacterium rhizogenes* infects the tissue system of plants where the T-DNA region harboring the gene encoding for auxin synthesis from the Ri-plasmid is transferred into the genome of the host cell/tissue. For the successful induction of hairy roots in *in vitro* conditions, there are several parameters/factors that need to be taken into account. The choice of the explant is of prime importance and, depending on the tissue/organ where high metabolite content is produced, is the right explant source for the induction of hairy roots. The strain of the *Agrobacterium rhizogenes* is also of importance, as some strains may not be virulent enough for the infection to take place. The infection time, condition, co-culture duration, incubation (light/dark) and the medium are some of the factors which influence the induction of hairy roots [56] Table **3**.

Table 3. Production of withanolides from hairy root cultures of *W. somnifera*.

S. NO.	Explant Source	Culture Medium	*Agrobacterium rhizogenes* Strain Used for Hairy Root Induction	Withanolides Content	References
1	Stem/leaf	Basal MS medium	LBA 9402	0.181mg/l/d	[57]

S. No.	Explant Source	Culture Medium	*Agrobacterium rhizogenes* Strain Used for Hairy Root Induction	Withanolides Content	References
2	Leaf	Basal MS medium	ATCC 15834	-	[58]
3	Petiole	Half strength MS medium	R1000	72.3 mg/g DW	[61].
4	Leaf	MS medium with 4% sucrose	R1601	157.4 µg/g DW	[59, 62].

Withanolides were produced in the highest amounts in hairy root cultures when compared to non-transformed roots. Hairy roots induced after transformation using *Agrobacterium rhizogenes* strain LBA 9402 produced 0.181mg/l/d of withanolide D, which is almost 7 fold increase when compared to non-transformed roots (0.026 mg/g/l/d [57]. Polymerase chain reaction (PCR) studies helped to confirm the transmission of T-DNA for the successful establishment of hairy roots. *A. rhizogenes* strain ATCC 15834 was employed for the induction of hairy roots from leaf explants of ashwagandha and it was seen that hairy roots showed increased production of secondary metabolites [58]. Withanolide-A production was increased by 2.7 fold (157.4 µg/g DW) in transformed hairy roots induced from leaf explants of *Withania* when compared to untransformed roots. *A. rhizogenes* strain R1601 was used for transformation and PCR studies confirmed its successful transformation efficiency [59]. *A. rhizogenes* strain R1000 was employed for hairy root establishment [60]. Secondary metabolites like withaferin-A were quantified from hairy roots induced from petiole explants using R1000. Hairy roots were well established on half strength MS medium and 72.3 mg/g DW of withaferin was accumulated in the hairy roots [61]. Optimization of carbon sources and pH for the production of the highest amounts of secondary metabolites from *Withania* was attempted by Praveen and Murthy. It is seen that sucrose at 4% was ideal for withanolide-A production, *i.e.*, 13.28 mg/g DW, and pH of 6 was suitable for the highest production of withanolide, *i.e.*, 13.84 mg/g DW [62]. A similar conclusion with respect to sucrose percentage was derived for withaferin-A and withanone production, which was carried out by Ganeshan Sivanandan *et al* . (2012). It is seen that half strength MS medium containing 4% sucrose was ideal for the production of 2.21 mg/g DW of withaferin A and 2.41 mg/g DW of withanone [63]. Praveen and Murthy optimized the media with respect to macroelements and nitrogen sources for the highest production of withanolide-A. It is observed that 2.0 X concentration of KNO_3 produced 15.27 mg g^{-1} DW, and NH_4^+/NO_3^- ratio of 0.00/18.80 mM showed the highest production of withanolide-A, *i.e.*, 14.68 mg/g DW [64]. Chandrasekaran *et al* . (2015) established a better approach for the *A. rhizogenes* mediated hairy root culture induction. It was suggested that transformation efficiency improved by 93.3%

when explants were subjected to sonication for 15 seconds, followed by heat treatment for 5 minutes [65].

ELICITATION STRATEGIES FOR IMPROVED PRODUCTION OF WITHANOLIDES

Elicitation is one of the most often used techniques in biotechnological approaches for increasing the production of plant secondary metabolites. Elicitors are compounds that are given in small concentrations for the production of secondary metabolites. The nature of the elicitor, time of exposure, and concentration of treatment are some of the important parameters that play a major role in elicitation. Many successful reports prove that elicitation helps in the production of phytochemicals in cell suspensions, shoots, and root cultures [66] Table **4**.

Table 4. Effect of elicitors on the production of withanolides from cell and organ cultures of *Withania somnifera*.

S. No.	Explant Source	Culture Type	Culture Medium with Elicitors	Withanolide Content		References
				Control	Elicited	
1	Hypocotyl	Cell suspension culture	MS medium + 100 µM Copper sulphate + 5% *Verticilium dahaliae* extracts.	Withaferin A- 2.65 mg/L	13.8-fold increase in production	[67]
2.	Leaf	Callus mediated adventitious roots	half strength MS medium + 0.5 mg/L IBA + 0.1 mg/L NAA + 100 mg/L Chitosan	Withanolide-A (17 fold less), withanolide-B (11 fold less), withaferin A (9 fold less), withanoside V (10 folds less), and withanoside IV (8 folds less)	Withanolide-A (323.85 mg/g DW), withanolide-B (0.275 mg/g DW), withaferin-A (4.347 mg/g DW), withanoside V (0.450 mg/g DW), and withanoside IV (0.528 mg/g DW)	[68]
3.	Leaf	Callus mediated adventitious roots	Half strength MS medium + 0.5 mg/L IBA + 0.1 mg/L NAA + 150 µM Salicylic acid.	withanolide-A (48 fold less), withanolide-B (29 fold less), withaferin-A (20 fold less), withanone (37 fold less)	Withanolide-A (64.65 mg/g DW), withanolide-B (33.74 mg/g DW), withaferin-A (17.47 mg/g DW), withanone (42.88 mg/g DW)	[53]

S. No.	Explant Source	Culture Type	Culture Medium with Elicitors	Withanolide Content		References
4.	Nodal explant	Shoot culture	MS + 0.6 mg/L BAP + 20 mg/L spermidine +100 µM salicylic acid	1.14 to 1.18 folds less.	Withnolde-A (8.48 mg/g DW), withnolde-B (15.47 mg/g DW), withaferin-A (29.55 mg/g DW) and withanone (23.44 mg/g DW).	[69]
5	Leaf	Hairy root culture induced using R1000	Half strength MS medium + 150 µM salicylic acid	Withanolide-A (58 fold less), withanone (46 fold less), and withaferin A (42 fold less)	Withanolide A (132.44 mg/g DW), withanone (84.35 mg/g DW), and withaferin A (70.72 mg/g DW)	[70]
6	Leaf	Cell suspension culture	MS + 1.5 mg/L 2,4-D + 0.2 mg/L kinetin + cell homogenate and culture filtrate of *Piriformospora indica*	2.6 mg/L	Withaferin-A, 3.8 mg/L	[71]

Initial studies using biotic and abiotic elicitors for withaferin-A production were attempted. Dual elicitation for cell suspension cultures using abiotic elicitor, copper sulphate at 100 µM concentration along with 5% (v/v) extracts of *Verticilium dahaliae* increased the withaferin-A production by 13.8 fold when compared to control cultures [67]. Chitosan at 100 mg/L concentration stimulated the highest production of withanolide-A (323.85 mg/g DW), withanolide-B (0.275 mg/g DW), withaferin-A (4.347 mg/g DW), withanoside-V (0.450 mg/g DW), and withanoside IV (0.528 mg/g DW) from the adventitious root culture induced from callus, when cultured on half strength medium supplemented with 0.5 mg/L IBA and 0.1 mg/L NAA [68]. Callus mediated adventitious roots induced in the previous study were treated with 150 µM salicylic acid, and produced increased production of withanolide-A (64.65 mg/g DW), withanolide-B (33.74 mg/g DW), withaferin-A (17.47 mg/g DW), withanone (42.88 mg/g DW) [53]. Shoot cultures established on MS medium supplemented with 0.6 mg/L BAP along with 20 mg/L spermidine and 100 µM salicylic acid produced the highest amounts of secondary metabolites like withanolide-A (8.48 mg/g DW), withanolide-B (15.47 mg/g DW), withaferin-A (29.55 mg/g DW) and withanone (23.44 mg/g DW) [69]. Salicylic acid of 150 µM concentration helped in the increased production of metabolites from the hairy root cultures of Withania. Metabolites like withanolide-A (132.44 mg/g DW), withanone (84.35 mg/g DW), and withaferin-

A (70.72 mg/g DW) were produced and showed better efficiency when compared to methyl jasmonate [70].

Studies revealed that biotic elicitors also play a major role in the production of plant secondary metabolites in Withania. 3% of cell homogenate and culture filtrate made from *Piriformospora indica* were employed as an elicitor. It is seen that withaferin-A production was increased by 1 to 2 times compared to control [71]. Hairy roots induced using R1000, when treated with 100 mg/L chitosan, showed increased production of withaferin-A by 4 fold when compared to control [72]. In Endophytic fungus, *Aspergillus terreus,* withanolide-A production was enhanced when treated with the root cell suspension culture [73]. Bioreactor scale production of withanolides was studied, and it was seen that 100 mg/L chitosan and 6.0 mM of squalene along with 1.0 mg/L picloram, 0.5 mg/L KN, 200 mg/L L-glutamine and 5% sucrose showed the increased production of metabolites [74].

METABOLIC ENGINEERING FOR INCREASED PRODUCTION OF WITHANOLIDES

Although the above-mentioned approaches have enhanced withanolide production, the amount of withanolide production for the commercialization of the process is insufficient, and hence there is a need for other approaches. Metabolic engineering is one such reliable approach for large-scale production of secondary metabolites. It can be used to increase withanolide yield by various approaches. The first approach is to obstruct the rate of approaching step or to block the competitive pathways by using antisense genes or antibodies. The other approach is to enhance the enzyme activity by using a sense gene needed from the same plant species, or a different plant species, or a totally different organism. Thus, all the above-mentioned metabolic engineering approaches necessitate a thorough understanding of the biosynthetic pathways and the genes involved in them. In the study conducted by [75], withanolide synthesis has been enhanced by diverting the metabolic flux from the isoprenoid pathway towards the withanolide biosynthetic pathway, which in turn was achieved by enhancing the enzyme squalene synthase by the introduction and overexpression of the squalene synthase gene (*WsSQS*) gene. This enzyme is involved in the reductive condensation of farnesyl diphosphate to squalene [76]. Also *AtSQS1* genes from *Arabidopsis thaliana* have been expressed in *W. somnifera,* and it has been observed that the withanolide content increased by 1.51-fold (330±0.87 µg/g) in comparison to control (218±0.17 µg/g) DW [77]. The withanolide content increased by 3.4 fold in *W. somnifera* overexpression was achieved by the overexpression of *WsWRKY1* gene that positively regulated the expression of triterpenoid biosynthetic genes [78]. The gene for Sterol Δ22-Desaturase, *WSCYP710A11*, has been isolated, characterized, and overexpressed in *W.*

somnifera. Withanolides have been overproduced under the influence of overexpression of *WSCYP710A11via* pathway channeling [79]. The other metabolic approach reported is the overexpression of the DXS gene, which is involved in catalyzing the MEP pathway's first step, thus regulating the expression of withanolide biosynthesis. The geranium gene (GrDXS), isolated from rose-scented geranium, was overexpressed in *W. somnifera*, which led to an increase in withanolide content [80].

GENETIC TRANSFORMATION STUDIES IN *WITHANIA SOMNIFERA*

One of the most important tools in plant biotechnology for metabolite enhancement is to genetically transform plants by manipulating the plant genome by introducing foreign genes. *Agrobacterium* mediated gene transfer, electroporation of protoplasts, gene gun bombardment, and microinjection are the most common methods employed in plant transformation. *Agrobacterium*-mediated plant transformation is the first method that initiated the transformation of plants to develop transgenic plants and is the most popular method of all the above-mentioned methods to date Table **5**. This method makes use of *Agrobacterium*'s inherent capacity to transform plants in order to complete its own life cycle. The *in vitro* tissue culture method is the most suitable biotechnological method for successfully carrying out *Agrobacterium*-mediated transformations. Thus, the regeneration ability of genotypes and explants plays a significant role in transformation efficiency [81]. Various gene transformation studies have been conducted to enhance withanolide in *Withania somnifera* using *Agrobacterium* strains. In the study conducted by [79] the gene *WSCYP710A11* that codes for sterol $\Delta 22$ -desaturase was cloned into *Agrobacterium tumefaciens* A4 and this strain was used to transform the leaf explants of *W. somnifera* by infecting the explant and the explant was cultured in half strength MS media for 2 days. The transformed plantlets exhibited the over-expression of WsCYP710A11 and also showed a substantial increase in withanolide content. The withanolide A content increased by 0.41 fold, withanone content increased by 0.37 fold and withaferin A content increased by 1.29 fold in comparison to control.

Table 5. Genetic transformation studies for enhanced withanolide content from *Withania somnifera*.

Strain	Media	Gene Inserted	Yield		References
		-	Control	Transformed	-
Agrobacterium tumefaciens	MS media with 0.1 mg/L kinetin and 0.2 mg/L 6-BAP	*WsSQS*	1.84, 2.24 and 2.25 mg/g dw of leaf, stem and root, respectively	3.55 (1.9 fold), 3.37 (1.5 fold) and 3.98 (1.8 fold) mg/g DW of leaf, stem and root, respectively	[75]

(Table 5) cont.....

Agrobacterium tumefaciens A4	MS media with kinetin (2 µL) and Indole--butyric acid (1 µL)	WSCYP710A11	-	0.41 fold, 0.37 fold and 1.29 fold increase in withanolide A, withanone, and withaferin A in comparison to control	[79]
Agrobacterium tumefaciens strain GV3101	Media containing MS salts and vitamins with 1.0 µM Thidiazuron (TDZ)	GrDXS	-	5 fold of total withanolide enhancement in WS-DXST8 lines and 2.8-3.3 fold enhancement in WS-DXST4, WS-DXST10 and WS-DXST11	[80]
Agrobacterium tumefaciens strains C58C1 (pRiA4) or C58C1 (pRiA4) (pBIs SQS1)	MS medium with 0.25 mg/L indole-3-butyric acid (IBA)	AtSQS1	218±0.17µg /g DW withaferin A	330±0.87µg/g DW (1.51-fold increase) withaferin A	[77]
Agrobacterium rhizogenes R1000	Half strength MS media supplemented with different concentrations of acetosyringone	WsSQS	-	2.82 mg/g DW withanolide A, 1.34 mg/g DW withanolide B, 1.83 mg/g DW withaferin A and 1.97 mg/g DW withanone	[82, 83]
Agrobacterium tumefaciens strain GV3101	MS agar medium 100 µM acetosyringone	WsGGPPS	-	Decrease in withanolide content	[84]
Agrobacterium tumefaciens strain GV3101	Liquid MS medium with 100 µM acetosyringone	WsSQS (WsSS)	2.05 mg/g DW of Withanolide A and Withaferin A was absent	6.04 mg/g DW (3 fold increment) of Withanolide A and Withaferin A was 0.20 mg/g DW	[85]
Agrobacterium tumefaciens strain GV3101	-	WsWRKY1	-	Withanamide production was observed	[86]

(Table 5) cont.....

Agrobacterium tumefaciens strain R1601	[Murashige and Skoog (MS), Chu (N6), Shenk and Hildebrandt (SH), Linsmaier and Skoog (LS)]	-	57.9 µg/g withanolide A	157.4 µg/g (2.7-fold increase)	[59]
Agrobacterium tumefaciens LBA 9402 and A4 strain	MS medium with 1.0 mg/L 6-benzylaminopurine (BA), 3% sucrose and solidified with 0.75% agar.	-	-	Maximum withaferin A of 0.44% DW withanolide D of 0.23% DW in the WSKHRL-1 and WSCHRL-22 respectively, induced by *A. rhizogenes* A4 strain	[87]
Agrobacterium rhizogenes strains R1000, MTCC 2364 and MTCC 532	MS medium	-	-	Maximum withaferin A of 6.17 mg/g DW and withanolide A (3.82 mg/g DW)	[88]

CONCLUSION AND PROSPECTS

Withania somnifera is one of the important medicinal plants with a high potential for therapeutic activities. The plant has been extensively used for general well-being and as one of the potential anti-cancer agents. It is also used to treat anxiety, stress related issues, antioxidants and many other such ailments. The principal bioactive molecules attributed to a large number of pharmacological activities are Withanolides, Withaferin-A, Withanone, withasomniferin and other bioactive molecules. These bioactive molecules with great medicinal importance have been used in formulating various drugs. Thus there is a need for the continuous supply of homogenous raw material in the form of micropropagated plants or even the cell lines, adventitious root cultures or hairy root cultures or any of the biotechnological approaches where good quality biomass can be obtained. Notable studies have shown the potential of obtaining plant biomass through micropropagation and also to obtain the biomass in the form of callus, adventitious roots, hairy roots or even development of elite germplasm with high content of bioactive molecules to meet the growing demand. Employment of elicitation strategies has also shown a considerable increase in the bioactive molecules concentration. Metabolic engineering approaches have helped in altering the biosynthetic pathway for enhanced production of some bioactive molecules. Deciphering the complete biosynthetic pathway of all the bioactive molecules and their pharmacological/therapeutic activities will boost the use of this plant in many more drug formulations. It also provides more opportunities to understand the events at the gene level and overexpress key genes using genetic engineering tools. The development of bioreactor scale production of secondary metabolites will help in meeting the homogenous supply of raw materials. There is also a need to perform studies in the field of genomics, proteomics, and metabolomics for enhanced productivity of a few bioactive molecules. With the advent of bioinformatics tools, we can easily understand the pharmacological efficiency of unexplored phytochemicals. Extensive clinical trials will also help us design remarkable formulations to treat various health problems.

REFERENCES

[1] Ahmad M, Dar NJ. 8 - *Withania somnifera*: Ethnobotany, pharmacology, and therapeutic functions. In: Bagchi D, Ed. Sustained Energy for Enhanced Human Functions and Activity. Academic Press 2017; pp. 137-54.
[http://dx.doi.org/10.1016/B978-0-12-805413-0.00008-9]

[2] Singh N, Bhalla M, De Jager P, Gilca M. An overview on ashwagandha: a Rasayana (rejuvenator) of Ayurveda. Afr J Tradit Complement Altern Med 2011; 8(5S) (Suppl.): 208-13.
[http://dx.doi.org/10.4314/ajtcam.v8i5S.9] [PMID: 22754076]

[3] Dar NJ, Hamid A, Ahmad M. Pharmacologic overview of *Withania somnifera*, the Indian Ginseng. Cell Mol Life Sci 2015; 72(23): 4445-60.
[http://dx.doi.org/10.1007/s00018-015-2012-1] [PMID: 26306935]

[4] Mayola E, Gallerne C, Esposti DD, *et al*. Withaferin A induces apoptosis in human melanoma cells through generation of reactive oxygen species and down-regulation of Bcl-2. Apoptosis 2011; 16(10): 1014-27.
[http://dx.doi.org/10.1007/s10495-011-0625-x] [PMID: 21710254]

[5] Yang ES, Choi MJ, Kim JH, Choi KS, Kwon TK. Combination of withaferin A and X-ray irradiation enhances apoptosis in U937 cells. Toxicol in vitro 2011; 25(8): 1803-10.
[http://dx.doi.org/10.1016/j.tiv.2011.09.016] [PMID: 21964475]

[6] Gupta A, Singh S. Evaluation of anti-inflammatory effect of *Withania somnifera* root on collagen-induced arthritis in rats. Pharm Biol 2014; 52(3): 308-20.
[http://dx.doi.org/10.3109/13880209.2013.835325] [PMID: 24188460]

[7] Khan MA, Subramaneyaan M, Arora VK, Banerjee BD, Ahmed RS. Effect of *Withania somnifera* (Ashwagandha) root extract on amelioration of oxidative stress and autoantibodies production in collagen-induced arthritic rats. J Complement Integr Med 2015; 12(2): 117-25.
[http://dx.doi.org/10.1515/jcim-2014-0075] [PMID: 25803089]

[8] Mohanty I, Arya DS, Dinda A, Talwar KK, Joshi S, Gupta SK. Mechanisms of cardioprotective effect of *Withania somnifera* in experimentally induced myocardial infarction. Pharmacol Toxicol 2004; 94(4): 184-90.
[http://dx.doi.org/10.1111/j.1742-7843.2004.pto940405.x] [PMID: 15078343]

[9] Rawat V, Bisht P. Antibacterial activity of *Withania somnifera* against Gram-positive isolates from pus samples. Ayu 2014; 35(3): 330-2.
[http://dx.doi.org/10.4103/0974-8520.153757] [PMID: 25972723]

[10] Singh G, Kumar P. Evaluation of antimicrobial efficacy of flavonoids of *withania somnifera* L. Indian J Pharm Sci 2011; 73(4): 473-8.
[http://dx.doi.org/10.4103/0250-474X.95656] [PMID: 22707839]

[11] Alam N, Hossain M, Mottalib MA, Sulaiman SA, Gan SH, Khalil MI. Methanolic extracts of *Withania somnifera* leaves, fruits and roots possess antioxidant properties and antibacterial activities. BMC Complement Altern Med 2012; 12(1): 175.
[http://dx.doi.org/10.1186/1472-6882-12-175] [PMID: 23039061]

[12] Girish KS, Machiah KD, Ushanandini S, *et al.* Antimicrobial properties of a non-toxic glycoprotein (WSG) from *Withania somnifera* (Ashwagandha). J Basic Microbiol 2006; 46(5): 365-74.
[http://dx.doi.org/10.1002/jobm.200510108] [PMID: 17009292]

[13] Saggam A, Limgaokar K, Borse S, *et al.* Withania somnifera (L.) Dunal: Opportunity for clinical repurposing in COVID-19 management. Front Pharmacol 2021; 12: 623795.
[http://dx.doi.org/10.3389/fphar.2021.623795] [PMID: 34012390]

[14] Khanal P, Chikhale R, Dey YN, *et al.* Withanolides from *Withania somnifera* as an immunity booster and their therapeutic options against COVID-19. J Biomol Struct Dyn 2022; 40(12): 5295-308.
[http://dx.doi.org/10.1080/07391102.2020.1869588] [PMID: 33459174]

[15] Rani G, Grover IS. *In vitro* callus induction and regeneration studies in *Withania somnifera*. Plant Cell Tissue Organ Cult 1999; 57(1): 23-7.
[http://dx.doi.org/10.1023/A:1006329532561]

[16] Samadi AK. Chapter Three : Potential anticancer properties and mechanisms of action of withanolides.The Enzymes. Academic Press 2015; 37: pp. 73-94.
[http://dx.doi.org/10.1016/bs.enz.2015.05.002]

[17] Thirugnanasambantham SK. *In vitro* and omics technologies opens a new avenue for deciphering withanolide metabolism in *Withania somnifera*. Int J High Risk Behav Addict 2016; 8(7): 17-26.

[18] Kaur K, Singh P, Guleri R, Singh B, Kaur K, Singh V. Biotechnological approaches in propagation and improvement of *Withania somnifera* (L.) Dunal. In: Kaul SC, Wadhwa R, Eds. Science of Ashwagandha: Preventive and therapeutic potentials. Cham: Springer International Publishing 2017; pp. 459-78.
[http://dx.doi.org/10.1007/978-3-319-59192-6_22]

[19] Yokota T. Chapter 12 - Brassinosteroids.New Comprehensive Biochemistry. Elsevier 1999; 33: pp. 277-93.

[20] Rahier A, Taton M, Benveniste P. Inhibition of sterol biosynthesis enzymes *in vitro* by analogues of high-energy carbocationic intermediates. Biochem Soc Trans 1990; 18(1): 48-52.
[http://dx.doi.org/10.1042/bst0180048] [PMID: 2185086]

[21] Knoch E, Sugawara S, Mori T, *et al.* Third DWF1 paralog in Solanaceae, sterol Δ^{24}-isomerase, branches withanolide biosynthesis from the general phytosterol pathway. Proc Natl Acad Sci 2018; 115(34): E8096-103.
[http://dx.doi.org/10.1073/pnas.1807482115] [PMID: 30082386]

[22] Shasmita , Rai MK, Naik SK. Exploring plant tissue culture in *Withania somnifera* (L.) Dunal: *in vitro* propagation and secondary metabolite production. Crit Rev Biotechnol 2018; 38(6): 836-50.
[http://dx.doi.org/10.1080/07388551.2017.1416453] [PMID: 29278928]

[23] Kulkarni AA, Thengane SR, Krishnamurthy KV. Direct *in vitro* regeneration of leaf explants of *Withania somnifera* (L.) Dunal. Plant Sci 1996; 119(1-2): 163-8.
[http://dx.doi.org/10.1016/0168-9452(96)04462-7]

[24] Bimal KG, Eun SS, Eun HK, Kabir L, Chang YY, Ill MC. Direct shoot organogenesis from petiole and leaf discs of *Withania somnifera* (L.) Dunal. Afr J Biotechnol 2010; 9(44): 7453-61.
[http://dx.doi.org/10.5897/AJB10.1250]

[25] Kulkarni AA, Thengane SR, Krishnamurthy KV. Direct shoot regeneration from node, internode, hypocotyl and embryo explants of *Withania somnifera*. Plant Cell Tissue Organ Cult 2000; 62(3): 203-9.
[http://dx.doi.org/10.1023/A:1006413523677]

[26] Fatima N, Anis M. Role of growth regulators on *in vitro* regeneration and histological analysis in Indian ginseng (*Withania somnifera* L.) Dunal. Physiol Mol Biol Plants 2012; 18(1): 59-67.
[http://dx.doi.org/10.1007/s12298-011-0099-x] [PMID: 23573041]

[27] Mishra T. An efficient regeneration of *Withania somnifera* (L.) Dunal through direct organogenesis. Medicinal Plants - International Journal of Phytomedicines and Related Industries 2014; 6(2): 143-6.
[http://dx.doi.org/10.5958/0975-6892.2014.00483.3]

[28] Saema A. Misra. Rapid *in vitro* plant regeneration from nodal explants of *Withania somnifera* (L.) Dunal: A valuable medicinal plant. Int J Sci Res 2015; 4(6): 1649-52.

[29] Kumar OA, Jyothirmayee G, Tata SS. Multiple shoot regeneration from nodal explants of Ashwagandha (L.) Dunal *Withania somnifera*. Asian J Exp Biol Sci 2011; 2(4): 636-40.

[30] Jhankare A, Tripathi MK, Tiwari G, Pandey A, Patel RP, Patidar H. Efficient plantlet regeneration from cultured nodal segment of *Withania somnifera* (L.) Dunal. Plant Cell Biotechnol Mol Biol 2013; 14(3-4): 99-110.

[31] Goswami B, Khan S, Banu TA, Akter S, Islam M, Habib A. *in vitro* mass propagation of *Withania somnifera* (L.) Dunal an important medicinal plant of Bangladesh. Bangladesh J Bot 2022; 51(2): 191-7.
[http://dx.doi.org/10.3329/bjb.v51i2.60414]

[32] Kaur K, Kaur K, Bhandawat A, Pati PK. *In vitro* shoot multiplication using meta-Topolin and leaf-based regeneration of a withaferin A rich accession of *Withania somnifera* (L.) Dunal. Ind Crops Prod 2021; 171: 113872.
[http://dx.doi.org/10.1016/j.indcrop.2021.113872]

[33] Udayakumar CK. *In vitro* plant regeneration from epicotyl explant of *Withania somnifera* (L.) Dunal. J Med Arom Plant Sci 2013; 7: 43-52.

[34] Nayak SA, Kumar S, Satapathy K, *et al*. *In vitro* plant regeneration from cotyledonary nodes of *Withania somnifera* (L.) Dunal and assessment of clonal fidelity using RAPD and ISSR markers. Acta Physiol Plant 2013; 35(1): 195-203.
[http://dx.doi.org/10.1007/s11738-012-1063-2]

[35] Sivanesan I. Direct regeneration from apical bud explants of *Withania somnifera* Dunal. Indian J Biotechnol 2007; 6: 125-7.

[36] Kanungo S. Direct organogenesis of *Withania somnifera* L. from apical bud. Int Res J Biotechnol 2011; 2: 58-61.

[37] Tata SS, Jyothirmayee G, Kumar OA. *In vitro* plant regeneration from mature seed explants of *Withania somnifera* (L.) Dunal, an important, rare and endangered medicinal plant. Not Sci Biol 2019; 11(4): 387-91.
[http://dx.doi.org/10.15835/nsb11410512]

[38] Rani G, Virk GS, Nagpal A. Callus induction and plantlet regeneration in *Withania somnifera* (L.) dunal. in vitro Cell Dev Biol Plant 2003; 39(5): 468-74.
[http://dx.doi.org/10.1079/IVP2003449]

[39] Dewir YH, Chakrabarty D, Lee SH, Hahn EJ, Paek KY. Indirect regeneration of *Withania somnifera* and comparative analysis of withanolides in *in vitro* and greenhouse grown plants. Biol Plant 2010; 54(2): 357-60.
[http://dx.doi.org/10.1007/s10535-010-0063-6]

[40] Chakraborty N, Banerjee D, Ghosh M, *et al*. Influence of plant growth regulators on callus mediated regeneration and secondary metabolites synthesis in *Withania somnifera* (L.) Dunal. Physiol Mol Biol Plants 2013; 19(1): 117-25.
[http://dx.doi.org/10.1007/s12298-012-0146-2] [PMID: 24381443]

[41] Rani A, Kumar M, Kumar S. *In vitro* callus induction and shoot regeneration from leaf explant of

Withania somnifera. Advances in Applied Research 2014; 6(1): 62.
[http://dx.doi.org/10.5958/j.2349-2104.6.1.011]

[42] Mishra. An efficient regeneration of *Withania somnifera* (L.) Dunal through direct organogenesis. Med. Plants - Int. J Phytomed Relat Ind 2014; 6(2): 143-6.

[43] Sabir F, Sangwan NS, Chaurasiya ND, Misra LN, Sangwan RS. *In vitro* withanolide production by *Withania somnifera* L. cultures. Z Naturforsch C J Biosci 2008; 63(5-6): 409-12.
[http://dx.doi.org/10.1515/znc-2008-5-616] [PMID: 18669028]

[44] Nagella P, Murthy HN. Establishment of cell suspension cultures of *Withania somnifera* for the production of withanolide A. Bioresour Technol 2010; 101(17): 6735-9.
[http://dx.doi.org/10.1016/j.biortech.2010.03.078] [PMID: 20371175]

[45] Nagella P, Murthy HN. Effects of macroelements and nitrogen source on biomass accumulation and withanolide-A production from cell suspension cultures of *Withania somnifera* (L.) Dunal. Plant Cell Tissue Organ Cult 2011; 104(1): 119-24.
[http://dx.doi.org/10.1007/s11240-010-9799-0]

[46] Ray S, Jha S. Production of withaferin A in shoot cultures of *Withania somnifera.* Planta Med 2001; 67(5): 432-6.
[http://dx.doi.org/10.1055/s-2001-15811] [PMID: 11488457]

[47] Sangwan RS, Chaurasiya ND, Lal P, *et al.* Withanolide A biogeneration in *in vitro* shoot cultures of ashwagandha (*Withania somnifera* DUNAL), a main medicinal plant in Ayurveda. Chem Pharm Bull 2007; 55(9): 1371-5.
[http://dx.doi.org/10.1248/cpb.55.1371] [PMID: 17827764]

[48] Sharada M, Ahuja A, Suri KA, *et al.* Withanolide production by *in vitro* cultures of *Withania somnifera* and its association with differentiation. Biol Plant 2007; 51(1): 161-4.
[http://dx.doi.org/10.1007/s10535-007-0031-y]

[49] Sivanandhan G, Mariashibu TS, Arun M, *et al.* The effect of polyamines on the efficiency of multiplication and rooting of *Withania somnifera* (L.) Dunal and content of some withanolides in obtained plants. Acta Physiol Plant 2011; 33(6): 2279-88.
[http://dx.doi.org/10.1007/s11738-011-0768-y]

[50] Sivanandhan G, Selvaraj N, Ganapathi A, Manickavasagam M. Effect of nitrogen and carbon sources on *in vitro* shoot multiplication, root induction and withanolides content in *Withania somnifera* (L.) Dunal. Acta Physiol Plant 2015; 37(2): 12.
[http://dx.doi.org/10.1007/s11738-014-1758-7]

[51] Khanam MN, Mohammad A, Saad Bin J, Mottaghipisheh J, Dezs"o C. Adventitious root culture—an alternative strategy for secondary metabolite production: A review. Agronomy 2022; 12(5): 1178.
[http://dx.doi.org/10.3390/agronomy12051178]

[52] Praveen N, Murthy HN. Production of withanolide-A from adventitious root cultures of *Withania somnifera.* Acta Physiol Plant 2010; 32(5): 1017-22.
[http://dx.doi.org/10.1007/s11738-010-0489-7]

[53] Sivanandhan G, Arun M, Mayavan S, *et al.* Optimization of elicitation conditions with methyl jasmonate and salicylic acid to improve the productivity of withanolides in the adventitious root culture of *Withania somnifera* (L.) Dunal. Appl Biochem Biotechnol 2012; 168(3): 681-96.
[http://dx.doi.org/10.1007/s12010-012-9809-2] [PMID: 22843063]

[54] Murthy HN, Praveen N. Influence of macro elements and nitrogen source on adventitious root growth and withanolide-A production in *Withania somnifera* (L.) Dunal. Nat Prod Res 2012; 26(5): 466-73.
[http://dx.doi.org/10.1080/14786419.2010.490914] [PMID: 21644171]

[55] Murthy HN, Praveen N. Carbon sources and medium pH affects the growth of *Withania somnifera* (L.) Dunal adventitious roots and withanolide A production. Nat Prod Res 2013; 27(2): 185-9.
[http://dx.doi.org/10.1080/14786419.2012.660691] [PMID: 22394118]

[56] Shajahan A, Thilip C, Faizal K, Mehaboob VM, Raja P, Aslam A. An efficient hairy root system for withanolide production in *Withania somnifera* (L.) dunal. In: Malik S, Ed. Production of plant derived natural compounds through hairy root culture. Cham: Springer International Publishing 2017; pp. 133-43.
[http://dx.doi.org/10.1007/978-3-319-69769-7_7]

[57] Ray S, Ghosh B, Sen S, Jha S. Withanolide production by root cultures of *Withania somnifera* transformed with *Agrobacterium rhizogenes*. Planta Med 1996; 62(6): 571-3.
[http://dx.doi.org/10.1055/s-2006-957977] [PMID: 17252504]

[58] Kumar V, Murthy KNC, Bhamid S, Sudha CG, Ravishankar GA. Genetically modified hairy roots of *Withania somnifera* Dunal: A potent source of rejuvenating principles. Rejuvenation Res 2005; 8(1): 37-45.
[http://dx.doi.org/10.1089/rej.2005.8.37] [PMID: 15798373]

[59] Murthy HN, Dijkstra C, Anthony P, *et al.* Establishment of *Withania somnifera* hairy root cultures for the production of withanolide A. J Integr Plant Biol 2008; 50(8): 975-81.
[http://dx.doi.org/10.1111/j.1744-7909.2008.00680.x] [PMID: 18713347]

[60] Sivanandhan G, Selvaraj N, Ganapathi A, Manickavasagam M. An efficient hairy root culture system for *Withania somnifera* (L.) Dunal. Afr J Biotechnol 2014; 13.
[http://dx.doi.org/10.4314/ajb.v13i43]

[61] Saravanakumar A, Aslam A, Shajahan A. Development and optimization of hairy root culture systems in *Withania somnifera* (L.) Dunal for withaferin-A production. Afr J Biotechnol 2012; 11.
[http://dx.doi.org/10.4314/ajb.v11i98]

[62] Praveen N, Murthy HN. Synthesis of withanolide A depends on carbon source and medium pH in hairy root cultures of *Withania somnifera*. Ind Crops Prod 2012; 35(1): 241-3.
[http://dx.doi.org/10.1016/j.indcrop.2011.07.009]

[63] Sivanandhan G, Rajesh M, Arun M, *et al.* Optimization of carbon source for hairy root growth and withaferin A and withanone production in *Withania somnifera*. Nat Prod Commun 2012; 7(10): 1934578X1200701.
[http://dx.doi.org/10.1177/1934578X1200701005] [PMID: 23156987]

[64] Praveen N, Murthy HN. Withanolide A production from *Withania somnifera* hairy root cultures with improved growth by altering the concentrations of macro elements and nitrogen source in the medium. Acta Physiol Plant 2013; 35(3): 811-6.
[http://dx.doi.org/10.1007/s11738-012-1125-5]

[65] Thilip C, Soundar Raju C, Varutharaju K, Aslam A, Shajahan A. Improved Agrobacterium rhizogenes-mediated hairy root culture system of *Withania somnifera* (L.) Dunal using sonication and heat treatment. 3 Biotech 2015; 5: 949-56.

[66] Ramirez-Estrada K, Vidal-Limon H, Hidalgo D, *et al.* Elicitation, an effective strategy for the biotechnological production of bioactive high-added value compounds in plant cell factories. Molecules 2016; 21(2): 182.
[http://dx.doi.org/10.3390/molecules21020182] [PMID: 26848649]

[67] Baldi A, Singh D, Dixit VK. Dual elicitation for improved production of withaferin A by cell suspension cultures of *Withania somnifera*. Appl Biochem Biotechnol 2008; 151(2-3): 556-64.
[http://dx.doi.org/10.1007/s12010-008-8231-2] [PMID: 18449479]

[68] Sivanandhan G, Arun M, Mayavan S, *et al.* Chitosan enhances withanolides production in adventitious root cultures of *Withania somnifera* (L.) Dunal. Ind Crops Prod 2012; 37(1): 124-9.
[http://dx.doi.org/10.1016/j.indcrop.2011.11.022]

[69] Sivanandhan G, Rajesh M, Arun M, *et al.* Effect of culture conditions, cytokinins, methyl jasmonate and salicylic acid on the biomass accumulation and production of withanolides in multiple shoot culture of *Withania somnifera* (L.) Dunal using liquid culture. Acta Physiol Plant 2013; 35(3): 715-28.

[http://dx.doi.org/10.1007/s11738-012-1112-x]

[70] Sivanandhan G, Kapil Dev G, Jeyaraj M, *et al.* Increased production of withanolide A, withanone, and withaferin A in hairy root cultures of *Withania somnifera* (L.) Dunal elicited with methyl jasmonate and salicylic acid. Plant Cell Tissue Organ Cult 2013; 114(1): 121-9.
[http://dx.doi.org/10.1007/s11240-013-0297-z]

[71] Ahlawat S, Saxena P, Ali A, Abdin MZ. *Piriformospora indica* elicitation of withaferin A biosynthesis and biomass accumulation in cell suspension cultures of *Withania somnifera.* Symbiosis 2016; 69(1): 37-46.
[http://dx.doi.org/10.1007/s13199-015-0364-9]

[72] Thilip C, Mehaboob VM, Varutharaju K, *et al.* Elicitation of withaferin-A in hairy root culture of *Withania somnifera* (L.) Dunal using natural polysaccharides. Biologia 2019; 74(8): 961-8.
[http://dx.doi.org/10.2478/s11756-019-00236-9]

[73] Kushwaha RK, Singh S, Pandey SS, Kalra A, Vivek Babu CS. Innate endophytic fungus, *Aspergillus terreus* as biotic elicitor of withanolide A in root cell suspension cultures of *Withania somnifera.* Mol Biol Rep 2019; 46(2): 1895-908.
[http://dx.doi.org/10.1007/s11033-019-04641-w] [PMID: 30706360]

[74] Sivanandhan G, Selvaraj N, Ganapathi A, Manickavasagam M. Enhanced biosynthesis of withanolides by elicitation and precursor feeding in cell suspension culture of *Withania somnifera* (L.) Dunal in shake-flask culture and bioreactor. PLoS One 2014; 9(8): e104005.
[http://dx.doi.org/10.1371/journal.pone.0104005] [PMID: 25089711]

[75] Patel N, Patel P, Kendurkar SV, Thulasiram HV, Khan BM. Overexpression of squalene synthase in *Withania somnifera* leads to enhanced withanolide biosynthesis. Plant Cell Tissue Organ Cult 2015; 122(2): 409-20.
[http://dx.doi.org/10.1007/s11240-015-0778-3]

[76] Patel N, Patel P, Khan BM. Metabolic Engineering: Achieving new insights to ameliorate metabolic profiles in Withania somnifera.Medicinal Plants - Recent Advances in Research and Development. Singapore: Springer Singapore 2016; pp. 191-214.
[http://dx.doi.org/10.1007/978-981-10-1085-9_7]

[77] Yousefian Z, Hosseini B, Rezadoost H, Palazón J, Mirjalili MH. Production of the anticancer compound withaferin a from genetically transformed hairy root cultures of *Withania Somnifera.* Natural Product Communications 2018; 13(8): 943-8.
[http://dx.doi.org/10.1177/1934578X1801300806]

[78] Singh AK, Kumar SR, Dwivedi V, *et al.* A WRKY transcription factor from *Withania somnifera* regulates triterpenoid withanolide accumulation and biotic stress tolerance through modulation of phytosterol and defense pathways. New Phytol 2017; 215(3): 1115-31.
[http://dx.doi.org/10.1111/nph.14663] [PMID: 28649699]

[79] Sharma A, Rana S, Rather GA, Misra P, Dhar MK, Lattoo SK. Characterization and overexpression of sterol Δ^{22}-desaturase, a key enzyme modulates the biosyntheses of stigmasterol and withanolides in *Withania somnifera* (L.) Dunal. Plant Sci 2020; 301: 110642.
[http://dx.doi.org/10.1016/j.plantsci.2020.110642] [PMID: 33218619]

[80] Jadaun JS, Sangwan NS, Narnoliya LK, *et al.* Over-expression of *DXS* gene enhances terpenoidal secondary metabolite accumulation in rose-scented geranium and *Withania somnifera* : Active involvement of plastid isoprenogenic pathway in their biosynthesis. Physiol Plant 2017; 159(4): 381-400.
[http://dx.doi.org/10.1111/ppl.12507] [PMID: 27580641]

[81] Ishnava P. Chauhan. Study of genetic transformation of medicinal plants, *Withania somnifera* (L.) Dunal by *Agrobacterium tumefaciens* (MTCC-431). Asian J Exp Biol Sci 2012; 3.

[82] Sivanandhan G, Selvaraj N, Ganapathi A, Lim YP. Up-regulation of Squalene synthase in hairy root culture of *Withania somnifera* (L.) Dunal yields higher quantities of withanolides. Ind Crops Prod

2020; 154: 112706.
[http://dx.doi.org/10.1016/j.indcrop.2020.112706]

[83] Ganeshan S, Natesan S, Andy G, Markandan M. An efficient hairy root culture system for *Withania somnifera* (L.) Dunal. Afr J Biotechnol 2014; 13(43): 4141-7.
[http://dx.doi.org/10.5897/AJB2014.14128]

[84] Srivastava Y, Tripathi S, Mishra B, Sangwan NS. Cloning and homologous characterization of geranylgeranyl pyrophosphate synthase (GGPPS) from *Withania somnifera* revealed alterations in metabolic flux towards gibberellic acid biosynthesis. Planta 2022; 256(1): 4.
[http://dx.doi.org/10.1007/s00425-022-03912-4] [PMID: 35648276]

[85] Grover A, Samuel G, Bisaria VS, Sundar D. Enhanced withanolide production by overexpression of squalene synthase in *Withania somnifera*. J Biosci Bioeng 2013; 115(6): 680-5.
[http://dx.doi.org/10.1016/j.jbiosc.2012.12.011] [PMID: 23313565]

[86] Jadaun JS, Kushwaha AK, Sangwan NS, Narnoliya LK, Mishra S, Sangwan RS. WRKY1-mediated regulation of tryptophan decarboxylase in tryptamine generation for withanamide production in *Withania somnifera* (Ashwagandha). Plant Cell Rep 2020; 39(11): 1443-65.
[http://dx.doi.org/10.1007/s00299-020-02574-4] [PMID: 32789542]

[87] Bandyopadhyay M, Jha S, Tepfer D. Changes in morphological phenotypes and withanolide composition of Ri-transformed roots of *Withania somnifera*. Plant Cell Rep 2007; 26(5): 599-609.
[http://dx.doi.org/10.1007/s00299-006-0260-0] [PMID: 17103214]

[88] Thilip C, Raja P, Mohamed Rafi K, Faizal KP, Thiagu G, Aslam A. Genetic transformation using *Agrobacterium rhizogenes* for the production of valuable anti-cancer compound, Withaferin-A from *Withania somnifera* (L.) Dunal. JARJ 2020; 1: 1-5.
[http://dx.doi.org/10.46947/jarj1120201]

CHAPTER 4

In Vitro Propagation and Phytochemical Screening of Some Important Medicinal Plants of Northern India-A Review

Rafiq Lone[1,*], **Shakir Ahmad Mochi**[1], **Younis Ahmad Hajam**[2], **Ibraq Khurshid**[3] **and Azra N. Kamili**[1]

[1] *Department of Botany, Central University of Kashmir, Ganderbal Jammu and Kashmir, India*

[2] *Department of Life Sciences and Allied Health Sciences, Sant Baba Bhag Singh University, Khiala Jalandhar, Punjab, India*

[3] *Department of Zoology, Central University of Kashmir, Ganderbal, Jammu and Kashmir, India*

Abstract: Plants are indispensable for the preservation of human life. They supply us with oxygen, food, fuel, and shelter while also holding a crucial role in disease treatment, such as cancer, diabetes, and tumors. Medicinal plants are harnessed across various cultures and nations as medicinal precursors. In today's era, biotechnological methods like tissue culture are vital for selecting, multiplying, and conserving medicinal plant genotypes. Regeneration under *in vitro* conditions notably enhances the production of high-quality plant-based medicines. Plant tissue culture techniques offer a unified approach for producing standardized phytopharmaceuticals, yielding consistent plant material for physiological characterization and active phytoconstituent assessment. While many medicinal plants are successfully regenerated under *in vitro* conditions, there are certain species that continue to be cultivated in soil, with their large-scale development through micropropagation remaining uncommon. The micropropagation technique employed for cloning these medicinal plants involves the utilization of various concentrations of plant growth regulators within a media variant (MS 1962). The process of plant regeneration is achieved through both organogenesis and embryogenesis, facilitated by the supplementation of auxins and cytokinins. In this context, this chapter provides a concise overview of the integrated micropropagation culture system designed for the effective propagation of medicinally significant specimens.

Keywords: Diseases, Microoperation, Medicinal plants, Murashige and skoog media.

---

[*] **Corresponding author Rafiq Lone:** Department of Botany, Central University of Kashmir, Ganderbal Jammu and Kashmir, India; E-mail: rafiqlone@gmail.com

Mohammad Anis & Mehrun Nisha Khanam (Eds.)
All rights reserved-© 2024 Bentham Science Publishers

INTRODUCTION

Plants play an indispensable role in supporting life on Earth. They have historically and will continue to provide essential resources such as daily sustenance, animal feed, fuel, shelter, recreation, and medicinal herbs for a significant portion of the global human population. The significance of medicinal plants lies in their biologically active compounds, which serve as the true agents of healing in medical processes. Throughout history, medicinal plants have been utilized for addressing various diseases, a practice evident in traditional usage recorded in both Vedic and post-Vedic texts. However, over the past two centuries, there has been a notable decline in the utilization of plant-based therapeutic systems, owing to the increasing emphasis on synthetic drug production. The advancement of modern medicine, marked by the discovery of antibiotics and corticosteroids, led to a decline in the utilization of plant-based remedies. Nonetheless, the adverse effects associated with their usage within the body have sparked a renewed attraction towards herbal medicine [1]. The World Health Organization (WHO) reports a growing global interest in traditional medicine, fostering a widespread availability of plant-derived drugs in health food stores across the globe, including affluent nations [2]. Unfortunately, this trend has triggered extensive harvesting of diverse wild plants, giving rise to significant challenges such as resource depletion and the endangerment of rare and vulnerable species. The resurgence of public interest in phytomedicines has led to heightened demand, coupled with the rapid expansion of pharmaceutical enterprises. However, this has resulted in rushed, unsustainable harvesting of medicinal plants for commercial purposes [3, 4]. In the Indian industry, over 95% of medicinal herbs are sourced from the wild. Various factors contribute to the extinction of these medicinal plants, including habitat destruction due to increased human activities such as settlements, agriculture, and development projects. Other factors encompass the introduction of non-native weeds, ecosystem stress from pollution and poisoning, global warming, greenhouse effects, and inappropriate chemical and pesticide usage. The illicit trade of rare and unique plant species, coupled with the degradation of forest regeneration capacity, has accelerated species extinction. Consequently, it is crucial to take action by initiating an effort to conserve existing germplasm before it becomes irretrievable. The objective of this chapter is to provide a concise overview of significant medicinal plants in North India, complemented by an outline of their phytochemical alterations Table 1.

Table 1. List of some important medicinal plants occurring in Northern India.

S. No.	Botanical Name and Family	Common Name	Explants	Life from	Ailments Treated	*Ex-vitro* Cultivation
01.	*Acorus calamus*	Vai	Rhizome	Herb	Diarrhoea, stomach pain, cough and swellings.	Multiple shoot regeneration (Mass propagation)
02.	*Aconitum heterophyllum* Wall. ex Royle	Patris	Roots	Herb	Roots are dried and grinded, boiled in water, then taken orally or used to cook rice which is eaten to cure joint problems. Treatment of skin diseases and healing of wounds.	Callus culture
03	*Gentiana kurroo* Royle	Neelikant	Apical Meristem	Herb	Roots and rhizomes are bitter tonic. Antiperiodic.	Rapid Micropropagation
04.	*Ajuga bracteosa* Wall Ex Benth.	Jain-a-adam	Leaves	Herb	Diuretic, diarrhoea and treatment of wounds.	Callus induction and multiple shoot induction
05.	*Meconopsisaculeta* Royle L.	GuleNeelam	Seeds	Herb	N/A.	*In vitro* plantlet regeneration
06.	*Artemisia absinthium* L.	Teethwan	Leaves	Herb	Anthelmintic, abdominal pain, fever and indigestion.	*In vitro* Mass Multiplication\ Regeneration
07.	*Arnebia benthamii* Wall. ex G.Don	Kahzaban	Shoot tips	Herb	Enhances lactation in women, cough and throat infection, root extract is mixed with oil to control hair fall.	Clonal multiplication (Mass propagation)
08.	*Artemisia amygdalina* Decne	Virteethwan	Whole plant	Herb	Abdominal pain, anthelmintic, and high fever.	Suspension culture

(Table 1) cont.....

S. No.	Botanical Name and Family	Common Name	Explants	Life from	Ailments Treated	Ex-vitro Cultivation
09.	*Picrorhiza kurro* Royle ex Benth	Kutki	Whole plant	Herb	Extracts from leaves and flowers are used to cure liver disorders and upper respiratory disease.	Mass Micropropagation
10.	*Saussurea lappa* C.B. ClarkeL.	Kuath	Shoot tips	Herb	Flowers are effective in the treatment of mental illness.	Callus culture/Large scale propagation
11.	*Viola canescens* Wallex.Roxb.	Bunafsha	Leaf derived calli	Herb	The whole plant is used to cure cough and Flu.	Somatic embryogenesis
12.	*Taxus wallichiana* L.	Pastul	Shoot cuttings	Tree	Fruits and bark of the tree are used to cure high-grade fever and body pain.	Adventitious root formation /Clonal propagation
13.	*Crocus sativus* L.	Kuang	Corm	Herb	Stigma of crocus is used to cure neurological disorders and sleep problems.	*In vitro* plantlet propagation
14.	*Lilium polyphyllum* D.Don ex Royle	-	Shoot	Herb	The bark is used as an astringent, leaves for diarrhoea and is also applied to wounds.	Callus culture
15.	*Acorus calamus* Linn	Bach	Rhizomes	Herb	Rhizomes are bitter tonic diuretic.	Clonal Propagation
16.	*Morus nigra.* L.	Tull	Shoot	Tree	Ripened fruits are used to treat constipation.	Micropropagation
17.	*Strawberry cv. Senga sengana*	N/A	Leaves	Herb	Fruits are good for reducing high cholesterol.	Callus Culture
18.	*Abies pindrow* (Royle ex D.Don)	Bunder	Stem segments	Tree	Anti-inflammatory and anti-rheumatism.	Callus Culture
19.	*Ajuja bracteosa* Wall ex Benth	-	Leaves, Petiole, Internode	Herb	Bark and leaves are astringent and hypoglycaemic.	Callus induction

(Table 1) cont.....

S. No.	Botanical Name and Family	Common Name	Explants	Life from	Ailments Treated	*Ex-vitro* Cultivation
20.	*Gerbera jamesonii* H. Bolus ex Hook	-	Flower heads	Herb	Flowers and leaves are used to cure cough and bronchitis.	Micropropagation
21.	*Bergenia ciliate* (Haw.) Sternb.	Pulfort	Seeds and leaf segments	Herb	Leaves are used to stop wound bleeding.	*In vitro* Regeneration\Callus culture
22.	*Thymus serphyllum* L	Banajvain	Whole plant	Herb	It is used externally as an antiseptic.	*In vitro* propagation
23.	*Colchicum luteum* Baker	Vir-kum-poash	Leaves Corms perianth, Filaent and seeds	Herb	Fresh corm extract is mixed with ghee and then applied externally to cure back pain.	Callus induction and regeneration
24.	*Podophyllum hexandrum* (Royle) T.S Ying	Wanwangun	Roots	Herb	Extract from roots is used to treat tumours, warts, diarrhoea and constipation.	Germplasm conservation
25.	*Rheum webbianum* Royle	Pambchalan	Leaves mid rib petioles and seeds	Herb	Crushed roots are mixed with ash to cure skin problems also, roots are used in healing wounds and skin treatment.	Mass micropropagation
26.	*Sambucus wightiana* Wall.	Brand	Flowers	Herb	Extract of roots is a diuretic, extract of leaf and roots is applied externally on livestock to cure foot and mouth disease.	Antimicrobial studies
27.	*Saussurea costus* (Falc.) Lipsch.	Kouth	Rhizome segments	Herb	Dried roots are boiled and used to cook rice which is used to cure joint problems.	Large scale *Ex-situ* Conservation
28.	*Saussurea lappa* (Falc.) Lipsch.	Chock dawa	Leaves	Herb	Leaves are boiled and applied on burns.	-

(Table 1) cont.....

S. No.	Botanical Name and Family	Common Name	Explants	Life from	Ailments Treated	*Ex-vitro* Cultivation
29.	*Alcea rosea* L.	Sazpuash	Seeds	Herb	Flowers are used to cure Tonsillitis.	Mass Micropropagation
30.	*Solanum nigrum* L.	Kambai	Flowers	Herb	Fruits are eaten raw, while an extract from leaves is used to treat Abdominal pain and jaundice.	*In vitro* regeneration and flower induction
31.	*Lavatera cashmeriana* Camb.	Sazzpuash	Seeds	Herb	Flowers are used to treat sore throat	Callogenesis
32.	*Salvia moorcroftiana* Wall. ex Benth.	Gulkan	Roots	Herb	Dried roots are crushed and made into a powder. The powder is applied to the affected portion externally.	Extraction of Mono sesquiterpenoid in leaves
33.	*Taraxicumofficinales* (L.) Weber ex F.H.Wigg	Hand	Leaves	Herb	Decoction of leaves is used to cure stomach cramps, internal ulcer, back pain, and tightening of vessels.	*In vitro* Micropropagation
34.	*Thymus linearis* Benth.	Jiand	Whole Plant	Shrub	Thyme is known to be expectorant.	*In Vitro* proliferative Activity
35.	*Inula royleana* D.C	Gugiphool	flower	Herb	Blub is used as a tonic.	*In vitro* micro propagation
36.	*Viola canescens* Wall ex Roxb.	Gulnakash	leaves	Herb	Extract of flowers is mixed with oil and applied to cure neck and joint problems. Dried leaves are boiled in water to cure foot fever.	Somatic Embryogenesis
37.	*Atropa acuminata* Royle Ex Lindl.	-	Petiole and Nodal explant	Herb	It is used in the treatment of biliary colic and whooping cough.	Rapid Micropropagation
38.	*Heracleum candicans* Wall.	-	Hypocotyl	Herb	Roots are nerve tonic.	Direct shoot regeneration

(Table 1) cont.....

S. No.	Botanical Name and Family	Common Name	Explants	Life from	Ailments Treated	*Ex-vitro* Cultivation
39.	*Hyoscamus niger* L.	Baazir bhang	Shoot tips	Herb	Flowers and leaves are used to treat rheumatism.	Micropropagation
40.	*Jurinea dolomiaea* Boiss	Dhup	Rhizome	Herb	Tuberous roots are used to cure various ailments.	Propagation
41.	*Gymnema sylvestre* R. Br.	Gurmar	Whole plant	Herb	The whole plant is used for manufacturing drugs for diabetes.	*In vitro* propagation
42.	*Inula racemosa* Hook. F	Pushkarmool	Roots	Herb	The whole plant is used to stop hiccups.	*In vitro* Propagation
43.	*Adhatoda vasica* Nees	Arusa	Whole plant	Herb	The whole plant is used to cure whooping cough and cold.	*In vitro* Propagation
44.	*Cichorium intybus* L.	Kaasnihundh	Shoot tip	Herb	The whole plant has anti-hepatotoxic properties.	Callus mediated shoot organogenesis
45.	*Angelica glauca* Edgew	Chihur	Rhizome	Herb	Flowers and leaves help in the treatment of dyspepsia.	*In vitro* propagation direct organogenesis
46.	*Asparagus racemosus* Willd.	Shatavari	Nodal and internodal explants	Herb	The whole plant is used to treat upset stomach (dyspepsia)	Callus induction

The most evident approach to analyze the ongoing pursuit of therapeutically effective novel medications, including anticancer drugs, antibacterial agents, and antihepatotoxic compounds, is to examine plants that have been selected for medicinal purposes over millennia. According to the "World Health Organization (WHO)," medicinal plants serve as the primary source for formulating medicines. Traditional medicines, containing substances derived from these plants are utilized by over 80% of individuals in developed nations Table **2**. However, a comprehensive study of such plants is essential to enhance our understanding of their attributes, safety, and efficacy.

Table 2. Phytochemical constituents of some medicinal plants.

S. No	Scientific Name	Secondary Metabolites	Effects	Applications
1	*Rheum webbianum* Royle	Anthraquinones *viz.* rhein, emodin, aloe-emodin, physcion and chrysophanol. Stilbene glycosides, including rhaponticin and the metabolite rhapontigenin	Positive	Roots, stems, leaves and leaf-stalks are purgative and are beneficial in treating indigestion, abdominal diseases, astringent, boils, purgative, wounds and flatulence. The roots are diuretic, laxative, purgative, febrifuge; used against indigestion, wounds and, gastritis *etc*.
2	*Arnebia benthamii* Wall	Terpenoids, flavonoids, saponins, tannins, polyphenols, alkaloids, anthraquinones and cardiac glycosides.	-	Antioxidant and antidiabetic activities.
3	*Gentiana kurroo* Royle	Iridoids, xanthones, C-glucoxanthonemangiferin, and C-glucoflavones)	-	Antibacterial, antioxidant, anti-arthritic, anti-inflammatory, analgesic activities and anti-diabetic activities.
4	*Bergenia ciliata* (Haw.) Sternb	Terpenoids, flavonoids and saponins	Negative	*B. ciliata* has been used for centuries in herbal formulations for the dissolution of kidney and bladder stones.
5	*Podophyllum hexandrum* (Royle) T.S.Ying	Podophyllotoxin	-	The semisynthetic analogues of podophyllotoxin are also used in the treatment of lung cancer, testicular cancer, neuroblastoma, hepatoma and other tumor diseases.
6	*Meconopsis aculeata* Royle	Terpenoids and alkaloids, phlobatannins, flavonoids	Positive	Flowers are analgesic and febrifuge.
7	*Ajuga bracteosa* L	Alkaloids, phenolics, flavonoids, tannins, terpenoids and saponins	-	Antimicrobial, antioxidant and anticancerous.
8	*Ajuga parviflora* L	Alkaloids, phenolics, flavonoids, tannins, terpenoids and saponins	-	Antimicrobial, antioxidant and anticancerous.

(Table 2) cont.....

S. No	Scientific Name	Secondary Metabolites	Effects	Applications
9	*Rumex nepalensis* Spreng.	Phenols, flavonoids, tannins, saponins, alkaloids, fixed oil	Positive	It exhibits pharmacological activities, including anti-inflammatory, antioxidant, antibacterial, antifungal, antiviral, insecticidal, purgative, analgesic, antipyretic, anti-algal, central nervous system depressant, genotoxic, wound healing and skeletal muscle relaxant activities.
10	*Crocus sativus* L.	Crocin, picrocrocin and safranin	-	The stigma of crocus shows anti-bacterial and antioxidant activities.
11	*Artemisia amygdalina* Decne	Alkaloids, cardiac glycosides, phenolics, steroids, tannins, terpenes, and other essential oils, with 1,8-cineole, 3-carene, artemisinin, artesunate, camphene, ludartin, p-cymene, santonin, α-cadinol, α-pinene, and β-pinene	Positive	Analgesic, antibacterial, anticancerous, anticonvulsant, sedative, antidepressant, antidiabetic, antifungal, antihelminthic, anti-inflammatory, antimalarial, antioxidant, antispasmodic antiulcer, antiviral, anxiolytic, sedative, and immunomodulatory properties.
12	*Saussurea costus* (Falc.) Lipsch.	Sesquiterpenoid lactones, lignins, and phytosterols	-	Therapeutic importance as an anti-gastric, antiasthmatic, antispasmodic, antimicrobial and anti-inflammatory agent.
13	*Hypericum perforatum* L.	Flavones (luteolin), flavonols (quercetin, kaempferol), biflavones (biapi-genin), amentoflavone, glycosides (rutin, hyperside, and isoquercitrin), myricetin, hyperin, oligomeric proanthocyanadins and miquelianin	Negative for steroids and glycosides	It is reported to have anticancerous and antiviral properties.
14	*Arisaema jacquemontii* Blume	Flavonoids, including flavones, flavanols and condensed tannins	Positive	The root extract prevented the growth of both Gram-positive and Gram-negative bacteria. It has antifungal activity. The antimicrobial and antioxidant activities of the extracts were positively associated with the total phenolic and flavonoid contents of the extract.

(Table 2) cont.....

S. No	Scientific Name	Secondary Metabolites	Effects	Applications
15	*Rheum emodi* Wall ex. Meissn	Anthraquinone (emodin, aleo-emodin, rhein, chrysophanol, physcion), stilbene (piceatannol, resvertrol)	Positive	It is used in treating various types of cancers and other ailments like jaundice, headache, migraine, paralysis, sciatica, asthma, diarrhoea and liver disorders, *etc.*
16	*Sambucus wightiana* Wall. ex. Wight & Arn	Steroids, tannins, phenolics, saponins, alkaloids, flavonoids	Positive	*Sambucus wightiana* is used in folk medicine in the treatment of skin diseases. The roots, leaves and berries are reported to be used for purgative properties. It is anti-inflammatory (bronchial tubes), expectorant, diuretic, diaphoretic and hypotensive.
17	*Solanum nigrum* L.	Flavonoid luteolin, oleuropein glucoside, alkaloid veremivirine, Myristic acid, Trimethylsilyl glycolic acid.	-	It has been extensively used in traditional medicine in India and other parts of the world to cure liver disorders, chronic skin ailments (psoriasis and ringworm), inflammatory conditions, fever, eye diseases, hydrophobia, cough and dropsy.
18	*Salvia moorcroftiana* Wall ex. Benth.	Phenalenone	positive	Its seeds and roots are used as an emetic, against cough, and also for haemorrhoids.
19	*Hyocymus Niger* L.	Alkaloids (hyoscyamine and scopolamine)	-	Powder of seeds is used for curing toothache. The plant has been used for the treatment of motion sickness, asthma and serves as an anaesthetic agent.
20	*Cichorium intybus* Linn.	Tannins, polysaccharides, sesquiterpenes, coumarin glycosides, anthocyanins, sesquiterpene lactones, phytoalexin, flavonoids	Positive	The root extract combined with sugary water is given in the form of 2 spoonfuls daily at bedtime to cure typhoid.

Acorus calamus L

Acorus calamus L (Acoraceae) is a medicinal plant of significant market importance and holds potential as a viable export crop. The propagation of *Acorus calamus* primarily relies on rhizomes; however, this method is inefficient and limits its production on a large-scale commercial level. A recent *in vitro* propagation method has been developed for Acorus calamus L., utilizing (MS, 1962) medium supplemented with various combinations of specific concentrations. These combinations include: 2,4-D (2 mg/L), 2,4-D (5 mg/L) +

BAP (5 mg/L), and Kn (2 mg/L), which induce leaf proliferation (caulogenesis). Additionally, BAP (2 mg/L), BAP (5 mg/L) + 2,4-D (5 mg/L), Kn (5 mg/L) + 2,4-D (5 mg/L) + BAP (5 mg/L) + NAA (5 mg/L), IBA (5 mg/L) + Kn (5 mg/L), and Kn (1 mg/L) + BAP (5 mg/L) + NAA (2 mg/L) + BAP (5 mg/L) + NAA (2 mg/L) promote shoot growth and rhizogenesis [5]. The survival rate under outdoor conditions was determined to be 75%. Root-bearing plantlets were successfully acclimatized in pots containing a mixture of sterilized soil and sandy soil (3:1 ratio). In the first case, complete regeneration of plants was observed in MS medium supplemented with BAP (5 mg/L) + NAA (5 mg/L), while in the second case, regeneration occurred with the addition of 2,4-D (mg/L) and Kn (5 mg/L). Notably, it has been documented that explants cultured on MS media enriched with Kn (1 mg/L) + BAP (5 mg/L) + NAA (2 mg/L) exhibited rapid root regeneration and the formation of a single leaf primordium. Additionally, research has shown that MS medium fortified with 2,4D (5 mg/L) + BAP (5 mg/L) led to the proliferation of multiple leaves, and when these shoots were sub-cultured on MS medium supplemented with IBA (1.5 mg/L), rhizogenesis was observed [6]. When subject to half-strength MS medium supplemented with IBA (4.9 µM), the same plant exhibited comparable outcomes [7]. The obtained rhizome, approximately 0.5 cm in length, was cultivated on MS medium supplemented with BAP (2 mg per liter), resulting in the development of 10 initial leaf primordia. On the other hand, when the MS medium was supplemented with 2,4-D (2 mg/L), leaf primordia were generated, but rhizogenesis did not occur. Consequently, it can be inferred that the proliferation of leaves was initiated by MS medium supplemented with 2,4-D (2 mg/L), (5 mg/L) + BAP (5 mg/L), and Kn (2 mg/L) [8].

Alcea rosea L. (Malvaceae)

Hollyhock, also known as Althaea rosea, finds application in both cosmetics and therapeutics due to its abundant tannins, carbohydrates, cyanide, and mucilage content. A micro-propagation technique was employed using MS (1962) medium along with various combinations of plant hormones, including BAP, NAA, IBA, and IAA. Explants were cultured under *in vitro* conditions, focusing on shoot tips for direct and indirect multiplication. The concurrent application of BAP and NAA yielded a higher count of shoots. IBA was incorporated to induce root formation in isolated shoots. Notably, laboratory conditions yielded a 60% survival rate for transplanted plantlets. Micropropagation is a widely employed technique for commercial plant propagation worldwide; however, the capacity for somatic organogenesis and plant regeneration varies among species. Plant hormones play a crucial role in tissue culture-based plant regeneration, with cytokinins particularly influential in shoot organogenesis [9]. Seed germination was documented on basal medium at half salt intensity, resulting in an average germination rate of 50%. Subsequently, shoot tips were cultivated under *in vitro* conditions using MS media, with the introduction of diverse combinations of phytohormones such as BAP, NAA, and IBA. Increasing the concentration of BAP from 1 µM to 7 µM led to an augmentation in the number of shoots, accompanied by callus formation. Similar outcomes were observed in a tissue culture study conducted on A. rosea. In a separate investigation, the authors [10] explored the impact of plant growth regulators (PGRs) on callogenesis in A. rosea cotyledonary explants, evaluating IAA, NAA, 2,4-D, IBA, BAP, and Kinetin. Additionally, the researchers examined callogenesis in nodal explants of A. rosea, noting that BAP was effective in inducing callus formation. Their findings indicated that the supplementation of 2,4-D was optimal for callogenesis initiation [11]. Cotyledons represent one of the diverse explant types utilized for callogenesis and organogenesis. Shoot tip explants cultivated under *in vitro* conditions were employed to achieve both direct multiplication (without callus formation) and indirect multiplication (with callus formation). For elongation, shoots were cultured in MS medium, with maximum elongation observed within two weeks. Isolated shoots were effectively cultivated in MS media supplemented with IBA (10 µM), resulting in healthy and robust root development.

Hyoscyamus niger L.

Hyoscyamus niger L. (Family: Solanaceae; Vernacular name: Bazir Bangh) encompasses diverse bioactive constituents, notably tropane alkaloids like hyoscyamine and scopolamine, of high significance to the pharmaceutical industry. These alkaloids exhibit notable anticholinergic properties. Additionally, Scopolamine-N-butyl bromide demonstrates spasmolytic activity [12]. A sterilization approach employing various sterilants has been devised, with successful seed sterilization achieved using $HgCl_2$ (0.02%) for 2 min [13]. Furthermore, a technique for the propagation of *Hyoscyamus niger via* shoot tip culture has been established [14]. Through the cultivation of shoot tips on MS media, employing diverse concentrations, as well as a combination of TDZ and BAP, regeneration was achieved. After an 8-week period of cultivation, the most robust outcomes in shoot number, shoot length, and percentage of shooting were attained through the employment of TDZ.

Seeds underwent treatment with various concentrations of a chemical sterilant. Notably, the highest sterilization efficiency, reaching 90%, was achieved using HgCl (0.02%) over a 20-minute interval, leading to 100% explant survival [15]. Post-sterilization, the seeds were introduced onto MS full media, yielding intriguing outcomes such as 100% seed germination within 3 weeks of cultivation. Under aseptic conditions, shoot tips cultivated *in vitro* were extracted and introduced onto both MS basal media and MS media supplemented with varying concentrations of BAP (0-16 μM). When placed on MS basal media, the explants displayed no growth. However, across all BAP concentrations, there was a consistent occurrence of indirect shoot regeneration accompanied by the development of Brownish Nodular Callus (BNC). The number of shoots exhibited a gradual rise up to BAP 4 μM, where the most substantial increase in average shoot number (7.2 cm) was achieved, boasting a 100% response rate. Concurrently, shoot tips were cultured on MS media with varying quantities of TDZ [16]. At all TDZ concentrations, a Brownish Nodular Callus (BNC) emerged at the explant base, followed by the indirect generation of multiple shoots and elongation on the same medium. Notably, the highest count of typical shoots (9 shoots) displayed a 100% response rate. In comparison to BAP, TDZ demonstrated enhanced responsiveness in terms of shoot formation. The promotion of shoot development was facilitated by the addition of a composite blend of auxins and cytokinins. When cultivating shoot tips under *in vitro* conditions, the utilization of MS medium supplemented with BAP+NAA and TDZ+NAA yielded shoot development, although the latter exhibited superior responsiveness as compared to BAP+NAA [17]. TDZ holds a pivotal role as a regulator under *in vitro* conditions, attributed to its multifaceted functions. Exogenous TDZ supplementation can counter apical dominance, thereby stimulating the breakage of lateral buds and the initiation of adventitious shoot growth [18]. A TDZ concentration of 2 μM exhibited superior results in terms of shoot regeneration from shoot tip explants. This concentration led to a higher percentage of shoot formation, an increased number of shoots per explant, and elevated shoot height [19]. Earlier investigations also highlighted that varying amounts of TDZ (0-2 mg per liter) and IBA (0-0.5 mg per liter) could induce shoot development in Latifolia. However, in terms of shoot regeneration, TDZ outperformed BAP, displaying greater success in regenerating shoots from henbane hypocotyls, cotyledons, and stem explants. Instead of utilizing a mixture of kinetin and BAP with IBA, MS medium supplemented with a blend of 0.91 μM TDZ and 0.98 μM IBA was employed [20]. Similarly, TDZ (0.05-1.0 μM) was also effective in inducing shoot growth in pigeon pea (Cajanus cajan) [21]. Conversely, the greatest number of shoots was achieved through various concentrations of BAP, showcasing similar outcomes. Comparable results were observed in the root tips of Ratula aquatica and Hyoscyamus niger, where the number of shoots exhibited variation corresponding to BAP concentrations.

Gymnema sylvestre R. Br.

Gymnema sylvestre R. Br (known as Gurmar in Hindi) holds significance as a medicinal plant; however, its population is declining in its natural habitat. This species is facing endangerment due to its consistent harvesting as a raw material in various pharmaceutical industries for the production of alternative medicines targeting diabetes, asthma, and eye diseases. In a controlled environment, diverse processes, including shoot multiplication, root establishment, and acclimatization, transpire during *in vitro* propagation of an explant [22]. Through the use of nodal segments (1-1.5 cm) from a five-year-old plant collected at three distinct locations, MS medium (1962) supplemented with 10.0 µM BA (N6-Benzyladenine) and 0.5 µM NAA was employed for aseptic cultivation [23]. Five nutritional media were utilized, which encompassed MS Medium, WPM, B5 (Gamborg Medium), SH (Schenk and Hildebrandt Medium), and NN (Nitsch and Nitsch Medium). The explants inoculated on B5 medium supplemented with 0.5 µM NAA, displayed the development of an apical bud along with two nodal segments, or alternatively, three nodal segments without an apical bud. This configuration was found to be more conducive for the growth of adventitious roots, exhibiting an 85% success rate in root development. Among the explants, one displayed a remarkable 4.27 cm root length on the 35th day of post-inoculation [24]. The process of shoot multiplication involves the cultivation of the explant on both basal medium and medium supplemented with different concentrations of BA, as this interaction significantly impacts the number of shoots generated [25]. Before inoculation, over a period of 15 days, the number of shoots per explant demonstrated a notably higher count, particularly on media such as MS, B5, SH, and NN. By the 30-day mark before inoculation, a considerable increase in the number of shoots per explant was observed only on B5 and NN media. Conversely, explorations conducted on the WPM medium yielded adverse outcomes, with a decrease in the number of shoots per explant recorded at both sampling stages.

Upon comparison of the study's outcomes, WPM exhibited insignificance at both the 15-day and 30-day intervals. In contrast, explants cultivated on NN medium displayed a noteworthy increase in the number of shoots per explant, showing increments of 17.7% and 23.6%, respectively [26]. A comparison between WPM and NN medium revealed that the latter yielded a higher number of nodes per shoot, with increments of 59% and 84% observed at 15 and 30 days, respectively. Prior to the 15-day inoculation period, the basal medium notably influenced the number of nodes per explant. At the 30-day mark prior to inoculation, the basal medium, as well as various doses of BA and their combinations, significantly impacted the number of nodes per explant [27]. In both sampling stages, the NN medium exhibited the highest number of nodes per explant, with the B5 medium being statistically comparable at the 15-day mark post-inoculation. Conversely, the WPM resulted in the lowest number of nodes per explant during both sampling phases [28]. The NN medium yielded a notably higher number of nodes per shoot, recording increments of 59% at 15 days and 84% at 30 days [29]. Prior to the 15-day inoculation period, the basal medium exerted a significant influence on the number of nodes per explant. By the 30-day mark following inoculation, the basal medium, along with varying doses of BA and their combinations, exhibited a substantial effect on the node count for each plant [30].

The B5 medium exhibited statistical equivalence at the 15-day mark after inoculation. Moreover, it resulted in a significant reduction in the number of nodes per explant during both sampling phases. Research findings indicate that the WPM medium contains 34.1% more nodes per explant, while the NN medium boasts 30.4% more nodes [31]. In contrast, the MS medium facilitated the formation of certain roots on each plant, showing statistical similarity to WPM, SH, and NN mediums. Notably, at the intervals of 28, 35, and 42 days prior to inoculation, the B5 medium demonstrated a higher count of roots on each plant, specifically 80.0%, 26.7%, and 70.0%, respectively, in comparison to the MS medium [32]. IAA, IBA, NAA, and coumarin significantly influence rooting percent at all sampling stages. Notably, NAA treatment proved highly effective throughout the sampling process [33]. IBA consistently displayed the lowest rooting (%) in all sampling phases, while remaining equivalent to IAA and coumarin. The addition of NAA led to a considerable enhancement in rooting: by 10.8% and 23.5% compared to IAA, IBA, and coumarin at 21 days; by 89.0% and 14.41% compared to IAA and coumarin at 28 days; and by 78.5%, 20.82%, and 10.09% compared to IAA, IBA, and coumarin at 35 days. A treatment of 5.0 μM auxins demonstrated high effectiveness during the initial two stages of sampling (21 and 28 days), while 15.0 μM exerted a stronger influence during the final phase of the sampling (35 days).

Abies pindrow (Royle ex D.Don)

Young nodal and internodal stem segments of *Abies pindrow* were introduced onto MS media supplemented with varying doses of 2,4-D, NAA, and BAP, both individually and in combinations. These segments were then kept in the dark for a duration of 3 weeks [34]. The MS medium treated with NAA (3.0 mg l-1) exhibited the highest percentage (10.33%) of callus induction on internodal stem explants devoid of bark. Subsequently, callus cultures were developed and maintained in a growth chamber for 41 weeks using the same medium to facilitate multiplication [35]. In regeneration media containing different concentrations of TDZ, BAP alone, and a combination of IBA, multiple cultures were sustained. The rooting of regenerated microshoots was facilitated by the application of MS medium with IBA (4.00 mg|l) and activated charcoal(200 mg|L) [36]. Nodal and internodal stem segments were employed as explants to induce a mass production system on MS media enriched with various concentrations of cytokinins and auxins [37]. Callus induction was achieved within 21 days of incubation in darkness, and the development of callus morphology was observed within 3 weeks of incubation in a culture room under diffused light, all on the same medium.

The initiation and development of callus were enhanced through the supplementation of MS medium with different combinations and doses of growth regulators. Notably, the highest mean % of callus formation, 10.33%, was achieved with NAA (3.0 mg per litre) treatment, followed by 5.66% when explants were cultured on media supplemented with 2,4-D (3.0 mg per litre) [34]. Nodal explants exhibited a lower percentage response (2.69%) for callus generation compared to internodal stem segments (5.0%). Notably, when internodal stem segments were cultured on MS medium supplemented with NAA (3.00 mg per litre), the highest percentage of callus (12.66%) was observed. Conversely, the lowest callus formation (0.44%) occurred when nodal segments were cultured on MS medium supplemented with 2,4-D (0.25 mg per litre). A more effective approach involved the balanced application of 2,4-D and NAA treatments in conjunction with 0.5 mg per litre of BAP to enhance the callus induction percentage from internodal explants. When BAP was combined with NAA, a higher induction in callus (8.56%) was observed, and a similar trend was noted with the supplementation of BAP on 2,4-D, resulting in a greater induction in callus (8.0%) from internodal explants. However, this value remained lower than the 12.66% achieved with solitary NAA (3.00 mg per litre) treatment. The callus displayed initial slow development, followed by accelerated growth in the subsequent two weeks. The callus exhibited a crumbly texture and a greenish-white color. Callus induction from the explants (nodal and internodal stem segments) was achieved approximately 6 weeks after culture. The resulting callus was then transferred to an MS medium supplemented with cytokinins and auxins at various concentrations [38]. Furthermore, the culture treated with BAP (0.25 mg/litre) exhibited minimal shoot regeneration (3.32%). To enhance shoot regeneration, 0.50 mg per litre of IBA was introduced in combination with TDZ (3.00 mg per litre), and another 0.50 mg per litre of IBA was added to BAP (5.00 mg/litre). The highest shoot regeneration was observed on MS medium supplemented with 5.00 mg/litre of BAP + 0.50 mg/litre of IBA, resulting in a regeneration rate of 24.00%. NAA (3.00 mg/litre) demonstrated an effective role in inducing and multiplying callus cultures. Notably, previous studies [39 - 41] have also employed MS media for callus induction in conifers. It has been demonstrated that successful callus induction requires a period of darkness. In the case of loblolly pine, callus was generated by utilizing various ratios of NAA and BAP. These findings underscore that plant growth regulators can effectively induce callus at different concentrations and these variations can potentially affect secondary metabolites production. The role of darkness in callus induction is also crucial [42]. Among these regulators, NAA emerged as the most influential for callus induction and multiplication, resulting in dense and greenish-white calli. The color of calli generated from different combinations of NAA and 2,4-D ranged from creamy to brownish hues [43]. A study focused on the impact of auxins and cytokinins on callus formation in various explants demonstrated that the utilization of MS media supplemented with BAP at 5.00 mg/litre and IBA at 0.50 mg/litre resulted in a higher regeneration percentage and an increased average shoot number [44].

Heracleum candicans Wall.

Heracleum candicans of the family Apiaceae is a fragile, medicinal herb indigenous to the Himalayan regions. Characterized by its delicate nature, it holds substantial demand within the pharmaceutical industry, particularly in the global market, owing to its diverse chemical constituents. Notably, it is abundant in Xanthotoxin, recognized for its application in treating leucoderma and in the formulation of suntan creams. Hypocotyl explants were employed for shoot regeneration. The explants were cultured on MS (1962) medium supplemented with plant growth hormones, including Auxins (Indole acetic acid - IAA, 2,4-dichlorophenoxyacetic acid - 2,4-D), cytokinins (kinetin and Benzyladenine BA) which fosters the initiation of shoot bud regeneration [45]. In some cultures, the addition of 2,4-D to the medium has led to the development of friable callus [46]. The shoots regenerated from this callus were acclimatized in vermiculite under standard conditions before being transferred to MS basal medium containing auxin for root induction.

The addition of auxins and cytokinins, either alone or in combination, has resulted in an accelerated proliferation of shoot buds. The development of adventitious shoot buds was directly observed from hypocotyl explants within 15-30 days. Among the growth hormones, 2,4-D (3mg|L) exhibited a more favourable response in terms of shoot bud proliferation [47, 48]. The most significant response was noted when BAP(3mg|L) was supplemented, resulting in an average of 4.5 shoots per explant within 20 days of inoculation. However, it has been observed that the number of shoots formed from each explant decreased after the treatment with BAP at 3 mg|L. When inoculated on MS media supplemented with Kinetin, the number of shoots has been observed to increase steadily up to 4 mg per litre, showing a proportional relationship to the increasing concentration of kinetin. In contrast, BAP at 3 mg per litre and IAA at 2 mg per litre yielded better results, with an average of 9.4 shoots on each explant. This result was better than the outcome achieved when BAP, kinetin, and 2,4-D were supplied alone [49]. When regenerated shoots were cultivated on MS media containing NAA, it was observed that within a 10-day period, young roots developed. The rooted plantlets were transplanted into pots filled with vermiculite and were successfully acclimated in a greenhouse. Various factors influence the regeneration of plants under *in vitro* conditions, including environmental conditions, media composition, explant sources, plant growth hormones, and genotypes [50 - 52]. It has been demonstrated that hypocotyls yield better results as responsive explants for shoot organogenesis in *Cuminum cyminum L* [53]. Similarly, the use of MS medium supplemented with 8.8 μM BAP resulted in the direct development of adventitious buds from hypocotyl explants in *Millettia pinnata (L.)* [54].

Jurinea dolomiaea Boiss

Jurinea dolomiaea, is an endangered medicinal and aromatic shrub native to the Himalayan region, particularly Kashmir, due to its extensive exploitation [55]. Various propagation methods have been employed, including rhizome cutting cultivation, aimed at preserving the species by ensuring a steady supply of planting materials for cultivation. Scaling up the cultivation of this plant could potentially contribute to bolstering its population in the wild. Employing ex situ conditions involving factors like soil textures, moisture levels, and different concentrations of plant growth regulators such as Indole acetic acid (IAA), Indole butyric acid (IBA), and Gibberellic acid (GA3) supplementation could facilitate sprouting and enhance the survival rate of rhizome cuttings, thereby aiding in species preservation [56].

The successful propagation of *J. dolomiaea* through rhizome cuttings is influenced by soil texture and field capacity. The sprouting rate of *J. dolomiaea* rhizome cuttings varies based on the quality of soil and moisture content. The highest sprouting rate of 80.5±4.2% was observed in soil with a sand to soil ratio of 1:1, followed by 63.8±4.8% in soil with a 1:2 sand to soil ratio. Notably, no sprouting occurred on pure sand. Interestingly, the sprouting rate was found to be influenced more by field capacity than soil texture. The highest sprouting rate occurred at a field capacity of 12%, followed by full field capacity, and finally, at 14% field capacity. These findings highlight the significant impact of soil texture and moisture levels on the successful rhizome cutting propagation of *J. dolomiaea*. The application of phytohormones can have a negative impact on vegetative propagation. When rhizome cuttings of *J. dolomiaea* were treated with GA3 at 25 ppm, sprouting was observed within 8 days. Interestingly, supplementation with GA3 at 25 ppm resulted in the highest shoot sprouting and percentage survival (rooted) at 94.45±0.24% and 83.34±0.38%, respectively, compared to the control treatments, which exhibited 77.78±0.43% sprouting and 44.45±0.51% rooting. Conversely, the supplementation of IAA at 25 ppm, 50 ppm, and 100 ppm led to significantly lower sprouting and rooting compared to the control. The application of GA3 proved more effective in promoting increased sprouting and rooting in various treatments. On the other hand, the use of IBA did not yield successful results, as no shoot sprouting or rooting was observed.

Arnebia benthamii Wall. ex G. Don

Arenebi benthamii (Wall. Ex G. Don) belongs to the family Boraginaceae. Its local name is "Kahzaban". This plant holds significant value as an essential "endangered Himalayan medicinal plant" and occupies the second rank in the list of prioritized medicinal plants for the Western Himalaya region. The propagation of shoots from shoot tip explants was achieved through the utilization of half-strength (MS) media, supplemented with varying concentrations of 6-benzyladenine (BA). Notably, a superior response, specifically multiple shoot formation, was attained with the application of 5µM BA [57].

An additional investigation was conducted to assess the combined impact of 6-benzyladenine (BA) and 1µM indole-3-butyric acid (IBA). Notably, when half-strength Murashige and Skoog (MS) medium was supplemented with 4µM BA and 1µM IBA, a higher number of multiple shoots were observed to develop. For the induction of roots from these shoots, various concentrations of IBA, indole--acetic acid (IAA), and naphthalene acetic acid (NAA) were employed. Among these, half-strength MS medium supplemented with 4µM IBA yielded the most robust root growth, contributing to an 80% survival rate of transferred plantlets under field conditions. In contemporary times, the tissue culture technique is increasingly employed for clonal proliferation and *in vitro* conservation of valuable germplasm that teeters on the brink of extinction. Different concentrations of 6-benzyladenine (BA) supplementation alone have been observed to induce the formation of multiple shoots. The highest average number of multiple shoots, reaching 9.7 per explant, was achieved with the supplementation of 5µM BA. However, an increase in the concentrations beyond 5µM of BA led to a reduction in the number of shoots. Similar findings have been reported in prior studies involving *Rotula aquatica* [58], *Hyoscyamus muticus* [59], *Atropa belladonna* [60], *Swertia chirata* [61], and *Glycyrrhiza glabra* [62]. Incorporation of thidiazuron (TDZ), a non-purine derivative with cytokinin-like activity, has been shown to enhance shoot formation from cotyledon and hypocotyl explants in *Arnebia euchroma*. Conversely, it has been documented that the addition of BA actually hampers shoot multiplication from shoot tip cultures in *A. benthamii*. A significant outcome concerning shoot multiplication was noted when a combination of 4µM BA and 1µM IBA was applied, resulting in a remarkable 15.3±1.6 shoots per plant. Correspondingly, in *Arnebia euchroma*, there were observations of organ development for leaf callus [63], and in Rheum emodi leaf explants, the emergence of adventitious shoot buds was observed, which further grew upon supplementation with a combined dose of BA and IBA [64]. For achieving optimal shoot multiplication in *A. benthamii*, the exogenous addition of phytohormones, specifically 4 µM BA and 1 µM IBA, proved to be efficacious. On the contrary, the addition of IBA facilitated root development in isolated shoots, whereas IAA and NAA did not stimulate root growth. The influence of IBA on root development in *A. euchroma* has been documented in earlier studies.

Atropa acuminata Royle Ex Lindl.

The induction of callus formation and shoot regeneration in petiole explants was pursued using various plant hormones, including BAP, Kn, IAA, NAA, IBA, and 2,4-D, both individually and in different combinations [65]. Interestingly, the supplementation of MS medium with BAP (3mg/L), either solely or in conjunction with IAA (2mg/L), exhibited the most favorable outcomes in terms of inducing regeneration in the callus. Remarkably, the combined treatments (MS+BAP and MS+BAP+IAA) resulted in the development of compact, light green callus within 18 days (80% success) and 38 days (60% success) of culturing, respectively. Comparatively, when MS medium was enriched with BAP (3mg/L), petiole explants exhibited a higher frequency of callus formation in comparison to using BAP (3mg/L) along with IAA (2mg/L), both in terms of the percentage of culture response and the duration required for callus development. The callus derived from the petioles was subsequently sub-cultured on MS medium, supplemented either with BAP (5mg/L) or BAP (3mg/L) in combination with IAA (2mg/L). Notably, the regenerated shoots exhibited lengths of 2.2±0.19 and 1.99±0.19 cm within 40, 48, and 53 days, respectively.

Upon sub-culturing the callus obtained from nodal explants on MS media with the inclusion of BAP (3mg/L) or BAP (2mg/L), regenerated shoots displayed an average shoot length of 2.0±0.20 and 1.9±0.23 cm, achieving success rates of 80% and 70%, respectively, within 14 and 15 days. Notably, shoots produced on MS medium supplemented with BAP (3mg/L) exhibited better health and greater length when compared to those regenerated on BAP (2mg/L), with quicker development [66]. The addition of BAP (3mg/L) to the MS medium led to the observation of shoot regeneration in Solanum nigrum L. When the cultivated shoots were subjected to *in vitro* conditions, they were cultured on both full and half-strength MS medium, supplemented with auxins such as IAA and IBA, either individually or in combination with cytokinins like BAP, to initiate the process of rooting. However, the percentage of cultures exhibiting rooting and the average number of roots per culture demonstrated a reduction [67].

Meconopsis aculeta Royle L.

Meconopsis aculeta Royle is an exceedingly endangered perennial alpine endemic angiosperm, exclusively found in the Himalayas (Kashmir). It thrives in a brief growing season and remains snow-covered for around 3-4 months during winter. In its natural habitat, the seeds of this species undergo an extended period of prechilling during winter before germinating in the subsequent spring season. The supplementation of exogenous phytohormones plays a pivotal role in the successful regeneration of plantlets from seed explants of *Meconopsis aculeta* under *in vitro* conditions. The application of growth hormones to plantlets during the pre-sowing seed phase significantly contributes to enhancing germination rates and regulating overall vigor [68]. Plant growth regulators are employed to regulate seed germination and seedling growth. Growth hormones play a crucial role in governing germination and vigor in different parts of plants [69]. In the context of seed germination, auxins, gibberellins, and cytokinins are primarily utilized across various plant species [70].

Seeds subjected to inoculation on MS medium, supplemented with Kinetin (6mg/L) and BAP (1mg/L), displayed germination within 33 days, resulting in a 60 percent cultural response. Conversely, the inclusion of Zeatin (1mg/l) in conjunction with NAA (0.1mg/l) in the basal MS medium led to germination in 45 days, yielding a 50% culture response rate. Furthermore, when the MS medium was augmented with BAP (2mg/l) in combination with NAA (1mg/l), germination occurred with a 45 percent culture response [71]. The addition of Zeatin in combination with NAA enhances culture responsiveness by countering the inhibitory effects of the seed coat, which not only covers the seed but also diminishes the germination rate in Meconopsis aculeta seeds. Various germination responses are applied as treatments to accelerate germination and ensure uniform emergence of seedlings. To overcome dormancy and facilitate seed germination under optimal conditions, the supplementation of Zeatin, BAP, and IAA has been found effective. The delay in seed germination is primarily attributed to physiological dormancy [72], which can be alleviated through the use of germination-regulating compounds (GRCs), also known as germination mitigating agents. These include Gibberellic acid, Zeatin, BAP, and Kinetin [73 - 77]. Research has indicated that the application of kinetin (6-furfurylaminopurine) to Meconopsis latifolia substantially enhances the likelihood of germination. Consequently, it can be inferred that successful germination of Meconopsis latifolia seeds necessitates an extended chilling treatment followed by the supplementation of GA3 under alternating light/dark conditions [78]. For the development of shoots, MS basal medium enriched with Kn (1mg/l) + NAA (0.1mg/l) and Zeatin (4mg/l) + IAA (1mg/l) was employed, with culture responses of 25% and 35%, and average shoot lengths of (1.75±1.00cm) and (1.20±1.00cm), respectively, following 40 and 29 days of treatment. Additionally, the utilization of MS medium supplemented with a combination of BAP (2mg/l) + NAA (1mg/l) led to a 35% culture response and an average shoot length of 1.10±1.00 cm, after 35 days of cultivation. Contrastingly, diverse studies have highlighted that the transplantation of shoots into MS media supplemented with plant growth regulators (PGRs) or free PGRs medium represents a more effective approach for augmenting shoot length [79]. Furthermore, the utilization of MS medium containing 2 or 5 mg/l benzyl adenine, kinetin, or isopentenyladenine, in conjunction with 0.2mg/l 2,4-dichlorophenoxyacetic acid, has demonstrated effectiveness in promoting seed germination and facilitating shoot development [80].

SUMMARY AND CONCLUSION

The utilization of medicinal plants has been deeply rooted in all cultures as a foundational source of medicine. Approximately 800 medicinal remedies are employed for treating various ailments in animals globally, with a particular emphasis on developing countries. In the modern age, there is a heightened focus on plants as medicinal sources; for instance, widely-used medications like aspirin and digitalis are derived from plant origins. Currently, the trade and commercialization of medicinal plants are experiencing substantial growth. The global herbal market is expanding at a rate of 7% annually. In the realm of plant-derived medicines, *in vitro* propagation or tissue culture methods hold significant potential for producing high-quality plant-based remedies. Well-established procedures now exist for the clonal multiplication of diverse medicinal plant species. This advancement facilitates the propagation of plants with consistent therapeutic qualities. In the contemporary era, the utilization of cell culture techniques for plant development is on the rise owing to their superior quality and the ease with which they enable the exploration of the synthetic pathways of secondary metabolites in economically valuable plants. Advancements in tissue culture methodologies have the potential to revolutionize the commercial production of endangered or exotic plants, along with their corresponding cells and the bioactive compounds they synthesize, in an improved and expedited manner. Consequently, significant progress has been achieved in the standardization of protocols for cultivating various medicinal plants, a development that has found application in different countries, including India.

REFERENCES

[1] Mishra N, Gupta RK. Conservation of medicinal plants in India: Tradition and neology.Herbal Medicine: Traditional Practices. Aavishkar Publishers Distributors India 2006; pp. 113-21.

[2] Pushpangadan P, Narayanan N. Medicinal plants.The Natural Resource of Kerala, Balachandran Thampi. Trivandrum: World Wide Fund for Nature–India 1997; pp. 312-30.

[3] Renuka C. Distribution of canes in Kerala and the need for their conservation. Proceedings of 9th All India Botanical Conference, IX–21. 65.

[4] Himadri K, Rao SS. Quantitative assessment of medicinal plants–II, chittoor district, andra pradesh, manphar. Tablerefrences 1998.

[5] Babar PS, Deshmukh AV, Salunkhe SS, Chavan JJ. Micropropagation, polyphenol content and biological properties of Sweet Flag (Acorus calamus): A potent medicinal and aromatic herb. Vegetos 2020; 33(2): 296-303.

[http://dx.doi.org/10.1007/s42535-020-00107-8]

[6] Altaf A, Shashidhara S, Rajasekharan PE, Hareesh Kumar V, Honnesh NH. *In vitro* regeneration of Acorus calamus—an important medicinal plant. J Curr Pharm Res 2010; 2(1): 36-9.

[7] Jadhav SK. *In vitro* mid term conversation of acorus calamus l. *via* . cold storage of encapsulated microrhizome brazilian archives of biology and technology. Biological and Applied Sciences 2017; 60.

[8] Amoo SO, Van Staden J. Influence of plant growth regulators on shoot proliferation and secondary metabolite production in micropropagated Huernia hystrix. Plant Cell Tissue Organ Cult 2013; 112(2): 249-56.
[http://dx.doi.org/10.1007/s11240-012-0230-x]

[9] Hill K, Schaller GE. Enhancing plant regeneration in tissue culture. Plant Signal Behav 2013; 8(10): e25709-, 25709.
[http://dx.doi.org/10.4161/psb.25709] [PMID: 23887495]

[10] Munir M, Hussain A, Haq I, *et al.* Callogenesis potential of cotyledonary explants of Althaea rosea L. from Pakistan. Pak J Bot 2012; 44: 271-5.

[11] Mushtaq H, Hussain A, Akram M, Shah FH. Growth conditions for callogenesis in Althaea rosea (L.). Cav Pak J Agric Res 1994; 15(1): 37-42.

[12] Whitaker C, Berjak P, Kolberg H, Pammenter NW, Bornman CH. Responses to various manipulations, and storage potential, of seeds of the unique desert gymnosperm, Welwitschia mirabilis Hook. fil. S Afr J Bot 2004; 70(4): 622-30.
[http://dx.doi.org/10.1016/S0254-6299(15)30201-5]

[13] Hawkes JG. The diversity of crop plants. Harvard University Press 2013.

[14] Bader GN, Rashid R, Ali T, Hajam TA, Kareem O, Jan I. Medicinal plants and their contribution in socio-economic upliftment of the household in gurez valley (J&K).Edible Plants in Health and Diseases. Singapore: Springer 2022; pp. 107-36.
[http://dx.doi.org/10.1007/978-981-16-4880-9_5]

[15] Shah D, Kamili AN, Sajjad N, Nazir N, Tyub S. Micropropagation of Hyoscyamus niger L. by shoot tip culture: An important medicinal plant of solanaceae family. IntJ Adv Res Sci Eng 2018; 7(4): 1880-8.

[16] Rani G, Talwar D, Nagpal A, Virk GS. Micropropagation of Coleus blumei from nodal segments and shoot tips. Biol Plant 2006; 50(4): 496-500.
[http://dx.doi.org/10.1007/s10535-006-0078-1]

[17] Park HY, Kim DH, Saini RK, Gopal J, Keum YS, Sivanesan I. Micropropagation and quantification of bioactive compounds in Mertensia maritima (L.) Gray. Int J Mol Sci 2019; 20(9): 2141.
[http://dx.doi.org/10.3390/ijms20092141] [PMID: 31052234]

[18] Thiha S. Effects of explants and growth regulators on *in vitro* regeneration of dragon fruit (hylocereus undatus haworth) (doctoral dissertation). 2019.

[19] Debnath SC. A two-step procedure for adventitious shoot regeneration from *in vitro* derived lingonberry leaves: shoot induction with TDZ and shoot elongation using zeatin. HortScience 2005; 40(1): 189-92.
[http://dx.doi.org/10.21273/HORTSCI.40.1.189]

[20] Barupal M, Kataria V, Shekhawat NS. *In vitro* growth profile and comparative leaf anatomy of the C3–C4 intermediate plant Mollugo nudicaulis Lam. In Vitro Cell Dev Biol Plant 2018; 54(6): 689-700.
[http://dx.doi.org/10.1007/s11627-018-9945-7]

[21] Singh ND, Sahoo L, Sarin NB, Jaiwal PK. The effect of TDZ on organogenesis and somatic embryogenesis in pigeonpea (Cajanus cajan L. Millsp). Plant Sci 2003; 164(3): 341-7.

[http://dx.doi.org/10.1016/S0168-9452(02)00418-1]

[22] Singh HP, Uma S, Selvarajan R, Karihaloo JL. Micropropagation for production of quality banana planting material in Asia-Pacific. Asia-Pacific Consortium on Agricultural Biotechnology (APCoAB),. New Delhi, India, 92.2011.

[23] Sudha CG, Krishnan PN, Pushpangadan P, Seeni S. *In vitro* propagation of Decalepis arayalpathra, a critically endangered ethnomedicinal plant. In Vitro Cell Dev Biol Plant 2005; 41(5): 648-54.
[http://dx.doi.org/10.1079/IVP2005652]

[24] Shah SN, Amjad MH, Ansari SA. Micropropagation of Gymnema sylvestre R. Br. Sky J Med Plant Res 2013; 2(3): 18-28.

[25] Sharma P, Rajam MV. Genotype, explant and position effects on organogenesis and somatic embryogenesis in eggplant (*Solanum melongena* L.). J Exp Bot 1995; 46(1): 135-41.
[http://dx.doi.org/10.1093/jxb/46.1.135]

[26] Zhou C. Development of micro-propagation in bigleaf maple (Acer macrophyllum) and screening for early markers preceding figured wood formation. (Doctoral dissertation, Science: Biological Sciences Department) 2018.

[27] Puhan P, Rath SP. Induction, development and germination of somatic embryos from *in vitro* grown seedling explants in Desmodium gangeticum L.: A medicinal plant. Res J Med Plant 2012; 6(5): 346-69.
[http://dx.doi.org/10.3923/rjmp.2012.346.369]

[28] S.R G, Monthony AS, Jones AMP. Basal media optimization for the micropropagation and callogenesis of Cannabis sativa L. BioRxiv 2020; 1-23.

[29] Gentile A, Jàquez Gutiérrez M, Martinez J, Frattarelli A, Nota P, Caboni E. Effect of meta-Topolin on micropropagation and adventitious shoot regeneration in Prunus rootstocks. Plant Cell Tissue Organ Cult 2014; 118(3): 373-81.
[http://dx.doi.org/10.1007/s11240-014-0489-1]

[30] Swain SS, Ray DK, Chand PK. ED-XRF spectrometry-based trace element composition of genetically engineered rhizoclones vis-à-vis natural roots of a multi-medicinal plant, butterfly pea (Clitoria ternatea L.). J Radioanal Nucl Chem 2012; 293(2): 443-53.
[http://dx.doi.org/10.1007/s10967-012-1796-9]

[31] Shi D. Effects of culture media and plant growth regulators on micropropagation of willow (Salix matsudana 'Golden Spiral') and hazelnut (Corylus colurna 'Te Terra Red). . Theses, Dissertations, and Student Research in Agronomy and Horticulture. 79. 2014.

[32] Kulkarni AA. Micropropagation and secondary metabolite studies in taxus spp. and *withania somnifera* (l.) dunal. Plant tissue culture division national chemical laboratory pune 411 008 2000.

[33] Sevik H, Guney K. Effects of IAA, IBA, NAA, and GA3 on rooting and morphological features of Melissa officinalis L. stem cuttings. ScientWorldJ 2013; 2013: 1-5.
[http://dx.doi.org/10.1155/2013/909507] [PMID: 23818834]

[34] Shah SN, Amjad MH, Ansari SA. Micropropagation of Gymnema sylvestre R. Br. Sky J Med Plant Res 2013; 2(3): 18-28.

[35] Sharma VK, Hänsch R, Mendel RR, Schulze J. A highly efficient plant regeneration system through multiple shoot differentiation from commercial cultivars of barley (Hordeum vulgare L.) using meristematic shoot segments excised from germinated mature embryos. Plant Cell Rep 2004; 23(1-2): 9-16.
[http://dx.doi.org/10.1007/s00299-004-0800-4] [PMID: 15221277]

[36] Biswas MK, Hossain M. Callus culture from leaf blade, nodal, and runner segments of three strawberry (Fragaria sp.) clones. Turk J Biol 2010; 34(1): 75-80.
[http://dx.doi.org/10.3906/biy-0708-7]

[37] Manoharan R, Tripathi JN, Tripathi L. Plant regeneration from axillary bud derived callus in white yam (Dioscorea rotundata). Plant Cell Tissue Organ Cult 2016; 126(3): 481-97.
[http://dx.doi.org/10.1007/s11240-016-1017-2]

[38] Bhat SJA, Gangoo SA, Geelani SM, Qasba SS, Parray AA. Callus culture and organogenesis in fir (Abies pindrow Royle). JCell Tis Res 2014; 14(3): 4653.

[39] Bari MA, Ferdaus KMKB, Hossain MJ. Callus induction and plantlet regeneration from *in vivo* nodal and internodal segments and shoot tip of dalbergia sissoo roxb. J Biosci 1970; 16: 41-8.
[http://dx.doi.org/10.3329/jbs.v16i0.3740]

[40] Nakazawa Y, Toda Y. Eucommiaulmoides Oliv. (Eucommiaceae): In-vitro culture and the production of iridoids, lignans, and other secondarymetabolites.Medicinal and Aromatic PlantsVIII. Berlin: Springer-Verlag 1995; pp. 215-331.

[41] Tang W, Ouyang F, Guo Z. Plant regeneration through organogenesis from callus induced from mature zygotic embryos of loblolly pine. Plant Cell Rep 1998; 17(6-7): 557-60.
[http://dx.doi.org/10.1007/s002990050441] [PMID: 30736635]

[42] Tang W. Micropropagation of loblolly pine by somatic organogenesis and RAPD analysis of regenerated plantlets. J For Res 2000; 11: 1-6.
[http://dx.doi.org/10.1007/BF02855486]

[43] Kaushal K, Nath AK, Kaundal P, Sharma DR. Studies on somaclonal variation in strawberry (fragaria x ananassa duch.) cultivars. Acta Hortic 2004; (662): 269-75.
[http://dx.doi.org/10.17660/ActaHortic.2004.662.39]

[44] Biswas MK, Roy UK, Islam R, Hossain M. Callus culture from leaf blade, nodal, and runner segments of three strawberry (Fragaria sp.) clones. Turk J Biol 2010; 34: 75-80.

[45] Bhattacharyya P, Kumaria S, Tandon P. High frequency regeneration protocol for Dendrobium nobile : A model tissue culture approach for propagation of medicinally important orchid species. S Afr J Bot 2016; 104: 232-43.
[http://dx.doi.org/10.1016/j.sajb.2015.11.013]

[46] Fu SF, Wei JY, Chen HW, Liu YY, Lu HY, Chou JY. Indole-3-acetic acid: A widespread physiological code in interactions of fungi with other organisms. Plant Signal Behav 2015; 10(8): e1048052.
[http://dx.doi.org/10.1080/15592324.2015.1048052] [PMID: 26179718]

[47] Sharma RK, Wakhlu AK. Regeneration of Heracleum candicans wall plants from callus cultures through organogenesis. J Plant Biochem Biotechnol 2003; 12(1): 71-2.
[http://dx.doi.org/10.1007/BF03263164]

[48] Sandhya B. Studies on Genetic Transformation in Arachis hypogea. Doctoral dissertation, Swami Ramanand Teerth Marthawada University 1999.

[49] Mahroofa J, Seema S, Farhana M, Irshad AN. Direct shoot regeneration from hypocotyl explants of Heracleum candicans Wall: A vulnerable high value medicinal herb of Kashmir Himalaya. Afr J Agric Res 2018; 13(28): 1419-24.
[http://dx.doi.org/10.5897/AJAR2017.12726]

[50] Bano R, Khan MH, Khan RS, Rashid H, Swat ZA. Development of an efficient regeneration protocol for three genotypes of Brassica juncea. Pak J Bot 2010; 42: 963-9.

[51] Zhang FL, Takahata Y, Xu JB. Medium and genotype factors influencing shoot regeneration from cotyledonary explants of Chinese cabbage (Brassica campestris L. ssp. pekinensis). Plant Cell Rep 1998; 17(10): 780-6.
[http://dx.doi.org/10.1007/s002990050482] [PMID: 30736591]

[52] Jana S, Shekhawat GS. Plant growth regulators, adenine sulfate and carbohydrates regulate organogenesis and *in vitro* flowering of Anethum graveolens. Acta Physiol Plant 2011; 33(2): 305-11.

[53] Dhir R, Shekhawat GS. *In vitro* propagation using transverse thin cell layer culture and homogeneity assessment in Ceropegia bulbosa Roxb. J Plant Growth Regul 2014; 33(4): 820-30.
[http://dx.doi.org/10.1007/s00344-014-9432-2]

[54] Tawfik AA, Noga G. Cumin regeneration from seedling derived embryogenic callus in response to amended kinetin. Plant Cell Tissue Organ Cult 2002; 69(1): 35-40.
[http://dx.doi.org/10.1023/A:1015078409682]

[55] Nagar DS, Jha SK, Jani J. Direct adventitious shoot bud formation on hypocotyls explants in Millettia pinnata (L.) Panigrahi : A biodiesel producing medicinal tree species. Physiol Mol Biol Plants 2015; 21(2): 287-92.
[http://dx.doi.org/10.1007/s12298-015-0293-3] [PMID: 25964721]

[56] Aryal S, Poudyal BH, Kunwar RM, Bussmann RW, Paniagua-Zambrana NY. Jurinea dolomitica Galushko Asteraceae.Ethnobotany of the Himalayas Ethnobotany of Mountain Regions. Cham: Springer 2021.
[http://dx.doi.org/10.1007/978-3-030-45597-2_135-1]

[57] Asma B, Irshad AN, Zahoor AK, Peerzada AS, Ali AR. Efficient propagation of an endangered medicinal plant Jurinea dolomiaea Boiss in the North Western Himalaya using rhizome cuttings under *ex situ* conditions. J Plant Breed Crop Sci 2014; 6(9): 114-8.
[http://dx.doi.org/10.5897/JPBCS2014.0473]

[58] Quadri RR, Kamili AN, Shah AM, Da Silva AJ. *In vitro* multiplication of Arnebia benthamii Wall. A critically endangered medicinal herb of the Western Himalaya. Funct Plant Sci Biotechnol 2012; 6(1): 54-7.

[59] Martin K. Rapid *in vitro* multiplication and *ex vitro* rooting of Rotula aquatica Lour., a rare rhoeophytic woody medicinal plant. Plant Cell Rep 2003; 21(5): 415-20.
[http://dx.doi.org/10.1007/s00299-002-0547-8] [PMID: 12789443]

[60] Grewal S, Koul S, Ahuja A, Atal CK. Hormonal control of growth, organogenesis & alkaloid production in *in vitro* cultures of Hyoscyamus muticus Linn. Indian J Exp Biol 1979.

[61] Stateva S, Desheva G. Adaptation of the Species Atropa belladonna L. Grown *in vitro* to the Environment. UARD Jubilee International Scientific Conference.

[62] Balaraju K, Agastian P, Ignacimuthu S. Micropropagation of Swertia chirata Buch.-Hams. ex Wall.: A critically endangered medicinal herb. Acta Physiol Plant 2009; 31(3): 487-94.
[http://dx.doi.org/10.1007/s11738-008-0257-0]

[63] Sharma S, Rathi N, Kamal B, Pundir D, Kaur B, Arya S. Conservation of biodiversity of highly important medicinal plants of India through tissue culture technology : A review. Agric Biol J N Am 2010; 1(5): 827-33.
[http://dx.doi.org/10.5251/abjna.2010.1.5.827.833]

[64] Manjkhola S, Dhar U, Joshi M. Organogenesis, embryogenesis, and synthetic seed production in Arnebia euchroma : A critically endangered medicinal plant of the Himalaya. *In Vitro*. Cell Dev Biol Plant 2005; 41(3): 244-8.
[http://dx.doi.org/10.1079/IVP2004612]

[65] Lal N, Ahuja PS. Propagation of indian rhubarh (rheum emodi wall.) using shoot-tip and leaf explant culture. Plant Cell Rep 1989; 8(8): 493-6.
[http://dx.doi.org/10.1007/BF00269057] [PMID: 24233537]

[66] Ahmed MR, Anis M. Role of TDZ in the quick regeneration of multiple shoots from nodal explant of Vitex trifolia L. : An important medicinal plant. Appl Biochem Biotechnol 2012; 168(5): 957-66.
[http://dx.doi.org/10.1007/s12010-012-9799-0] [PMID: 23065400]

[67] Rajput S, Agrawal V. Micropropagation of Atropa acuminata Royle ex Lindl. (a critically endangered medicinal herb) through root callus and evaluation of genetic fidelity, enzymatic and non-enzymatic

antioxidant activity of regenerants. Acta Physiol Plant 2020; 42(11): 160.
[http://dx.doi.org/10.1007/s11738-020-03145-6]

[68] Ahmad M, Wani TA, Kaloo ZA, Ganai BA, Yaqoob U, Ganaie HA. Germination studies of critically endangered medicinal angiosperm plant species meconopsis aculeta royle endemic to kashmir himalaya, India: A multipurpose species. Med Aromat Plants 2018; 7(1): 2167-0412.
[http://dx.doi.org/10.4172/2167-0412.1000315]

[69] Duggleby RG, McCourt JA, Guddat LW. Structure and mechanism of inhibition of plant acetohydroxyacid synthase. Plant Physiol Biochem 2008; 46(3): 309-24.
[http://dx.doi.org/10.1016/j.plaphy.2007.12.004] [PMID: 18234503]

[70] Ghane SG, Lokhande VH, Ahire ML, Nikam TD. Indigofera glandulosa Wendl. (Barbada) a potential source of nutritious food: Underutilized and neglected legume in India. Genet Resour Crop Evol 2010; 57(1): 147-53.
[http://dx.doi.org/10.1007/s10722-009-9496-1]

[71] Raghav A, Kasera PK. Seed germination behaviour of Asparagus racemosus (Shatavari) under *in-vivo* and *in-vitro* conditions. Asian J Plant Sci Res 2012; 2: 409-13.

[72] Malik K, Saxena P. Somatic embryogenesis and shoot regeneration from intact seedlings of Phaseolus acutifolius A., P. aureus (L.) Wilczek, P. coccineus L., and P. wrightii L. Plant Cell Rep 1992; 11(3): 163-8.
[http://dx.doi.org/10.1007/BF00232172] [PMID: 24213552]

[73] Peleg Z, Blumwald E. Hormone balance and abiotic stress tolerance in crop plants. Curr Opin Plant Biol 2011; 14(3): 290-5.
[http://dx.doi.org/10.1016/j.pbi.2011.02.001] [PMID: 21377404]

[74] Baskin JM, Baskin CC. Seeds: Ecology, biogeography and evaluation of dormancy and germination. New York: Academic Press 1998.

[75] Bewley JD, Black BM. Physiology and Biochemistry of Seed Germination, Part-II. New York: Springer Verlag 1982; pp. 32-4.
[http://dx.doi.org/10.1007/978-3-642-68643-6]

[76] Ungar IA. Alleviation of seed dormancy in Spergularia marina. Bot Gaz 1984; 145(1): 33-6.
[http://dx.doi.org/10.1086/337422]

[77] Kaber K, Beltepe S. Effect of kinetin and gibberellic acid in overcoming high temperature and salinity (NaCl) stresses on the germination of barley and lettuce seeds. Phyton 1989; 30: 65-74.

[78] Nautiyal AR. Removal of seed dormancy in Corylus colurna Linn. J Tree Sci 1993; 12: 103-6.

[79] Dar AR, Reshi Z, Dar GH. Germination studies on three critically endangered endemic angiosperm species of the Kashmir Himalaya, India. Plant Ecol 2009; 200(1): 105-15.
[http://dx.doi.org/10.1007/s11258-008-9436-8]

[80] Sivanesan I, Song JY, Hwang SJ, Jeong BR. Micropropagation of Cotoneaster wilsonii Nakai : A rare endemic ornamental plant. Plant Cell Tissue Organ Cult 2011; 105(1): 55-63.
[http://dx.doi.org/10.1007/s11240-010-9841-2]

CHAPTER 5

Phytochemistry, Antioxidants, Antimicrobial Activities and Edible Coating Application of *Aloe Vera*

Awad Y. Shala[1], Hayam M. Elmenofy[2], Eman Abd El-Hakim Eisa[3,4] and Jameel M. Al-Khayri[5,*]

[1] *Medicinal and Aromatic Plants Research Department, Horticulture Research Institute, Agricultural Research Center, Giza-12619, Egypt*

[2] *Fruit Handling Research Department, Horticulture Research Institute, Agricultural Research Center, Giza-12619, Egypt*

[3] *Department of Floriculture and Dendrology, Hungarian University of Agriculture and Life Science (MATE), 1118-Budapest, Hungary*

[4] *Botanical Gardens Research Department, Horticulture Research Institute, Agricultural Research Center (ARC), Giza-12619, Egypt*

[5] *Department of Agricultural Biotechnology, College of Agriculture and Food Sciences, King Faisal University, Al-Ahsa-31982, Saudi Arabia*

Abstract: *Aloe vera* (L.) Burm. f. is a medicinal plant that has gained widespread interest due to the distinctive biological activities associated with its biologically active phytocomponents. To combat the difficulties caused by microbe resistance, it is urgently necessary to investigate potent antimicrobials as a natural alternative to synthetic chemicals. This challenging task is attracting a lot of interest from the scientific community worldwide. The previous antimicrobial results of *A. vera* indicated its broad spectrum to treat a variety of infectious diseases, which will support the development of new herbal antimicrobial agents and avoid the side effects of conventional antibiotics as well as preserve the fruit quality and extend the shelf-life of various vegetables and fruits To take advantage of the prospective uses of this plant, the current review offers insight into the phytochemical composition, and its production-limiting factors, antimicrobial and antioxidant properties, as well as the promising use of *A. vera* in postharvest fruit-coating.

Keywords: *Aloe vera*, Antimicrobial, Anthraquinone, Antioxidant, Fruit coating, Polysaccharide.

---

* **Corresponding author Jameel M. Al-Khayri:** Department of Agricultural Biotechnology, College of Agriculture and Food Sciences, King Faisal University, Al-Ahsa-31982, Saudi Arabia; E-mail: jkhayri@kfu.edu.sa

Mohammad Anis & Mehrun Nisha Khanam (Eds.)
All rights reserved-© 2024 Bentham Science Publishers

INTRODUCTION

Aloe vera (L.) Burm. f. (also known as *Aloe barbadensis* Miller) is a medicinal plant belonging to the family Xanthorrhoeaceae that is assumed to have its origins in the arid regions of Southern Europe, Africa and Asia [1]. *A. vera* has acquired popularity due to its beneficial phytochemicals, which have potent therapeutic medicinal properties. It is frequently used in traditional medicine to treat wounds, minor burns, and skin irritations, as well as internally to treat numerous ailments, including constipation, coughs, ulcers and diabetes.

The name *A. vera* is derived from "Alloeh" (an Arabic word that means "shining bitter substances") and "vera" (Latin word for "true"). Moreover, because of its medical properties, the plant has earned the names "survival plant", "medicine plant" and "lily of the desert" [2]. *A. vera* (Fig. **1a**) is a flowering succulent Crassulacean acid metabolism (CAM) xerophyte, that develops water-storage tissue in the leaves inside the parenchymatous tissues to enable the plants to survive in dry environments. The plant has a bright yellow tubular flower. It is a perennial plant with whorled-shaped, green fleshy leaves along the stem.

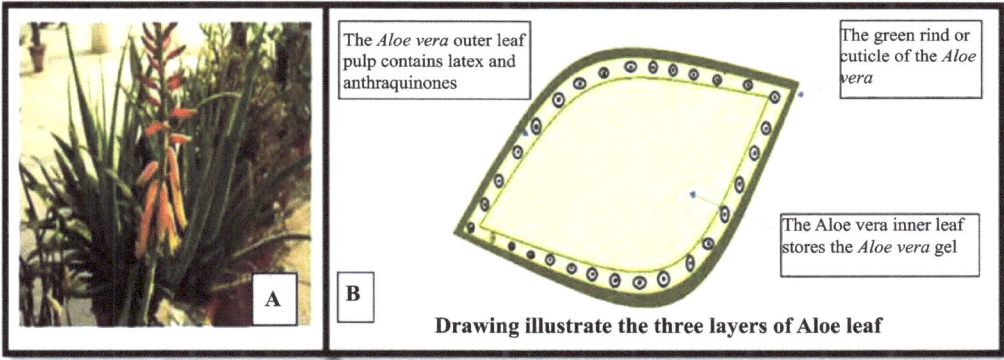

Fig. (1). (**A**) *Aloe vera* plant and (**B**) diagram of leaf layers. Source: This figure was reconstructed based on Boudreau and Beland [9].

The elongated and pointed leaves of the plant, which resembles a cactus and has common green leaves, have a thick outer rind and an inner full of pulpy substance [3]. It can be cultivated in saline soils as well as those irrigated with saline water [4 - 6] due to its salt tolerance capacity that is ascribed to its high K/Na ratio. Additionally, the plant can survive under severe as well as prolonged drought due to its storage-succulent leaves and drought-avoidance mechanisms [7]. The green dagger-shaped leaves of *A. vera* are the most used portion of the plant, and each leaf has three distinct layers (Fig. **1b**). Acemannan, the primary bioactive polysaccharide found in leaves, appears to be the greatest bioactive polysaccharide. The mucilaginous layer or inner gel contains 98% of water and is responsible for the majority of the plant's pharmacological purposes. The middle layer is made up of latex, an anthraquinone- and glycoside-rich bitter yellow sap that has laxative properties [8]. Finally, the outer, thick green covering or rind, which is composed of 15–20 cells and serves as protection, is also photosynthetically active; thus, synthesizing carbohydrates and proteins.

A. vera phytochemicals have been shown to have a variety of biological properties, including antioxidant [10, 11], anti-inflammatory [12], antimalarial [13], antiviral [14], anti-fungal and antimycoplasmic activity. This plant is used in a variety of industries, including cosmetics, pharmaceuticals, healthcare and food processing, and is particularly effective at inhibiting bacteria resistance as a promising natural antimicrobial alternative agent. Similarly, the anti-oxidant properties of its gel are related to the activities of glutathione peroxidase, superoxide dismutase enzymes and phenolic derivatives [15]. Furthermore, the gel encourages cell formation, improves skin healing, increases water retaining capacity and provides a cooling effect; additionally, as a drink, it maintains mucous membranes [16]. Contrary to the high incidence of fresh fruit and vegetable food-borne diseases, awareness of the health benefits and high nutritional value of fresh fruits and vegetables expands their use, leading to selection of different approaches for the microbiological protection of fruits and vegetables at the highest levels. Edible coatings, modified atmosphere packaging and radiation preservatives are the most significant innovative technologies [17]. Coatings, such as *A. vera*, which recently has attracted much attention in reducing microorganism proliferation and early germ count in fresh products, are regarded as a pleasant environment that are economical and easy to apply after harvest [18]. It is interesting to note that *A. vera* gel (AVG), which is crucial to maintaining the consistency of all fruit's bioactive components, can be used as a natural substitute for synthetic fungicides. Proteins, soluble sugars, polysaccharides, minerals and vitamins make up the majority of the AVG components, though *Aloe* species have relatively low levels of lipids (0.07 - 0.42%) [19, 20]. Compared to a lipid-based coating, a polysaccharide-based coating has a low water vapor barrier. Thus, the hydrophobic characteristics of AVG can be improved by inserting a lipid source in the composite and subsequently improving the barrier performance of the coating [21]. Additionally, the strong, orderly configuration of the polysaccharides is an excellent mechanical and gas barrier. Despite various findings on the constituents of polysaccharides in Aloe pulp, it is generally agreed that acetylated glucomannan molecules are essential for the dense mucilage-like properties of crude aloe gel. When applied to fruits and vegetables *A. vera* forms a semi-permeable coating which creates a barrier around the treated fruits and can alter the environment. This barrier restricts exchange of gas and water vapor, which slows down the metabolic processes and delays fruit ripening [22].

It naturally possesses and antimicrobial properties [23]. These operations would eventually extend the longevity of the treated product [24, 25]. Therefore, this review highlights *A. vera* bioactive compounds, an overview of their antimicrobial potency and discusses the impact of environmental factors on the accumulation of phytochemical constituents for optimizing pharmaceutical production, as well as, their applications in preserving fruits and vegetables as a promising edible coating.

BIOACTIVE COMPONENTS

Table **1** shows multiple phytocomponents in the *A. vera* plant parts. The majority of therapeutic activities of plants are strongly correlated to their polysaccharides content [26]. Nevertheless, the biological properties of *A. vera* were ascribed to synergistic effects between a variety of components [27]. "The polysaccharides are composed of linear chains of glucose and mannose molecules. Cellulose, hemicellulose, glucomannans, mannose derivative, and acetylated compounds are the most common polysaccharides. Acemannan and glucomannan are thought to be the two most important functional components of *A. vera*". Aloe possesses two categories of sugars, mannose and glucose as monosaccharides (simple molecules), and cellulose and acemannan as polysaccharides (complex molecules) [28]. *A. vera* contains anthraquinones. Among the anthraquinones, aloins A and B (together recognized as barbaloin) are situated in the outer green rind, aloin and aloe-emodin anthraquinones. The roots are rich in anthraquinones which have antiviral properties against the human influenza. The insoluble polysaccharides of cell wall polymers consist mainly of pectic substances as the major type of polysaccharides in the *A. vera* parenchyma, cellulose and also hemicelluloses [29 - 31], while the leaf skin also contains large amounts of polysaccharides-containing xylose.

Screening of different plant parts, particularly leaves and flowers, showed an abundance of various chemical components present: fatty acids, alkanes, phenolic acids/polyphenols, sterols, pyrimidines, alkaloids, dicarboxylic acids, indoles, aldehydes, alcohols, ketones, organic acids, carbohydrates, flavonoids, tannins, terpenoids and saponins in aloe leaves. The presence of 26 bioactive components in the ethanolic extract of *A. vera* leaves by the GC-MS analysis, which include 1-tetradecyne, N-hexadecanoic acid, Tridecanoic acid methyl ester, Hexadecanoic acid ethyl ester, Oleic acid, Tetracontane, 3,5,24-trimethyl-,(9,12,15-Octadecatrienoic acid methyl ester, Oxalic acid allyl pentadecyl ester, 9-Ocatadecenal, Oxalic acid, allyl hexadecyl ester, 1-Octanol 2-butyl-, Eicosane, Didodecyl phthalate, 1-Octadecyne, Squalene, Sulfurous acid hexyl pentadecyl ester, Phytol, 1-Iodo-2-methylundecane, Octadecane, 2-methyl, α-Tocopherol,

Table 1. Phytochemical compounds in the different plant parts of *A. vera*.

Plant part	Phytochemicals	References
Leaf	Isoaloeresin D and aloins	[32]
	Anthraquinones	[33]
	Tannin, Saponins, flavonoids, terpenoids	[2]
	Terpenoids, flavonoids, carbohydrates, Tannins, and alkaloids	[34]
	Aloin and aloe-emodin	[13]
	Phenolic compounds, alkaloids, steroids, glycosides, reducing sugars, terpenoids, tannins, flavonoids and saponin glycosides	[35]
	Alkaloids, saponins, tannins, sterols, flavonoids, phenol	[36]
	Alkaloids, tannins, saponins, flavonoids, glycosides, steroids.	[37]
	Flavonoids, steroids, terpenoids, proteins, phenols, carbohydrates, reducing sugar, starch, tannins, glycosides	[38]
Leaf gel	Phenolic acids/polyphenols, alkaloids, alkanes, sterols, indoles, dicarboxylic acids, fatty acids, pyrimidines, aldehydes, organic acids, alcohols and ketones	[39]
	Protein	[12]
	Polysaccharide	[40]
	Polyphenols, flavonoids, aloeresin A, aloesin, aloin, Aloe-emodin and aloenin	[11]
	Acemannan, Free sugars (glucose, fructose, mannose), organic acid (malic acid, citric, lactic and succinic acids), maltodextrin	[29]
	Phenolic compounds (anthrones, chromones) and luteolin glucosides	[41]
	Polysaccharides (glucose, mannose and galactose)	[42]
Green rind	Aloin A and B, aloinoside A and B, and 5-hydroxyaloin	[29]
	Polyphenol	[1]
	Flavonoids, tannins, sterols, alkaloids, oses, holosides, mucilages, triterpenes and reducing compounds metabolites	[43]
	Phenolic compounds (anthrones, chromones), luteolin and apigenin glucosides, p-coumaroylquinic acid	[41]
	Cinnamic acid, chlorogenic acid, aloe-emodin, orientin, aloeresin D, aloin A, aloeresin and aloin B	[44]
Fillet	Phenolic compounds, protein, fat, carbohydrates, oxalic acid, quinic acid, malic acid, tocopherols, fibers	[41]
Cell wall fibers	Galacturonic acid, total carbohydrate, uronic acid	[26]

Flower	Glucuronic acid, galactose, glucose and mannose	[42]
	Polyphenol	[1]
	Phenolics (caffeic acid, coumarin, GA, D-catechin, naringenin, resveratrol, vanillic acid, thymol, cinnamic acid, quercetin, and naringin), flavonoid	[10]
	Apigenin-2"-O-pentoxide-C-hexoside, flavonoids apigenin-6,8-C-diglucoside, apigenin-6-C-glucoside, and traces of luteolin glucoside derivatives	[41]
Root	Anthraquinones, aloesaponarin-I and aloesaponarin-II	[14]

Nonadecane, 2-methyl, Sulfurous acid, butyl heptadecyl ester, Vitamin E, 9,12-Octadecadienoic acid (Z,Z), phenylmethyl ester, Lupeol and Sitosterol. Additionally, eleven (11) phenolic compounds from flowers by high performance liquid chromatography, which include gallic acid, cinnamic acid, coumarin, vanillic acid, caffeic acid, D-catechin, resveratrol, naringenin, thymol, naringin and quercetin, and flavonoid were identified [42 - 44]. Also, eighteen (18) phenolic compounds in the leaf skin and flowers by using reversed-phase high performance liquid chromatography, including quercetin, sinapic acid, apigenin, gallic acid, kaempferol, chlorogenic acid, syringic acid, protocatechuic, vanillic acid, caffeic acid, coumaric acid, catechin, gentisic acid, rutin, ferulic acid, epicatechin, myricetin and quercetin that confirms both plant organs as natural sources of antioxidants. *A. vera* is said to include a wide range of organic and inorganic elements. Some inorganic minerals, including calcium, magnesium, zinc, iron and copper ions, may have therapeutic properties [45]. The increased acidity of *A. vera* gel may be attributed to the accumulation of organic acids, such as malic acid, in the pulp cells.

A. vera latex is found in the pericyclic cells at the leaf edges [46], and its secondary metabolites play no direct role in plant growth and development but function as a defensive mechanism to prevent it from being consume. Furthermore, latex is often used in beverage production due to its aromatic characteristics and bitter taste, as well as in pharmaceuticals due to its purgative properties attributed to the existence of hydroxyanthracene derivatives.

Today, several countries manufacture *A. vera* gel from the leaf pulp after removing the outer green rind, so aloe skin is treated as the main byproduct of gel manufacturing. However, the possibility of industrial utilization of *A. vera* skin instead of discarding it because it has a similar quantity of polysaccharides to the gel juice. Furthermore, the aloin content of the leaf skin is significantly higher than that of the gel. The gel after purification from unrequired components can be sold fresh or in powdered concentrate and used in various preparations for food, therapeutic, health and cosmetic purposes. The distinction between natural gel and commercial aloe gel powders is more relevant since adulteration is a considerable concern for the *A. vera* global market, especially given the high price of raw materials. Maltodextrin is still a common adulterant in *A. vera* gel powders. Malic acid, glucose and polysaccharide acemannan are natural constituents of aloe gel. Malic acid, in particular, was the sole organic acid in the fresh *A. vera* gel and plays a vital role in plant photosynthesis as a storage of carbon dioxide. On the contrary, commercial aloe gel powders contain large amounts of organic acids other than malic. Lactic, citric and succinic acids are the most abundant organic acids. Citric acid is a naturally occurring preservative that is often added to foods to enhance flavor and prevent oxidation. Whereas lactic acid as well as succinic acid, as indicators of bacterial fermentation and enzymatic degradation, should not be present in a high-quality aloe gel concentrate, and lactic acid, which is not a natural constituent of *A. vera* exists due to lactobacillus fermentation. Therefore, it is an undesirable attribute of aloe in its unprocessed form. Consequently, inadequate processing or storage can lead to chemical deterioration, insufficient thermal treatment, persistent enzyme activity, increased organic acid content in the gel throughout the citric acid cycle, and persistent enzyme activity.

Acetic acid is produced when acemannan is deacetylated during such chemical degradation, indicating that other chemical degradation has also taken place. Even at low concentrations, the presence of acetic acid is an indicator that some chemical degradation has occurred.

FACTORS AFFECTING PLANT CHEMICAL COMPOSITION

The growing conditions, plant age, geography, light intensity, harvest season, climate, water quality and water deficiency all have an impact on the chemical composition of *A. vera* plants. Additionally, plant species, which play an important role in the accumulation of bioactive compounds in plants during their growth and development should be taken into consideration, especially for aloe mass-production for pharmaceutical purposes as this will also reflect on plant efficiency as an antimicrobial agent.

The major bioactive components of β-polysaccharide depend extremely on the species hence, *A. vera* plants have a higher content of β-polysaccharide than *A. arborescens* [47]. According to information on plant component concentration and its active ingredients, anthraquinone aloin is primarily located in the outer green rind and not the inner parenchyma. Additionally, it was discovered that there was a relationship between aloe active constituents and leaf portions as well as plant age. Aloin anthraquinone, for example, was primarily found in the leaf skin and was higher in leaves from 3-year-old plants as compared with leaves from 5-year-old plants, while β-polysaccharides were concentrated in the leaf pulp only, particularly in the vascular bundle sheath and unaffected by plant age. It was also reported that aloin and β-polysaccharide contents were significantly reduced in plants under reduced light intensities. There was no significant variation in aloin concentration between plants exposed to the sun and shade when aloe plants were cultivated in partial shade (30% full sunlight) which increased the primary solute concentrations (glucose, galactose and quinic acid) [48]. Similarly, the highest concentration of aloin, total lipids, total proteins, total phenolics, polyamines (putrescine and spermidine) and total antioxidants in *A. vera* gel is obtained from plants harvested in the summer season as compared to plants harvested in spring and winter seasons.

Agro-climatic conditions lead to pronounced influences on the phytochemical diversity of *A. vera* plants collected from colder regions in Northern India exhibited higher antioxidant activity, as well as, total phenolic content as compared to plants procured from Southern India because of the sensitivity of *A. vera* to cold conditions. These are stress conditions for plants, causing them to generate additional phytochemicals to withstand adverse conditions. Furthermore, the lower temperature has a significant impact on the aloe phytochemical content; therefore, the quantity of aloin and aloe-emodin showed variability across climatic zones; their quantity declined with increasing temperature as negatively correlated; hence, colder climatic sites displayed higher amounts of both compounds than *A. vera* samples collected from warmer climate regions [49]. It was stated that aloe-emodin production was more abundant from plants collected from Indian (humid-subtropical and semi-arid zone) than from plants collected from an Indian tropical zone; thus anthraquinones production is significantly enhanced from plants grown in subtropical environment conditions.

Salt stress have a different osmotic adjustment in *A. vera* plants. The main osmotic modification of *A. vera* occurs when inorganic cations accumulate in the roots under salt stress and does not combine organic solutes of low molecular weight such as soluble sugars [50]. Furthermore, the salinity decreased total soluble sugars and glucose percentage while increasing sucrose percentage, which plays a vital role in reducing osmotic potential, implying that osmotic alteration in *A. vera* is not dependent on the accumulation of organic solutes under salinity stress conditions. Consequently, *A. vera* can be cultivated in saline soil irrigated with saline water with little or no membrane damage because of increased K^+/Na^+ ratio, decreased Na^+/Ca^{2+} ratio and improved osmotic adjustment brought on by salt. In this regard, The aloin and β-polysaccharides concentrations increased by irrigation with saline water. Additionally, during seawater stress, aloin concentration increased in the top and middle leaves without significant changes in the base leaves, and following a 42% seawater treatment polysaccharides decreased only in the base leaves without any variations in the top and middle leaves.

A. vera plants grown under water and salt stress recorded higher carbohydrates, glycosides, amino acids and proline content; on the other hand, aloin content was low [51]. As a result, amino acids and proline content increased in plants exposed to water and salt stress, demonstrating the role of the proline as an adaptive mechanism for water and salt stresses. The total sugars, soluble sugars, oligofructans, and polyfructans were significantly increased in the leaf tip and leaf base in plants exposed to water stress; however, all of these components are considered osmolytes which can protect aloe plants from water shortage [52]. Plants subjected to water deficit conditions showed remarkable variations in their active constituents. On the other hand, when water deficit increases the concentration of total protein in the plant leaves decreases. Interestingly. *A. vera* mucilage, acemannan bioactive polymer, was most affected by the application of water deficit, and as a result, mannose and glucose levels reduced as the water deficit increased from 40 to 60%. Mucilage dry matter, and pectic substances were also significantly enhanced by water deficit applied at 60%. *A. vera*, a crassulacean acid metabolism enables the preservation of water within the tissue because the innermost part of the leaf involves large thin-walled parenchyma cells in which water is retained in the form of a viscous mucilage [53]. Thus, the stomatal opening is maintained by resistance to high water stress in addition to the fact that the parenchyma's high-water content keeps stomata open at night and closed during the day, causing all exogenous gas exchanges to occur at night. Low temperature is another important factor for stomatal opening in CAM plants of arid and semi-arid lands that encourages the exchange of gas and fixes CO_2 [54].

ANTIOXIDANT ACTIVITY OF *A. VERA*

Reactive oxygen species, such as hydrogen peroxide, superoxide anion radical, and hydroxyl radical may contribute to cytotoxicity and metabolic changes. Lipid peroxidation is a free radical-related process that might take place in biological systems either enzymatically or non-enzymatically, and frequently results in cellular damage as a result of oxidative stress. Different toxic by-products of peroxidation can destruct other biomolecules, involving DNA mutation, membrane damage, and decrease membrane fluidity. Natural non-toxic antioxidants, on the other hand, play a vital role in the prevention of these damages [55]. As a result, the scientific community is becoming more interested in researching natural antioxidants as a possible replacement for synthetic antioxidants. Their main therapeutic effect is protecting human bodies against chronic diseases caused by oxidative stress. This is accomplished through scavenging free radicals. To prevent lipid peroxidation, natural antioxidants are preferred over synthetic antioxidants in the food industry to prevent lipid peroxidation for extending the shelf life of fruits and vegetables. Additionally, synthetic antioxidants have potential toxic and carcinogenic health effects [56].

The antioxidant activity of various parts of the aloe plant is influenced by plant part, development stage and plant age.

Aloesin derivatives displayed effective DPPH radical and superoxide anion scavenging activities, isorabaichromone together with p-coumaroylaloesin and feruloylaloesin indicated that the powerful isorabaichromone superoxide anion scavenging efficacy may have been attributed to its caffeoyl group [57]. The antioxidant effect of *A. vera* leaf ethanolic extract was related to its development stage. A three-year-old *A. vera* had higher amounts of polysaccharides and flavonoids in comparison to 2- and 4-year-old *A. vera*. Additionally, the extract from a three-year-old demonstrated higher radical scavenging capacity in comparison to BHT and α-tocopherol. The extracts of *A. vera* rind produced considerably higher free radical-scavenging activity than extracts of ethanol-extracted *A. vera* gel. The extracts of *A. vera* skin, by supercritical extraction of carbon dioxide and ethanol, exhibited significant antioxidant activities as compared to α-tocopherol and BHT, and the highest antioxidant activity of *A. vera* extracts was due to the rind chemical components [58]. According to Chunhui *et al.*, the antioxidant efficacy of extracted polysaccharides from the gel and leaf skin of *Aloe barbadensis* Miller showed significant superoxide radical scavenging activity and moderate ferrous chelating, hydroxyl radical scavenging, reductive and lipid peroxidation inhibition. Similarly, it was observed that leaf skin extract was more effective on DPPH scavenging activity and ferric reducing capacity than flower extract which is also associated with abundance of phenolics and its relation to antioxidant activity. *In vivo* and *in vitro* zebrafish models, a pure polysaccharide extracted from aloe gel demonstrated effective radical scavenging capacity.

Significant differences in radical scavenging activity were found in aloe leaf parts as well as plant age, with the outer green rind having the highest DPPH radical scavenging activity and oxygen-radical absorption efficacy compared to the inner parenchyma and whole leaf. Furthermore, DPPH radical scavenging was significantly higher in five-year-old plants than three-year-old plants which might be correlated to the older plants having a distinct secondary metabolite profile. Additionally, the ability to counteract free radicals can be ascribed to the existence of total phenolics, flavonoids or flavonols [59]. The methanolic extract of leaves displayed elevated antioxidant potential "in species gathered from Northern India than in Southern India" due to raised content of phenolic compounds, alkaloids, glycosides, flavonoids, and saponin glycosides. Moreover, because climatic fluctuations, especially temperature has a considerable impact on antioxidant activity, the maximum antioxidant activity was correlated with plant extracts collected from Indian highland and semi-arid zones; on the contrary, the lowest antioxidant efficacy resulted from accessions collected from Tropical

Indian zones. lowers. The ethanol extract of flowers suppressed DPPH free radicals with an inhibition constant (IC_{50}) of 0.25 mg/ml and showed increased Fe^{3+} reducing power with an observed (EC_{50}) 2.1 mg/ml and increased nitrite scavenging activity with an (IC_{50}) value of 0.92 mg/ml, reduced both hydroxyl and superoxide radicals with the IC_{50} values of 0.92 and 0.85 mg/ml, respectively. Ethanol extracts had a protective effect against free radical-induced DNA damage. When lipid peroxidation is excluded, the best antioxidant activities of the floral extract are attributed to the highest largest amount of phenols, particularly vanillic acid. Isolated compounds from *A. vera* resin exhibited potent anti-lipid peroxidation efficacy with IC_{50} values ranging from 201.7 ± 0.9 to 868.1 ± 0.9 µmol L^{-1} as compared with butyl hydroxy toluene (BHT $IC_{50} = 58.4 \pm 0.8$ µmol L^{-1}), which is due to the existence of -OH, $-OCH_3$ and methyl groups in isolated components that is considered prominent features for suppression of free radicals [60]. The tannin extract from Aloe leaves green rind and its fractions have moderate antiradical potential that might be due to the existence of fatty acids such as phytol and palmitic acids. However, there is no flavonoid that supports the antioxidant potency due to the synergistic action of complete bioactive components of *A. vera* gel extract rather than the individual component. Fillet extract demonstrated the maximum antioxidant capacity. The methanolic extract of *A. vera* demonstrated significant antioxidant efficacy for DPPH radical scavenging and Fe^{3+} reduction, which was correlated to the presence of phenolic compounds in plant extracts.

ANTIMICROBIAL ACTIVITY

Antibacterial Activity

According to Table 2 *A. vera* showed significant inhibitory effects against dangerous microorganisms that are responsible for causing numerous diseases in humans. This information will help to formulate antibacterial drugs from aloe plants. The whole-leaf components of *A. vera* plants are thought to have direct antibacterial effects due to anthraquinones and saponins, while polysaccharides have been linked to indirect bactericidal action through activation of phagocytic leukocytes to damage bacteria [61, 62]. Moreover, because of its anthraquinone glycoside concentration, extracts, whether gel or leaf, have antimicrobial properties. Its reputation as a plant that heals burns is supported by the fact that the aloe leaf had a larger inhibitory impact than the gel on the growth of the skin infection-causing *P. aeruginosa*.

Table 2. Antibacterial activity of *A. vera* extracts.

Plant Part used	The Solvent used for Extraction	Method used	Tested Microorganisms	References
Leaves	Inner leaf gel	Agar diffusion	*Shigella flexneri* *Streptococcus progenies*	[64]
Leaves	Leaves ethanol extract and gel	NA	*Staphylococcus aureus,* *Pseudomonas aeruginosa*	[16]
Inner leaves gel	Hexane, chloroform, methanol, water	Microtitre assay	*Shigella flexneri* *Staphylococcus aureus* *Enterobacter cloacae* *Enterococcus bovis*	[65]
Leaves	Aqueous, ethanol and acetone	Disc diffusion technique	*Escherichia coli,* *Staphylococcus aureus,* *Pseudomonas aeruginosa,* *Streptococcus pyogenes*	[2]
Leaves	Ethanolic extract	Agar diffusion method	*Pseudomonas aeruginosa,* *Proteus mirabilis,* *Escherichia coli,* *Morganella morganii,* *Klebsiella pneumoniae,* *Enterococcus bovis,* *Staphylococcus aureus*	[63]
Leaves	Epidermis methanol/ethyl acetate extract and gel water extract	Diffusion disc method	*Escherichia coli,* *Bacillus cereus,* *Bacillus licheniformis,* *Staphylococcus epidermidis* and *Saccharomyces boulardii*	[27]
Leaves	Distilled water, ethanol, and acetone solution	Disc diffusion technique	*Escherichia coli,* *Streptococcus pyogenes,* *Staphylococcus aureus,* *Pseudomonas aeruginosa*	[39]
Leaves	Ethanol, methanol and aqueous (hot) extracts	Agar well diffusion method	*Escherichia coli* *Staphylococcus aureus*	[66]

(Table 2) cont.....

Plant Part used	The Solvent used for Extraction	Method used	Tested Microorganisms	References
Leaves	Methanolic extract	Agar well diffusion method	*Proteus mirabilis* ATCC 43071, *Shigella flexneri* ATCC 12022, *Serratia marcescens* ATCC 27137, *Salmonella typhi* ATCC 1331, *Enterococcus faecalis* ATCC 29212, *Pseudomonas aeruginosa* ATCC 27853, *Klebsiella pneumonia* ATCC 700603, *Escherichia coli* ATCC 25922 *Staphylococcus aureus* ATCC 259323	[49]
Leaves	Hexane, methanol, ethyl acetate and distilled water	Agar well diffusion method	*Serratia marcescens* (MTCC No. 3124), *Pseudomonas aeruginosa* (MTCC No. 7436), *Bacillus cereus* (MTCC No. 135), *Escherichia coli* (MTCC No. 448)	[36]
Leaves	Hydroethanolic Extracts	NA	*Staphylococcus epidermidis* (clinical isolate Ibis 2999), *Staphylococcus aureus* (ATCC 11632), *Staphylococcus lugdunensis* (clinical isolate Ibis 2996), *Micrococcus flavus* (ATCC10240), *Listeria monocytogenes* (NCTC 7973), *Salmonella typhimurium* (ATCC 13311), *Pseudomonas aeruginosa* (ATCC 27853), *Escherichia coli* (ATCC 25922)	[41]
Leaves and roots	Ethanol extract	Disc diffusion method	*Acinetobacter baumannii, Escherichia coli, Proteus vulgaris Proteus mirabilis, Bacillus cereus, Enterococcus faecalis, Pseudomonas aeruginosa, Agrobacterium tumefacins, Salmonella typhii, Bacillus megaterium, Bacillus subtitis, Staphylococcus aureus, Streptococcus pyogenes*	[67]

(Table 2) cont.....

Plant Part used	The Solvent used for Extraction	Method used	Tested Microorganisms	References
Leaves and roots	Ethanol extract	Agar disc diffusion	*Acinetobacter baumanni, Proteus vulgaris, Escherichia coli, Pseudomonas aeruginosa, Bacillus cereus, Staphylococcus aureus, Bacillus Subtitis, Enterococcus*	[68]
Leaves	Acetone, methanol extracts and n-butanol, chloroform, aqueous, ethyl acetate fractions	Aromatogram by diffusion method	*Bacillus cereus* (ATCC108 76), *Staphylococcus aureus* (ATCC43300), *Staphylococcus aureus* (A TCC291), *Acinetobacter baumannii* (ATCC19606), *Escherichia coli* (ATCC259 22), *Pseudomonas aeruginosa* (ATCC278 53)	[69]

NA: Not available.

Additionally, anthraquinones, which are structural analogues of tetracycline and thus behave like tetracycline, as well as, inhibit bacterial protein synthesis by blocking the ribosomal pathway, are made by acemannan to create the mucilaginous layer that captures the microbial flora and effectively prevents them from invading the system. The bacteria cannot survive in the media, including *A. vera* extract, and they also observed a potent antibacterial effect with ethanolic extract, whereas no inhibition effect was noticed for aqueous extract because "Anthraquinones are alcohol, acetone, *etc.* , soluble but poorly or insoluble in water" [63].

Aloe leaf gel extracts had higher antibacterial potentials on Gram-negative bacteria than on Gram-positive bacteria; also, ethanol extracts had the least minimum inhibitory concentration when compared to methanol and aqueous extracts [64]. A recent study found that *A. vera* flower extract has potent antibacterial activity against the multidrug-resistant bacteria *P. aeruginosa* as antibiotic streptomycin, suggesting that the flower is an underutilized plant part as a natural agent for infection therapy caused by this opportunistic pathogen and the urgent need for new antibiotics by WHO to overcome this threatening bacterium.

Antifungal Activity

Recently, there has been a growing demand for new therapeutic alternatives of plant herbal origin, such as *A. vera,* which contains diverse components with distinct efficacies against a lot of antibiotics resistant fungi, such as *Candida spp.,* as shown in Table 3, which reports on *A. vera*'s vital role against plant and human pathogenic fungi. Previous research found that *A. vera* extracts displayed antifungal activity against a wide range of fungi, but the degree of effectiveness was influenced by the presence of different active constituents in the plant extracts, the extraction solvent, solvent temperature, the selected part of the plant and geographical site, as well as climatic conditions. The aloe liquid fraction has an effective role against plant pathogenic fungi, reducing colony growth rate at a concentration of 105 $\mu l l^{-1}$ in *Fusarium oxysporum, Colletotrichum coccodes* and *Rhizoctonia solani* with a wider range of antifungal potency than *A. vera* pulp. Accordingly, both the pulp and liquid fraction of *A. vera* were proposed as natural products to combat fungi that attack industrial crops, as well as to prevent fungicides application [70].

Table 3. Antifungal activity of *A. vera*.

Plant Part used	Solvent used for Extraction	Method used	Tested Microorganisms	References
Leaves	Leaves ethanol extract and gel	NA	*Trichophyton mentagrophytes Trichophyton schoeleinii Microsporum canis Candida albicans*	[16]
Leaves	Manually pulp, Mechanically pulp and liquid fraction	NA	*Rhizoctonia solani, Colletotrichum coccodes, Fusarium oxysporum*	[70]
Leaves	Hydroalcoholic plant extract	Agar-dilution method	*Botrytis gladiolorum, Fusarium oxysporum* f.sp. *gladioli, Heterosporium pruneti Penicillium gladioli*	[33]
Leaves	Aqueous, ethanol and acetone extracts	Disc diffusion technique	*Aspergillus flavus Aspergillus niger*	[2]
Leaves	Methanol, ethanol and ethyl acetate	Agar-well diffusion method	*Crytococcus neoformans, Candida albicans, Penicillium marneffei Phythium sp., Aspergillus niger, Fusarium oxysporum, Rhizoctonia solani*	[71]
Leaves	Aloe protein	NA	*Candida krusei, Candida paraprilosis Candida albicans*	[12]
Leaves	Ethanol, methanol and aqueous (hot) extracts	Agar well diffusion method	*Candida albicans*	[66]

(Table 3) cont.....

Plant Part used	Solvent used for Extraction	Method used	Tested Microorganisms	References
Leaves	Methanolic extract	Agar well diffusion method	*Aspergillus niger* ATCC 282 *Candida albicans* ATCC 3018	[49]
Leaves	Alcohol and aqueous extracts	Well-diffusion method	*Aspergillus niger, Candida albicans*	[72]
Leaves	Hexane, methanol ethyl acetate, and distilled water	Agar well diffusion method	*Aspergillus oryzae, Aspergillus niger, Penicillium chrysogenum, Trichoderma viridae, Aspergillus flavus*	[36]
Leaves and flower	Hydroethanolic extracts	NA	*Candida albicans* (clinical isolate Ibis 475/15), *Aspergillus niger* (ATCC 6275), *Aspergillus flavus* (ATCC 9643), *Penicillium funiculosum* (ATCC 36839), *Trichophyton tonsurans* (clinical isolate Ibis16/17), *Trichophyton mentagrophytes* (clinical isolate Ibis 2979/18), *Microsporum canis* (clinical isolate Ibis 2990/18), *Microsporum gypseum* (clinical isolate Ibis 3277/18)	[41]
Leaves	Methanolic extract	Poisoned food method	*Aspergillus niger Sclerotium rolfsii*	[37]
Leaves and roots	Ethanol extract	Disc diffusion method	*Candida albicans, Aspergillus fumigatus, fusarium oxysporum, Aspergillus niger*	[67]
Dried *Aloe vera*	Ethanol extract	Disc diffusion method	*Candida albicans*	[73]

NA: Not available.

Furthermore, the hydroalcoholic extracts derived from *A. vera* fresh leaves had antifungal efficacy versus the mycelial production of *Heterosporium pruneti, Botrytis gladiolorum, Penicillium gladioli,* and *Fusarium oxysporum* f.sp. *gladioli* and have proven that the minimum fungicidal concentration varied between 80 and 100 µl/ml, depending on the fungal species. The specific plant extract has direct antifungal action, as compared with other extracts hence, the highest antifungal effect was recorded in acetone extracts of Aloe leaves against human clinical *Aspergillus flavus* and *Aspergillus niger* when compared to aqueous and ethanol extracts of Aloe leaves. The methanol and ethanol extracts displayed the highest antifungal potency as compared to ethyl acetate extract, which may be due to the existence of polar constituents extracted from whole plant leaves [71]. Additionally, the antifungal activity of Malaysian *A. vera* leaf depends on the solvent used for extraction and the extraction method; therefore, the antifungal activity of alcohol extract against *A. niger* with MIC value 4.4 g/ml was more effective than the aqueous extract with MIC value 5.1 g/ml; however, no inhibition zone was observed for both extracts against *C. albicans* that might be correlated with the impact of geographical and climatic conditions on the phytochemical composition of the plants, as well as its antifungal efficacy as compared with other investigations from other regions [72]. Furthermore, the high temperature used in the Soxhlet extraction method for extract preparation may influence the active components leading to the ineffectiveness against *C. albicans*. However, the ability of ethanol extract of *A. vera* to *C. albicans* growth inhibition is attributed to the distinct chemical components like flavonoids, alkaloids, tannins, saponins and steroids/terpenoids included in the extract and its inhibitory activity was greater than or comparable with fluconazole, supporting the use of these extracts as natural antifungal agents to overcome superficial infections and life-threatening systemic infections caused by *C. albicans* [73]. Additionally, it has been described that the antifungal potency of *A. vera* depends on the inhibition of germination and suppression of mycelial growth.

ALOE VERA AS AN EDIBLE COATING FOR VEGETABLES AND FRUITS

The properties of fruits should not be altered by the coating since many fruits, including guava and stone fruit, are consumed with their exocarps. When used alone, *A. vera* gel is recorded as an edible coating to preserve fruit quality during storage of various fruits [74 - 77], but most experiments combined *A. vera* to extend shelf life with another substance such as salicyclic acid, ascorbic acid, oils and chitosan on stone fruits [78 - 80], guava [81, 82], mango, table grapes [83, 84], papaya [85], apple [86], tomatoes [87 - 89], apple slices [90, 91], pomegranate arils [92], sour cherry [93], strawberry [94, 95], orange and sapodilla fruits [96], persimmon [97] and cucumber [98].

Effect of *A. vera* Coating on Microbial Decay and Physiological Disorders of Fruits

Microbial Activity

Both *in-vivo* and *in-vitro* studies on the antifungal action of *A. vera* indicated that it reduced microbial spoilage [99]. Microbial populations have significantly decreased in Aloe-treated fruit [100, 101], and *A. niger*, *C. herbarum*, *F. moniliform*, *P. expansum*, *P. digitatum*, *p. cinerea* and *Alternaria alternata*, as well as *Curvularia hawaiiensis*, *Penicillium italicum*, *R. solani*, *Verticillium dahlia*, *Botryotinia fuckeliana*, under "*in vitro*" conditions have been identified as postharvest fruit pathogens [102], while the higher the concentration of Aloe gel in the medium, the stronger the inhibition of the growth of certain fungal species, as well as when applied to blueberries. *Aloe vera* gel prevented the growth of *V. dahliae*. Aloe's inhibitory impact, depends on the type of fungus and was based on germination suppression and mycelial growth inhibition and could be mainly related to the presence of active compounds like aloe-emodin and aloenin, as well as anthraquinones, which are considered important antimicrobial agents by inhibiting solute transportation of *E. coli* and *S. aureus* membranes [103]; the environmental, agricultural growth conditions and procedures for acquiring gel may be crucial factors in deciding antifungal activity for the *A. vera* plants [104].

A. vera gel inhibited the growth of seventeen (17) bacteria [105]. Many components of *A. vera* gel, such as saponin, anthraquinone and acemannan, are considered to have antibacterial properties that are more potent against gram-positive compared to gram-negative bacteria. On the other hand, reported that *A. vera* gel inhibited the growth of both gram-positive, as well as gram-negative bacteria. Furthermore, *A. vera* gel coatings decreased the use of O_2 and the development of Co_2 to avoid anaerobic conditions.

Additionally, it has been suggested that *A. vera* may be used to preserve food and reduce the likelihood of deterioration in table grapes [106] and sweet cherries. The way for two forms of *A. vera* extract (fresh and grade), as a possible edible coating for post-harvest fungal control of *Colletotrichum gloeosporioides*, *A. niger*, and *L. theobromae*, as well as both fresh *A. vera* and grade gel-filtered equally delayed mycelial growth of fungi on papaya fruits.

A. vera is effective as an antifungal coating for most fungal pathogens in stone fruits, papaya and [107] mandarins. It was proposed that *A. vera* could regulate the decrease of green and blue mold, as well as enhance fruit resistance to decay agents [108]. Researchers noted the presence of antimicrobial properties against postharvest pathogens for fruit coated with *A. vera* combined with chitosan, such as avocado [109], blueberries, bell peppers [110], tomato, mango [111] and table grapes [112].

Additionally, cucumber fruit coated with *A. vera* + chitosan- is more effective in fighting bacteria, and tends to reduce microbial load and thus minimize cucumber fruit spoilage. An infrared study revealed the dislocation experienced between starch and chitosan molecules gel to connect phenolic compounds from *A. vera* to starch, thus improving edible coating properties [113], and the implementation of *A. vera* gel with chitosan improves all mechanical properties [114].

The *A. vera* gel-coated salak fruit was able to reduce decay percent during storage, as well as disease severity for Aloe- gel-coated papaya fruit; *A. vera* combined with salicylic acid-treated table grapes, strawberry and orange fruits were more efficient in reducing decay, as well as a decreasing microbial load [115]. The mesophilic aerobic and yeast mold populations declined in storage when *A. vera* gel-coated strawberry fruits were stored alone or in combination with ascorbic acid. Similar results regarding antimicrobial efficacy against *B. subtilis* and *E. coli* by using *A. vera* oil combined with cellulose acetate membrane [116]. The *A. vera* gel combined with modified atmosphere packaging treatment effectively reduced the rate of decay of cherry laurel fruit [117]. Along with *F. cretica* or ginger extract, *A. vera* gel coating significantly reduced banana anthracnose and drastically reduced deterioration [118]. Strawberries coated in a mixture of *A. vera* gel and basil oil also showed the same pattern. Furthermore, *A. vera*, which relies on starch, proved effective in controlling fungal decrease in cherry tomatoes [119]. A similar trend where the addition of *A. vera* enhanced antagonistic yeasts' ability to inhibit disease-causing organisms while maintaining the quality of mandarin fruits [120].

Physiological Disorders

Many mechanisms, such as the increasing spread of gas through the fruit's surface and the emission of ethylene, are possible for the weight reduction of vegetables and fruits. *A. vera* gel coating was successful in forming a physical barrier that prevents humidity loss and, consequently, dehydration. Additionally, weight loss in the treated fruits has been reduced and is likely to be prevented by hydrogen bonding between the hydroxy groups in polysaccharide-based coatings and hydrophilic constituents such as phenols [121, 122].

In line with these advantages, concurrent use of *A. vera* combined with salicylic acid treatments was more effective in lowering weight loss [123]. Similarly, showed that *A. vera* gel delayed the weight loss of papaya and salak fruits compared to controls during the storage period, and cherry laurel fruits weight reductions with *A. vera* and modified atmosphere packaging treatments were delayed during the process of storing. Additionally, this was shown that *A. vera* is affected in sour cherry fruit through the inhibition of water diffusion, thereby delaying stem browning and maintaining fruit quality. Furthermore, *A. vera* preserved apricot fruit when combined with the seed mucilage of basil plants, and strawberry fruit when mixed with ascorbic acid or basil oil. *A. vera* and rosehip oil were found to significantly reduce weight loss in coated plum fruit. Additionally, *A. vera*, a starch-based organism, also reduced the weight loss in tomato cherry fruit. The same trend was observed in nectarine fruit.

Table grapes, strawberries, mangoes and sapodilla fruit, respectively, were considerably improved by guar gum, salicylic acid, or *F. cretica* plant extract in combination with an aloe gel coating as compared to the control. The use of lotus root slices coated in a combination of *A. vera* and ascorbic acid resulted in a decrease in relative electrolyte leakage and malondialdehyde [124]. Furthermore, the integration of *A. vera* with salicylic acid may improve the orange fruit shelf life by lowering malondialdehyde, electrolyte leakage and chilling injury. In contrast to the untreated control, the *A. vera* gel-treated guava fruits showed reduced increases in malondialdehyde as an indicator of lipid peroxidation, as well as in persimmon.

Impact of *Aloe vera* Coatings on Fruit and Vegetables' Physical-chemical Qualities

Fruit Firmness

Due to the hygroscopic properties of the edible coating, such as *A. vera*, water and gas can form barriers between fruit and the atmosphere, reducing transpiration and respiration rates and thereby delaying metabolic activity, as well as ripening. Similar outcomes were detected in table grapes, peaches and plums and strawberries. Aloe gel-coated fruit was firmer during fruit ripening, while the weight loss in nectarine and blueberry aloe gel-coated fruit was substantially reduced.

The acceptability of fresh fruits is influenced by fruit firmness. Furthermore, the texture is a crucial feature in the evaluation of the freshness of fruit and vegetables for consumers, as well as industry. The porosity of the fruit surface was said to be reduced because *A. vera* molecules were attached to the fruit surface. Moreover, the coating could affect the reduction of degrading enzymes such as pectin methylesterase and polygalacturonase enzymes during fruit ripening by inhibition of their activity [125].

In this regard, the impact of *A. vera* coating on firmness preservation mainly focused on slowing metabolism, respiration rate, and enzyme activity; which will delay the ripening process [126]. Furthermore, the polysaccharide nature and pectin of *A. vera* can prevent water transmission, thus delaying the loss of firmness and probably producing hydrophobic compounds in the aloe fraction, which will also provide an effective barrier to the lost humidity.

Similarly, *A. vera* merged with gum Arabic was successfully applied to maintain the firmness of avocado fruit during cold storage. When combined with basil seed mucilage and salicylic acid, a high degree of strength for strawberry and apricot fruits during storage as compared to controls was obtained. Additionally, the firmness of table grapes and papaya fruit was maintained during storage. Also, aloe gel combined with *F. cretica* plant extract or modified atmosphere packaging coating retains high firmness in sapodilla and cherry fruits *A. vera* gel-coated nectarine fruit remained firm as it ripened.

Soluble Solids Content and Acidity

In most stored fruits, soluble solids content (SSC) was increased while acidity decreased as glycoside is broken up into subunits, insoluble polysaccharide is hydrolyzed into simple sugars and organic acid is utilized in respiratory function. *A. vera* coating can change the inner fruit atmosphere, act as a gas barrier and reduce the absorption of oxygen, which in turn restricts physiological processes such as respiration, resulting in reduced increments in SSC and acidity in coated fruit, and thereby retarded ripening and fruit senescence. Aloe gel-coated sapodilla fruit significantly reduces loss in soluble solid contents and preserves acidity when combined with *F. cretica* plant extract. Guava juice's highest pH increased dramatically for *A. vera* + gum Arabic coated fruit, due to a limited supply of O_2 and decreased oxidation of organic acids, thereby maintaining acidity levels [127, 128] and their potential conversion to sugars [129].

Aloe gel-treated guava fruits reduced the increase in total sugar and high levels of total soluble solids, while Hassanpour in raspberry fruit and in cherries demonstrated that coated-fruit with aloe gel maintained SSC and acidity over storage time, however, the SSC and ripening index were considerably lower in Aloe gel-coated fruits. There were no significant differences in SSC or acidity in fruit coated with aloe gel. The decrease in the acidity values of *A. vera*-coated fruit after storage while maintaining pH levels, it was discovered that the SSC of tomato fruit and table grapes coated with *A. vera* + chitosan increased progressively while the total acidity reduced gradually. Contrarily, *A. vera* particularly when combined with basil seed mucilage coatings, results in decreased SSC in coated fruits, which may be due to a decrease in the respiration rate.

Respiration and Ethylene Production

Recent studies showed that *A. vera* affected respiration and ripening of fruits; these findings had been detected in table grapes, peaches and plums, as well as strawberries and blueberries and recorded that *A. vera* gel + guar gum coating significantly slows the respiration rate of mango fruit [130]. The effectiveness of chitosan and aloe gel are potent in reducing respiration rate that delayed the ripening of cucumber and table grapes fruits and an increase in their shelf life was shown to follow a similar pattern. After storage, aloe gel combined with basil seed mucilage on apricots cherry fruit wrapped in a modified atmosphere, or strawberries combined with basil oil reduced respiration and ethylene production, changing the internal atmosphere of the fruits acting as a gas barrier and reducing oxygen consumption, thus delaying fruit ripening and senescence [131].

Sensory Attributes and Color

Fresh quality is an essential factor for the satisfaction of customers and is linked to the product's color, texture, taste and aromas. As observed with a hue angle, fruits coated with *A. vera* will maintain color change during storage. Aloe-treated fruit also achieved the highest scores for the fruit visual appearance, as demonstrated in a negative value by preventing oxidation or enzyme browning in green pulp Kiwi fruit.

reported that the best coating employed to increase the shelf-life of fruit was *A. vera*, as well as to preserve the product's sensory characteristics throughout storage. Additionally, showed that the rheological and optical characteristics of edible coatings were significantly affected by *A. vera* gel, primarily due to the high concentrations of Aloe gel solids. In this regard, it is shown that using *A. vera* will minimize the color change in date fruits, papaya and salak fruits during storage. Sensory qualities have reportedly been preserved over longer periods of storage by using *A. vera* and plant extracts. attained that treated fruit with Aloe gel blended with basil seed mucilage was highly valued to maintain the sensory parameters of apricot fruit, except for sweetness. However, the color change of mango fruit and apple slices during storage can be reduced by combining aloe gel with guar gum or cysteine. Furthermore, the highest scores were given to pomegranate arils preserved with the incorporation of aloe gel and ascorbic + citric acids, and in sweet cherry. Similar results have also been noted in conserving visual consistency and minimizing browning in lotus root slices by coating *A. vera* gel with ascorbic acid. Additionally, noticed remarkably less browning in banana fruits treated with starch-chitosan edible treated banana fruits and the same trend was noted by combining with basil oil.

Total Phenolic Content and Antioxidant Activity

Fruits and vegetables have elevated amounts of polyphenol, and ascorbic acid that contributes to dietary activity as natural antioxidants. *A. vera* gel coating was used by numerous researchers for the preservation of total antioxidant activity, ascorbic acid and total phenol in many fruits according to these advantages. Monosaccharides in the polysaccharide structure in *A. vera* gel, rhamnose and arabinose, in particular, have strongly radical scavenging activity both *in vitro* and *in vivo*. However, applying edible coatings on fruits allows both phenolic and ascorbic acid to accumulate and enhances antioxidant activity, limiting changes in tissue browned by the polyphenol oxidase enzyme, which supports the statement that there is a strong relationship between all phytochemicals and antioxidant activities.

The content of flavonoid increases in response to the use of aloe gel coating during postharvest storage attributed to the enhancement of phenylalanine ammonia-lyase enzyme (the polyphenolic biosynthesis enzyme and the flavonoid compounds in plants); flavonoids are a class of secondary plant phenolics, and act as potent antioxidants [132]. Furthermore, ascorbate peroxidase, catalase and superoxide dismutase activity improved in aloe gel-treated guava fruits compared to untreated fruits. Aloe gel treatment of guava fruit reduced lipid peroxidation, which is compatible with the maintenance of non-enzymatic and enzymatic antioxidant activities, thereby delaying ripening and retaining the quality of guava fruit during post-harvest storage. Additionally, the total phenolic content as an antifungal and antioxidant agent is higher in the *A. vera* liquid fraction than in the pulp fraction, and, consequently, the liquid fraction has the least value of IC_{50} and thus has higher antioxidant activity, and direct associations between scavenging efficacy and phenolic and flavonoid concentrations in the skin, pulp and ethanol extracts of *A. vera* as earlier documented [133]. Furthermore, various compounds such as aloe-emodin, a major component that helps the antioxidant function of *A. vera* indicated that *A. vera* coated apricot fruits contributed the highest level of antioxidant potential, which was also achieved in mango, blueberry and raspberry fruits. The concentration of phenolics increases generally with ripening and then decreases due to oxidation and senescence. Total phenolics to increased when *A. vera* coatings were treated with garlic essential oil in banana fruit and gum arabic in guava fruit during storage; additionally, the total phenolic content of the *A. vera* + chitosan-coated tomato fruits lasted longer than the control fruits, as did pectate lyase activity, and antioxidant activity induced in mango fruit.

Applying *A. vera* gel coating, with *F. cretica* extract, significantly influenced the maintenance of the total phenol, flavonoid, high ascorbic acid and radical scavenging activity during the storage of sapodilla fruits, the same trend was noticed with ascorbic acid-combined with *A. vera* in strawberries fruits. Furthermore, the free radical scavenging activity increased, and a great potential for producing an efficient membrane for food packaging comprising *A. vera* oil and cellulose acetate was attained. Additionally, coating lotus root slices with *A. vera* and ascorbic acid has maintained the increasing activity of antioxidants, increased total phenolics and the activity of radical scavenging and referred to *A. vera* combined with ascorbic acid as the most powerful way to postpone changes in ripening and preserve the consistency of strawberry fruits when stored [134]. Ascorbic acid is a potent antioxidant that is soluble in water and is intended to mitigate or avoid damage to the fruit by scavenging reactive oxygen species, and its losses occur during ripening due to the oxidation process, so the surface of the fruit coated by *A. vera* acts as a conservative layer and inhibits the absorption of O_2 and the autoxidation of ascorbic acid; a comparable trend that was also observed by as a result of the combination with rosehip oil. However, stated that *A. vera* gel and gum Arabic restrict O_2 accessibility. Furthermore, ascorbic acid levels continuously increased, resulting in higher total antioxidants in the coated fruits than in the control during storage due to slower rises in SSC and reduction in acidity as a result of semi-permeable coatings of *A. vera*, as consistent with chitosan coating for tomatoes and mango fruit. The ability of mango and strawberry fruits' to keep their ascorbic acid content may also be significantly increased by coating them, thus having a major impact on the shelf life of mango and strawberry fruits. The coating of papaya and persimmon with *A. vera* increased tissue strength by enhancing the antioxidants and free radical scavenging potential. According to earlier research, *A. vera* is effective at increasing ascorbic acid, overall anthocyanin, phenolic content, total flavonoids, and antioxidant efficiency in treated fruits. It was claimed that strawberry fruit containing salicylic acid and aloe reduced increases in total carotene. Furthermore, it was discovered that table grapes treated with aloe and salicylic acid had increased total phenol content and antioxidant capacity, while cherry fruit treated with aloe and salicylic acid and packaged in a modified atmosphere followed a similar pattern [135]. Additionally, the combination of *A. vera* gel with chitosan-induced antioxidant activity, which delayed tomato fruit ripening and increased the shelf life of table grapes and tomatoes, was observed. As suggested by Rastegar and Atrash, *A. vera* and *Spirulina platensis* had the greatest favorable impact on the maximal retention on the maximum preservation of phenol, flavonoid and antioxidants in mango fruit during storage. On the contrary, Aloe gel-coated 'Arctic Snow' nectarine fruit did not show any effect on ascorbic acid and total antioxidant levels.

IN VITRO PROPAGATION RESPONSES OF *ALOE VERA*

A. vera is naturally propagated through axillary shoots, which is a relatively slow method of multiplication [136]. Male sterility is another obstacle to rapid propagation. As a result, *in-vitro* propagation techniques, also known as plant cell and tissue culture, are required to regenerate and propagate plants from single cells, tissues, and organs under sterile and controlled environmental conditions to meet the growing demand. Shoot proliferation in various genotypes of the same species is significantly influenced by explant source, size, age, genotype, media composition, culture conditions, and explant phenolic content as well as media discoloration. Advances in tissue culture methodology have made numerous recalcitrant plants amenable to *in vitro* regeneration, as well as the evolution of haploids, somatic hybrids, and pathogen-free plants. Biotechnological approaches for improving *in situ* and *ex-situ* conservation programs are becoming increasingly important for scarce medicinal plants [137]. Micropropagation of *A. vera* using traditional methods or offshoots has many obstacles, including a slow propagation rate (a single plant yields three to four offshoots per year) and male sterility, which makes it difficult to use the plant's seed [138]. To solve this problem, plant tissue culture propagation has been used for rapid clonal propagation of many plants, particularly, for plants with a slow rate of multiplication, to deliver elite clones for commercial production using several concentrations and combinations of Plant Growth Regulators (PGR) in an aseptic environment [139]. Implementation of this strategy helps to propagate thousands of *A. vera* plants from a few initial mother plants yearly. The major factors influencing plant growth, leaf shape and morphology, microstructural and biochemical characterization, photosynthetic potency, and additional physiological activities are controlled *in vitro* physicochemical parameters [140, 141]. Various explants were assessed for understanding *in vitro* response in the nutrient media, such as apical bud, nodal segments, axillary bud, stem cutting, leaves, lateral shoots, shoot discs, adventitious shoot and shoot tips [142, 143]. Despite this, several reports [144] proposed that the shoot tip is the ideal explant for *A. vera* micropropagation; on the contrary, it observed that utilizing a shoot tip as an explant was not ideal for proliferation and the apical bud explants provided the best outcomes [145]. Several previous studies targeted to standardize ideal *in vitro* growth conditions for *Aloe vera* (L.) Burm. f. Various plant growth regulators singly or in combination such as auxin and cytokinin have a considerable influence on the shoot proliferation in the tissue culture of *A. vera*. Additionally, cytokinins (Kinetin, BA, 2-iP and zeatin) play a role in breaking apical dominance in buds and inducing subsidiary meristems that grow into shoots [146]. BAP is the most credible for shoot proliferation of *Aloe polyphylla* and *A. vera* [147] stimulates axillary bud growth and RNA synthesis, accompanied by increases in intracellular proteins and enzymes that increase bud growth [148]; however, shoot elongation is smothered at high concentrations [149, 150], that may due to the presence of sufficient levels of exogenous growth regulators [151]. On the other hand, auxins, namely (IAA (Indole-3-acetic acid), NAA (α-Naphthalene acetic acid), 2, 4-D (2, 4-Dichloro phenoxy acetic acid), IBA (Indole-3-butyric acid) play a significant role in adventitious root formation by increasing cell division and elongation. For instance, 4.0 mg/l BA + 0.2 mg/l IAA + NAA (0.5-1.0 mg/l) are recommended for rooting in half-strength MS medium [152]. The best-performing combination for shoot induction is 4.0 mg/l BAB + 0.2 mg/l NAA with the MS medium, while the best-performing combination for root induction, is 2.0 mg/l IBA + 1 mg/l NAA with the MS medium [153]. In other studies, media containing 0.5-1.0 mg/l NAA and IBA resulted in a greater rate of rooting and a high number of roots per shoot, however, when the shoots were grown on half-strength MS medium devoid of auxin, no roots were observed

[154]. On the contrary, aloe rooting is accomplished on a hormone-free media [155, 156]. For the maximum rate of shoot multiplication, previous studies recommended utilizing a micropropagation strategy that combines 4 mg/l BAP + 1 mg/l IAA in addition to MS which resulted in a 100% survival rate and appeared to be healthy and morphologically identical to the mother plant [157].

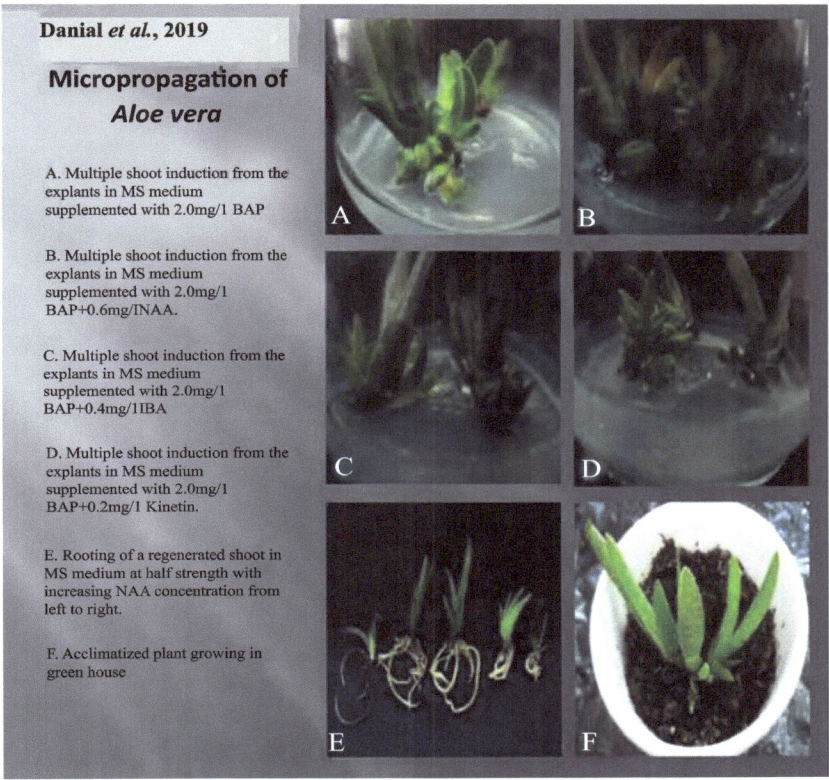

Fig. (2). *In vitro* propagation of *Aloe vera*. (Source: Danial *et al.*, 2019).

Moreover, it was stated that the medium's moderately low salt concentrations are known to improve root establishment of small shoots [158]. IBA promotes more adventitious and axillary buds than NAA, furthermore, IAA did not form adventitious buds, but it did form axillary buds [159, 160]. The best callus formation medium was 1 mg/l (2,4-D) + 0.2 mg/l (KIN). The most common limitation is the browning of the explant, caused by phenolic compound accumulation in the media; however, this it was found that (MS + 0.2 mg/l NAA + 1 g/l PVP+10 mg/l citric acid and 0.5 g/l AC) was an effective strategy to control the browning and accomplish 100% of the survival of rooted plantlets after acclimatization [161]. The problem can be solved using polyvinylpyrrolidone and proline. The addition of adenine to the medium has a significant influence on shoot multiplication and shoot elongation. The addition of activated charcoal (AC) *in vitro* has an effective role in rooting, which reduces light and generates a favorable environment for rhizosphere growth [162] and increased the length of the plantlets, as well as, sucrose is superior to any other carbon source [163]. Additionally, Aloe gel is considered as a conventional rooting method resulting in 100% rooting and the maximum roots per culture [164]. Table **4** and Fig. **(2)** show some recent reports on *in vitro* propagation methods for *A. vera*.

Table. 4. *In vitro* propagation responses of *Aloe vera*.

Explant type	Medium Content	Responding	Survival Response	References
Shoot apical meristem	-MS+35.5 µM BAP + 9.8 µM IBA + 81.4 µM AS	Shoot bud induction	----	[164]
	Aloe gel	100% Rooting		
Shoot tip	MS+ BAP 2 mg/l	80% Shoot induction	75%	[137]
	MS+(2 and 2.5) IBA	70%Root induction		
Rhizomatous stem	MS+0.5 mg/l NAA + 1.5 mg (BAP)	Promoted earliest shoot induction	90%	[165]
	MS+2.5 mg/l BAP	Maximum shoot multiplication		
	MS + 0.5 mg/l IAA +2 g/l (AC)	Promoted root induction		
Adventitious shoot	MS+0.5 mg/l (NAA) + 0.2 mg/l (BA) + 4 g/LPVP	Adventitious root induction	60%	[166]
Shoot with a covering of leaf sheath	MS+ 0.2 mg/ l NAA + 4 mg/ l BA	Highest shoot induction	95%	[167]
	MS +4 mg/ l BA	Highest shoot proliferation		
	B5 + 2 mg/ l NAA	Optimal rooting response		
Apical bud	1/2 Strength MS+1.5 BAB+0.5 NAA	Root induction 70%	85%	[145]
	MS+0.5 NAA	Shoot induction 95%		
Lateral shoot	MS+ 4.0 mg/l BAP + 0.2 mg/l NAA	Shoot initiation and elongation	100%	[153]
	MS+2.0 mg/l IBA + 1.0 mg/l NAA	Root proliferation and elongation		

(Table 4) cont.....

Explant type	Medium Content	Responding	Survival Response	References
Rhizomatous stem and leaf segment	MS+(NAA)2.5 mg/l +2 mg/l (BAP)+0.5 mg/l (IBA)	Callus induction	85-90%	[168]
Rhizomatous stem	MS+2.5 mg/l BAP +2.5 NAA	Shoot proliferation		
Leaf segment	MS+(NAA)2 mg/l+IAA (1)mg/l	Root induction		
Shoot tip	MS medium + 2.0 mg/l BA + 0.5 mg/l NAA + 40.0 mg/l Ads	90% Explants produced multiple shoots	85%	[160]
	Half strength MS + 1.0 mg/l IBA	90% Shoot rooted		
Shoot tip	MS media + 1.0 mg/l BAP	Highest No. of shoots	93%	[139]
	MS media + 0.2 mg/l IBA+0.1BAB	Highest multiplication factor		
	MS-medium without any hormones	Highest No. of roots		
Shoot tip	-MS+0.2 BAB -MS+2.0 mg/l BAP+ 0.6mg/l NAA -MS+2.0 mg/l BAP+0.4 mg/l IBA -MS+2.0 mg/l BAP+0.2 mg/l Kinetin.	Multiple shoot induction	100%	[150]
	-MS ½ Strength +3.0 NAA	Root induction		
Axillary meristems	-MS+1.0 mg/l BAP + 0.15 mg/l (IAA)	Shoot proliferation	100%	[169]
	-1/2-Strength MS+200 mg/l (AC)	Root induction		

Microstructural and Histochemical Changes in *Aloe vera* from *In Vitro* to *In Vivo*

The leaves are extremely sensitive to changes in their environment, as a result, foliar ultra-structural assessment offers a potential solution for detecting anatomical and physiological problems in micropropagated plantlets which could aid in reducing the chance of mortality, boost greenhouse acclimatization, and ensure effective plants establishment in the soil [170].

A. vera growth and development during micropropagation may be better understood with a thorough evaluation of foliar anatomical and histochemical traits; in addition, it will help in the identification of potential secondary metabolites and structural localization for the acclimatization process. However, there is no data available regarding the structural transitions and secondary metabolites localization in tissues that are influenced by various physicochemical factors. Furthermore, when the same researchers compared the foliar developmental transitions of micropropagated *A. vera* plantlets at three diverse stages of growth and development from *in vitro*, *ex vitro*, and (*in vivo*) environments, they discovered that field transferred plantlets had enhanced phenotype, foliar microstructures and leaf pigmentation, as well as histochemical depositions of mucilage, starch tannins, lignin, polyphenol, cutin and suberin (Figs. **3** and **4**).

Fig. (3). *In vitro A. vera* regeneration (**A**) Direct regeneration of shoot buds from nodal meristem explants cultured in MS medium with BAP (3.0 mg/l) (culture period: 5 weeks). (**B**) Shoot proliferation and phenotypic traits achieved on MS medium supplemented with BAP (1.0 mg/l) and IAA (0.15 mg/l) (culture period: 4 weeks after first subculture). (**C**) Root induction from the shoots on half-strength MS medium containing 200 mg L-1AC (culture period: 2 weeks) (*indicates roots). (**D**) Greenhouse hardening of the rooted shoot in soilrite. (**E**) Acclimatization of *A. vera* in the earthen pot, including vermicompost and garden soil (**F**) *A. vera* plant transplanted to the field (Source: Manokari *et al.*, 2021).

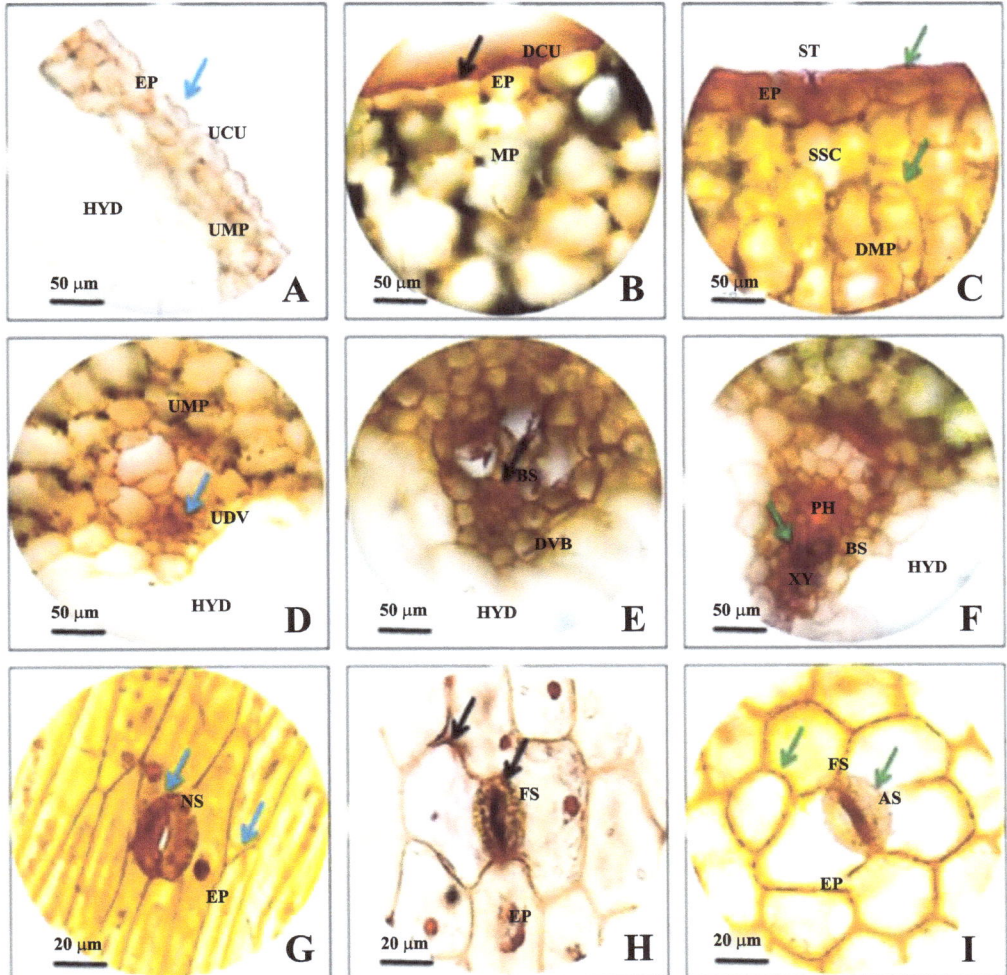

Fig. (4). Foliar-anatomical and histochemical developments in *A. vera* leaves that were *in vitro* regenerated at subsequent stages (stain- safranin). (**A**), (**D**) and (**G**) are initials. Under *in vitro* conditions, transverse leaf sections with underdeveloped structural and lignin accumulation in the cell walls were noticed under *in vitro* environments. (**B**), (**E**) and (**H**). T.S. of greenhouse acclimatized leaves show gradual developments in foliar anatomy and lignin contents in cell walls. (**C**), (**F**) and (**I**) Full lignin deposition was identified in the T.S. of 3 months old field adapted leaves (Source: Manokari *et al.*, 2021).

CONCLUSION

The promising antimicrobial effects of the previously mentioned results obtained by applying several extracts and Aloe plant parts, as well as, its components support the usage of *A. vera* as alternative antimicrobial agents, which requires in-depth clinical trials to confirm the potency and safety of *A. vera* extracts as eco-friendly candidate antimicrobial agents. Furthermore, the therapeutic compounds that are responsible for its antimicrobial activity should be identified and isolated from plant parts or its extracts and possible synergistic influences between them should be explored.

More research studies are required to understand the antioxidant mechanisms of *A. vera* constituents and the impact of the growth stage on antioxidants and antimicrobial properties. However, due to the growing demand and extensive application of it in alternative medicine and dietary supplements, analytical methods for detecting adulterants and undesired additions should be developed. Moreover, due to its nutritional quality and pathogens resistance, which can be obtained by *A. vera* as a safe alternative to postharvest chemical applications, maintaining the entire fruit and vegetable after harvest is vital. Further studies are required to explain the possible mechanisms underlying its effect on maintaining the quality of vegetables and fruits when applied individually or in combination with different plant extracts.

ACKNOWLEDGMENTS

The authors wish to express their gratitude to the Deanship of Scientific Research, Vice Presidency for Graduate Studies and Scientific Research, King Faisal University, Saudi Arabia, for their support [Project No. GRANT4668].

REFERENCES

[1] López A, De Tangil M, Vega-Orellana O, Ramírez A, Rico M. Phenolic constituents, antioxidant and preliminary antimycoplasmic activities of leaf skin and flowers of *Aloe vera* (L.) Burm. f. (syn. A. barbadensis Mill.) from the Canary Islands (Spain). Molecules 2013; 18(5): 4942-54.
[http://dx.doi.org/10.3390/molecules18054942] [PMID: 23624648]

[2] Arunkumar S, Muthuselvam M. Analysis of phytochemical constituents and antimicrobial activities of *Aloe vera* L. against clinical pathogens. World J Agric Sci 2009; 5(5): 572-6.

[3] Park Y, Jo TH. Perspective of industrial application of *Aloe vera*. New Perspect Aloe 2006; 191-200.

[4] Cardarelli M, Rouphael Y, Rea E, Lucini L, Pellizzoni M, Colla G. Effects of fertilization, arbuscular mycorrhiza, and salinity on growth, yield, and bioactive compounds of two Aloe species. HortScience 2013; 48(5): 568-75.
[http://dx.doi.org/10.21273/HORTSCI.48.5.568]

[5] Murillo-Amador B, Córdoba-Matson MV, Villegas-Espinoza JA, Hernández-Montiel LG, Troyo-Diéguez E, García-Hernández JL. Mineral content and biochemical variables of *Aloe vera* L. under salt stress. PLoS One 2014; 9(4): e94870.
[http://dx.doi.org/10.1371/journal.pone.0094870] [PMID: 24736276]

[6] Jiang CQ, Quan LT, Shi F, *et al*. Distribution of mineral nutrients and active ingredients in *Aloe vera* irrigated with diluted seawater. Pedosphere 2014; 24(6): 722-30.
[http://dx.doi.org/10.1016/S1002-0160(14)60059-X]

[7] Delatorre-Castillo JP, Delatorre-Herrera J, Lay KS, *et al*. Preconditioning to water deficit helps *Aloe vera* to overcome long-term drought during the driest season of Atacama desert. Plants 2022; 11(11): 1523.

[http://dx.doi.org/10.3390/plants11111523] [PMID: 35684295]

[8] Rahman S, Carter P, Bhattarai N. *Aloe vera* for tissue engineering applications. J Funct Biomater 2017; 8(1): 6.
[http://dx.doi.org/10.3390/jfb8010006] [PMID: 28216559]

[9] Boudreau MD, Beland FA. An evaluation of the biological and toxicological properties of *Aloe barbadensis* (miller), *Aloe vera*. J Environ Sci Health Part C Environ Carcinog Ecotoxicol Rev 2006; 24(1): 103-54.
[http://dx.doi.org/10.1080/10590500600614303] [PMID: 16690538]

[10] Debnath T, Ghosh M, Lee YM, Nath NCD, Lee KG, Lim BO. Identification of phenolic constituents and antioxidant activity of *Aloe barbadensis* flower extracts. Food Agric Immunol 2018; 29(1): 27-38.
[http://dx.doi.org/10.1080/09540105.2017.1358254]

[11] Cardarelli M, Rouphael Y, Pellizzoni M, Colla G, Lucini L. Profile of bioactive secondary metabolites and antioxidant capacity of leaf exudates from eighteen Aloe species. Ind Crops Prod 2017; 108: 44-51.
[http://dx.doi.org/10.1016/j.indcrop.2017.06.017]

[12] Das S, Mishra B, Gill K, *et al.* Isolation and characterization of novel protein with anti-fungal and anti-inflammatory properties from *Aloe vera* leaf gel. Int J Biol Macromol 2011; 48(1): 38-43.
[http://dx.doi.org/10.1016/j.ijbiomac.2010.09.010] [PMID: 20888359]

[13] Kumar S, Yadav M, Yadav A, Rohilla P, Yadav JP. Antiplasmodial potential and quantification of aloin and aloe-emodin in *Aloe vera* collected from different climatic regions of India. BMC Complement Altern Med 2017; 17(1): 369.
[http://dx.doi.org/10.1186/s12906-017-1883-0] [PMID: 28716028]

[14] Borges-Argáez R, Chan-Balan R, Cetina-Montejo L, *et al. In vitro* evaluation of anthraquinones from *Aloe vera* (*Aloe barbadensis* Miller) roots and several derivatives against strains of influenza virus. Ind Crops Prod 2019; 132: 468-75.
[http://dx.doi.org/10.1016/j.indcrop.2019.02.056] [PMID: 32288269]

[15] Hamman J. Composition and applications of *Aloe vera* leaf gel. Molecules 2008; 13(8): 1599-616.
[http://dx.doi.org/10.3390/molecules13081599] [PMID: 18794775]

[16] Agarry OO, Olaleye MT, Bello-Michael CO. Comparative antimicrobial activities of *Aloe vera* gel and leaf. Afr J Biotechnol 2005; 4(12): 1413-4.

[17] Mostafidi M, Sanjabi MR, Shirkhan F, Zahedi MT. A review of recent trends in the development of the microbial safety of fruits and vegetables. Trends Food Sci Technol 2020; 103: 321-32.
[http://dx.doi.org/10.1016/j.tifs.2020.07.009]

[18] Ebrahimi F, Rastegar S. Preservation of mango fruit with guar-based edible coatings enriched with *Spirulina platensis* and *Aloe vera* extract during storage at ambient temperature. Sci Hortic 2020; 265: 109258.
[http://dx.doi.org/10.1016/j.scienta.2020.109258]

[19] Paladines D, Valero D, Valverde JM, Díaz-Mula H, Serrano M, Martínez-Romero D. The addition of rosehip oil improves the beneficial effect of *Aloe vera* gel on delaying ripening and maintaining postharvest quality of several stonefruit. Postharvest Biol Technol 2014; 92: 23-8.
[http://dx.doi.org/10.1016/j.postharvbio.2014.01.014]

[20] Zapata PJ, Navarro D, Guillén F, *et al.* Characterisation of gels from different *Aloe spp.* as antifungal treatment: Potential crops for industrial applications. Ind Crops Prod 2013; 42: 223-30.
[http://dx.doi.org/10.1016/j.indcrop.2012.06.002]

[21] Guillén F, Díaz-Mula HM, Zapata PJ, *et al. Aloe arborescens* and *Aloe vera* gels as coatings in delaying postharvest ripening in peach and plum fruit. Postharvest Biol Technol 2013; 83: 54-7.
[http://dx.doi.org/10.1016/j.postharvbio.2013.03.011]

[22] Vieira JM, Flores-López ML, De Rodríguez DJ, Sousa MC, Vicente AA, Martins JT. Effect of

chitosan– *Aloe vera* coating on postharvest quality of blueberry (*Vaccinium corymbosum*) fruit. Postharvest Biol Technol 2016; 116: 88-97.
[http://dx.doi.org/10.1016/j.postharvbio.2016.01.011]

[23] Sogvar OB, Koushesh Saba M, Emamifar A. *Aloe vera* and ascorbic acid coatings maintain postharvest quality and reduce microbial load of strawberry fruit. Postharvest Biol Technol 2016; 114: 29-35.
[http://dx.doi.org/10.1016/j.postharvbio.2015.11.019]

[24] Rasouli M, Koushesh Saba M, Ramezanian A. Inhibitory effect of salicylic acid and *Aloe vera* gel edible coating on microbial load and chilling injury of orange fruit. Sci Hortic 2019; 247: 27-34.
[http://dx.doi.org/10.1016/j.scienta.2018.12.004]

[25] Tahir HE, Xiaobo Z, Mahunu GK, Arslan M, Abdalhai M, Zhihua L. Recent developments in gum edible coating applications for fruits and vegetables preservation: A review. Carbohydr Polym 2019; 224: 115141.
[http://dx.doi.org/10.1016/j.carbpol.2019.115141] [PMID: 31472839]

[26] Ni Y, Turner D, Yates KM, Tizard I. Isolation and characterization of structural components of *Aloe vera* L. leaf pulp. Int Immunopharmacol 2004; 4(14): 1745-55.
[http://dx.doi.org/10.1016/j.intimp.2004.07.006] [PMID: 15531291]

[27] Pellizzoni M, Ruzickova G, Kalhotka L, Lucini L. Antimicrobial activity of different *Aloe barbadensis* Mill. and *Aloe arborescens* Mill. leaf fractions. J Med Plants Res 2012; 6(10): 1975-81.

[28] Nema J, Shrivastava SK, Mitra NG. Physicochemical study of acemannan polysaccharide in Aloe species under the influence of soil reaction (pH) and moisture application. African J Pure Appl Chem 2012; 6(9): 132-6.

[29] Bozzi A, Perrin C, Austin S, Arce Vera F. Quality and authenticity of commercial *aloe vera* gel powders. Food Chem 2007; 103(1): 22-30.
[http://dx.doi.org/10.1016/j.foodchem.2006.05.061]

[30] Femenia A, García-Pascual P, Simal S, Rosselló C. Effects of heat treatment and dehydration on bioactive polysaccharide acemannan and cell wall polymers from *Aloe barbadensis* Miller. Carbohydr Polym 2003; 51(4): 397-405.
[http://dx.doi.org/10.1016/S0144-8617(02)00209-6]

[31] Minjares-Fuentes R, Medina-Torres L, González-Laredo RF, Rodríguez-González VM, Eim V, Femenia A. Influence of water deficit on the main polysaccharides and the rheological properties of *Aloe vera* (*Aloe barbadensis* Miller) mucilage. Ind Crops Prod 2017; 109(May): 644-53.
[http://dx.doi.org/10.1016/j.indcrop.2017.09.016]

[32] Saccù D, Bogoni P, Procida G. Aloe exudate: Characterization by reversed phase HPLC and headspace GC-MS. J Agric Food Chem 2001; 49(10): 4526-30.
[http://dx.doi.org/10.1021/jf010179c] [PMID: 11599983]

[33] Rosca-Casian O, Parvu M, Vlase L, Tamas M. Antifungal activity of *Aloe vera* leaves. Fitoterapia 2007; 78(3): 219-22.
[http://dx.doi.org/10.1016/j.fitote.2006.11.008] [PMID: 17336466]

[34] Raphael E. Phytochemical constituents of some leaves extract of *Aloe vera* and Azadirachta indica plant species. Glob Adv Res J Environ Sci Toxicol 2012; 1(2): 014–7.

[35] Kumar S, Yadav A, Yadav M, Yadav JP. Effect of climate change on phytochemical diversity, total phenolic content and *in vitro* antioxidant activity of *Aloe vera* (L.) Burm.f. BMC Res Notes 2017; 10(1): 60.
[http://dx.doi.org/10.1186/s13104-017-2385-3] [PMID: 28118858]

[36] Nalin Pagi DD, Payal Patel HJ, Jasani H, Patel P. Antimicrobial activity and phytochemical screening of *Aloe vera* (*Aloe barbadensis* Miller). Int J Curr Microbiol Appl Sci 2017; 6(3): 2152-62.
[http://dx.doi.org/10.20546/ijcmas.2017.603.246]

[37] Chinche AV, Gade RM, Shinde AN, Vairagade MT, Kendhale KV. Fractionation of secondary metabolites from Tulsi (*Ocimum sanctum*) and *Aloe vera* (Aloe barbadensis Mill.) and their antifungal activity against *Aspergillus niger* and *Sclerotium rolfsii*. Int J Curr Microbiol Appl Sci 2020; 9(5): 445-52.
[http://dx.doi.org/10.20546/ijcmas.2020.905.050]

[38] Bista R, Ghimire A, Subedi S. Phytochemicals and antioxidant activities of *Aloe vera* (*Aloe barbadensis*). J Nutrit Sci Heal Diet 2020; 1(1): 25-36.
[http://dx.doi.org/10.47890/JNSHD/2020/RBista/10243803]

[39] Nejatzadeh-Barandozi F. Antibacterial activities and antioxidant capacity of *Aloe vera*. Org Med Chem Lett 2013; 3(1): 5.
[http://dx.doi.org/10.1186/2191-2858-3-5] [PMID: 23870710]

[40] Kang MC, Kim SY, Kim YT, *et al*. In vitro and in vivo antioxidant activities of polysaccharide purified from aloe vera (*Aloe barbadensis*) gel. Carbohydr Polym 2014; 99: 365-71.
[http://dx.doi.org/10.1016/j.carbpol.2013.07.091] [PMID: 24274519]

[41] Añibarro-Ortega M, Pinela J, Barros L, *et al*. Compositional features and bioactive properties of *Aloe vera* leaf (fillet, mucilage, and rind) and flower. Antioxidants 2019; 8(10): 444.
[http://dx.doi.org/10.3390/antiox8100444] [PMID: 31581507]

[42] Chang XL, Chen BY, Feng YM. Water-soluble polysaccharides isolated from skin juice, gel juice and flower of *Aloe vera* Miller. J Taiwan Inst Chem Eng 2011; 42(2): 197-203.
[http://dx.doi.org/10.1016/j.jtice.2010.07.007]

[43] Benzidia B, Barbouchi M, Hammouch H, *et al*. Chemical composition and antioxidant activity of tannins extract from green rind of *Aloe vera* (L.) Burm. F. J King Saud Univ Sci 2019; 31(4): 1175-81.
[http://dx.doi.org/10.1016/j.jksus.2018.05.022]

[44] Solaberrieta I, Jiménez A, Garrigós MC. Valorisation of *Aloe vera* skin by-products to obtain bioactive compounds by microwave-assisted extraction: Antioxidant activity and chemical composition. Antioxidants 2022; 11(6): 1058.
[http://dx.doi.org/10.3390/antiox11061058] [PMID: 35739955]

[45] Dagne E, Bisrat D, Viljoen A, Van Wyk B-E. Chemistry of aloe species. Curr Org Chem 2000; 4(10): 1055-78.
[http://dx.doi.org/10.2174/1385272003375932]

[46] Steenkamp V, Stewart MJ. Medicinal applications and toxicological activities of Aloe products. Pharm Biol 2007; 45(5): 411-20.
[http://dx.doi.org/10.1080/13880200701215307]

[47] Lucini L, Pellizzoni M, Molinari GP. Anthraquinones and β-polysaccharides content and distribution in Aloe plants grown under different light intensities. Biochem Syst Ecol 2013; 51: 264-8.
[http://dx.doi.org/10.1016/j.bse.2013.09.007]

[48] Paez A, Michael Gebre G, Gonzalez ME, Tschaplinski TJ. Growth, soluble carbohydrates, and aloin concentration of *Aloe vera* plants exposed to three irradiance levels. Environ Exp Bot 2000; 44(2): 133-9.
[http://dx.doi.org/10.1016/S0098-8472(00)00062-9] [PMID: 10996366]

[49] Kumar S, Yadav M, Yadav A, Yadav JP. Comparative analysis of antimicrobial activity of methanolic extracts of *Aloe vera* and quantification of aloe-emodin collected from different climatic zones of India. Arch Clin Microbiol 2015; 6(2:1): 1-10.

[50] Jin ZM, Wang CH, Liu ZP, Gong WJ. Physiological and ecological characters studies on *Aloe vera* under soil salinity and seawater irrigation. Process Biochem 2007; 42(4): 710-4.
[http://dx.doi.org/10.1016/j.procbio.2006.11.002]

[51] Tawfik KM, Sheteawi SA, El-Gawad ZA. Growth and aloin production of *Aloe vera* and *Aloe eru* under different ecological conditions. Egypt J Biol 2001; 3: 149-59.

[52] Salinas C, Handford M, Pauly M, Dupree P, Cardemil L. Structural modifications of fructans in *Aloe barbadensis* miller (*Aloe vera*) grown under water stress. PLoS One 2016; 11(7): e0159819.
[http://dx.doi.org/10.1371/journal.pone.0159819] [PMID: 27454873]

[53] Newton LE. Aloes in habitat. Boca Raton: CRC Press 2004; pp. 3-36.

[54] Rodríguez-García R, Rodríguez DJ, Gil-Marín JA, Angulo-Sánchez JL, Lira-Saldivar RH. Growth, stomatal resistance, and transpiration of *Aloe vera* under different soil water potentials. Ind Crops Prod 2007; 25(2): 123-8.
[http://dx.doi.org/10.1016/j.indcrop.2006.08.005]

[55] Chun-hui L, Chang-hai W, Zhi-liang X, Yi W. Isolation, chemical characterization and antioxidant activities of two polysaccharides from the gel and the skin of *Aloe barbadensis* Miller irrigated with sea water. Process Biochem 2007; 42(6): 961-70.
[http://dx.doi.org/10.1016/j.procbio.2007.03.004]

[56] Hu Y, Xu J, Hu Q. Evaluation of antioxidant potential of *aloe vera* (*Aloe barbadensis* miller) extracts. J Agric Food Chem 2003; 51(26): 7788-91.
[http://dx.doi.org/10.1021/jf034255i] [PMID: 14664546]

[57] Yagi A, Kabash A, Okamura N, Haraguchi H, Moustafa SM, Khalifa TI. Antioxidant, free radical scavenging and anti-inflammatory effects of aloesin derivatives in *Aloe vera*. Planta Med 2002; 68(11): 957-60.
[http://dx.doi.org/10.1055/s-2002-35666] [PMID: 12451482]

[58] Hu Q, Hu Y, Xu J. Free radical-scavenging activity of *Aloe vera* (*Aloe barbadensis* Miller) extracts by supercritical carbon dioxide extraction. Food Chem 2005; 91(1): 85-90.
[http://dx.doi.org/10.1016/j.foodchem.2004.05.052]

[59] Lucini L, Pellizzoni M, Pellegrino R, Molinari GP, Colla G. Phytochemical constituents and *in vitro* radical scavenging activity of different Aloe species. Food Chem 2015; 170: 501-7.
[http://dx.doi.org/10.1016/j.foodchem.2014.08.034] [PMID: 25306376]

[60] Rehman NU, Al-Riyami SA, Hussain H, Ali A, Khan AL, Al-Harrasi A. Secondary metabolites from the resins of *Aloe vera* and *Commiphora mukul* mitigate lipid peroxidation. Acta Pharm 2019; 69(3): 433-41.
[http://dx.doi.org/10.2478/acph-2019-0027] [PMID: 31259740]

[61] Boateng JS. Analysis of commercial samples of aloe. Ph D thesis, Univ Strat 2000.

[62] Pugh N, Ross SA, ElSohly MA, Pasco DS. Characterization of Aloeride, a new high-molecular-weight polysaccharide from *Aloe vera* with potent immunostimulatory activity. J Agric Food Chem 2001; 49(2): 1030-4.
[http://dx.doi.org/10.1021/jf001036d] [PMID: 11262067]

[63] Pandey R, Mishra A. Antibacterial activities of crude extract of *Aloe barbadensis* to clinically isolated bacterial pathogens. Appl Biochem Biotechnol 2010; 160(5): 1356-61.
[http://dx.doi.org/10.1007/s12010-009-8577-0] [PMID: 19263248]

[64] Ferro VA, Bradbury F, Cameron P, Shakir E, Rahman SR, Stimson WH. *In vitro* susceptibilities of *Shigella flexneri* and *Streptococcus pyogenes* to inner gel of *Aloe barbadensis* Miller. Antimicrob Agents Chemother 2003; 47(3): 1137-9.
[http://dx.doi.org/10.1128/AAC.47.3.1137-1139.2003] [PMID: 12604556]

[65] Habeeb F, Shakir E, Bradbury F, *et al.* Screening methods used to determine the anti-microbial properties of *Aloe vera* inner gel. Methods 2007; 42(4): 315-20.
[http://dx.doi.org/10.1016/j.ymeth.2007.03.004] [PMID: 17560318]

[66] Stanley MC, Ifeanyi OE, Eziokwu OG. Antimicrobial effects of *Aloe vera* on some human pathogens. IntJCurrMicrobiolAppSci 2014; 3(3): 1022-8.

[67] Danish P, Ali Q, Mm H, Malik A. Antifungal and antibacterial activity of *Aloe vera* plant extract. Biol

Clin Sci Res J 2020; e004.

[68] Haq A, Ali Q, Rashid MS, Waheed F, Hayat S, Malik A. Antibacterial and antifungal activity of *Aloe vera* plant. Life Sci J 2020; 17(7): 76-82.

[69] Bendjedid S, Lekmine S, Tadjine A, Djelloul R, Bensouici C. Analysis of phytochemical constituents, antibacterial, antioxidant, photoprotective activities and cytotoxic effect of leaves extracts and fractions of *Aloe vera*. Biocatal Agric Biotechnol 2021; 33: 101991.
[http://dx.doi.org/10.1016/j.bcab.2021.101991]

[70] Jasso de Rodríguez D, Hernández-Castillo D, Rodríguez-García R, Angulo-Sánchez JL. Antifungal activity *in vitro* of *Aloe vera* pulp and liquid fraction against plant pathogenic fungi. Ind Crops Prod 2005; 21(1): 81-7.
[http://dx.doi.org/10.1016/j.indcrop.2004.01.002]

[71] Khaing TA. Evaluation of the antifungal and antioxidant activities of the leaf extract of *Aloe vera* (*Aloe barbadensis* Miller). World Acad Sci Eng Technol 2011; 75: 610-2.

[72] Saniasiay J, Salim R, Mohamad I, Harun A. Antifungal effect of Malaysian *Aloe vera* leaf extract on selected fungal species of pathogenic otomycosis species in *In vitro* culture medium. Oman Med J 2017; 32(1): 41-6.
[http://dx.doi.org/10.5001/omj.2017.08] [PMID: 28042402]

[73] Nabila VK, Putra IB. The effect of *Aloe vera* ethanol extract on the growth inhibition of *Candida albicans*. Med Glas 2020; 17(2): 485-9.
[PMID: 32662608]

[74] Hassanpour H. Effect of *Aloe vera* gel coating on antioxidant capacity, antioxidant enzyme activities and decay in raspberry fruit. Lebensm Wiss Technol 2015; 60(1): 495-501.
[http://dx.doi.org/10.1016/j.lwt.2014.07.049]

[75] Mendy TK, Misran A, Mahmud TMM, Ismail SI. Application of *Aloe vera* coating delays ripening and extend the shelf life of papaya fruit. Sci Hortic 2019; 246: 769-76.
[http://dx.doi.org/10.1016/j.scienta.2018.11.054]

[76] Mendy TK, Misran A, Mahmud TMM, Ismail SI. Antifungal properties of *Aloe vera* through *in vitro* and *in vivo* screening against postharvest pathogens of papaya fruit. Sci Hortic 2019; 257: 108767.
[http://dx.doi.org/10.1016/j.scienta.2019.108767]

[77] Parven A, Sarker MR, Megharaj M, Meftaul Md. Prolonging the shelf life of Papaya (*Carica papaya* L.) using *Aloe vera* gel at ambient temperature. Sci Hortic 2020; 265: 109228.
[http://dx.doi.org/10.1016/j.scienta.2020.109228]

[78] Ahmed MJ, Singh Z, Khan AS. Postharvest *Aloe vera* gel - coating modulates fruit ripening and quality of 'Arctic Snow' nectarine kept in ambient and cold storage. Int J Food Sci Technol 2009; 44(5): 1024-33.
[http://dx.doi.org/10.1111/j.1365-2621.2008.01873.x]

[79] Nourozi F, Sayyari M. Enrichment of *Aloe vera* gel with basil seed mucilage preserve bioactive compounds and postharvest quality of apricot fruits. Sci Hortic 2020; 262: 109041.
[http://dx.doi.org/10.1016/j.scienta.2019.109041]

[80] Martínez-Romero D, Zapata PJ, Guillén F, *et al.* The addition of rosehip oil to Aloe gels improves their properties as postharvest coatings for maintaining quality in plum. Food Chem 2017; 217: 585-92.
[http://dx.doi.org/10.1016/j.foodchem.2016.09.035] [PMID: 27664675]

[81] Anjum MA, Akram H, Zaidi M, Ali S. Effect of gum arabic and *Aloe vera* gel based edible coatings in combination with plant extracts on postharvest quality and storability of 'Gola' guava fruits. Sci Hortic 2020; 271: 109506.
[http://dx.doi.org/10.1016/j.scienta.2020.109506]

[82] Rehman MA, Asi MR, Hameed A, Bourquin LD. Effect of postharvest application of *Aloe vera* gel on

shelf life, activities of anti-oxidative enzymes, and quality of 'gola' guava fruit. Foods 2020; 9(10): 1361.
[http://dx.doi.org/10.3390/foods9101361] [PMID: 32992728]

[83] Serrano M, Valverde JM, Guillén F, Castillo S, Martínez-Romero D, Valero D. Use of *Aloe vera* gel coating preserves the functional properties of table grapes. J Agric Food Chem 2006; 54(11): 3882-6.
[http://dx.doi.org/10.1021/jf060168p] [PMID: 16719510]

[84] Castillo S, Navarro D, Zapata PJ, *et al.* Antifungal efficacy of *Aloe vera* in vitro and its use as a preharvest treatment to maintain postharvest table grape quality. Postharvest Biol Technol 2010; 57(3): 183-8.
[http://dx.doi.org/10.1016/j.postharvbio.2010.04.006]

[85] Marpudi SL, Abirami LSS, Pushkala R, Srividya N. Enhancement of storage life and quality maintenance of papaya fruits using *Aloe vera* based antimicrobial coating. Indian J Biotechnol 2011; 10(1): 83-9.

[86] Ergun M, Satici F. Use of *Aloe vera* gel as biopreservative for "Granny Smith" and "Red Chief" apples. J Anim Plant Sci 2012; 22(2): 363-8.

[87] Athmaselvi KA, Sumitha P, Revathy B. Development of *Aloe vera* based edible coating for tomato. Int Agrophys 2013; 27(4): 369-75.
[http://dx.doi.org/10.2478/intag-2013-0006]

[88] Chauhan OP, Nanjappa C, Ashok N, Ravi N, Roopa N, Raju PS. Shellac and *Aloe vera* gel based surface coating for shelf life extension of tomatoes. J Food Sci Technol 2015; 52(2): 1200-5.
[http://dx.doi.org/10.1007/s13197-013-1035-6] [PMID: 25694740]

[89] Khatri D, Panigrahi J, Prajapati A, Bariya H. Attributes of *Aloe vera* gel and chitosan treatments on the quality and biochemical traits of post-harvest tomatoes. Sci Hortic 2020; 259: 108837.
[http://dx.doi.org/10.1016/j.scienta.2019.108837]

[90] Chauhan OP, Raju PS, Singh A, Bawa AS. Shellac and aloe-gel-based surface coatings for maintaining keeping quality of apple slices. Food Chem 2011; 126(3): 961-6.
[http://dx.doi.org/10.1016/j.foodchem.2010.11.095]

[91] Song HY, Jo WS, Song NB, Min SC, Song KB. Quality change of apple slices coated with *Aloe vera* gel during storage. J Food Sci 2013; 78(6): C817-22.
[http://dx.doi.org/10.1111/1750-3841.12141] [PMID: 23647574]

[92] Martínez-Romero D, Castillo S, Guillén F, *et al. Aloe vera* gel coating maintains quality and safety of ready-to-eat pomegranate arils. Postharvest Biol Technol 2013; 86: 107-12.
[http://dx.doi.org/10.1016/j.postharvbio.2013.06.022]

[93] Ravanfar R, Niakousari M, Maftoonazad N. Postharvest sour cherry quality and safety maintenance by exposure to Hot- water or treatment with fresh *Aloe vera* gel. J Food Sci Technol 2014; 51(10): 2872-6.
[http://dx.doi.org/10.1007/s13197-012-0767-z] [PMID: 25328241]

[94] Mohammadi L, Tanaka F, Tanaka F. Preservation of strawberry fruit with an *Aloe vera* gel and basil (*Ocimum basilicum*) essential oil coating at ambient temperature. J Food Process Preserv 2021; 45(10): e15836.
[http://dx.doi.org/10.1111/jfpp.15836]

[95] Hosseinifarahi M, Jamshidi E, Amiri S, Kamyab F, Radi M. Quality, phenolic content, antioxidant activity, and the degradation kinetic of some quality parameters in strawberry fruit coated with salicylic acid and *Aloe vera* gel. J Food Process Preserv 2020; 44(9): 1-14.
[http://dx.doi.org/10.1111/jfpp.14647]

[96] Khaliq G, Ramzan M, Baloch AH. Effect of *Aloe vera* gel coating enriched with Fagonia indica plant extract on physicochemical and antioxidant activity of sapodilla fruit during postharvest storage. Food Chem 2019; 286: 346-53.

[97] Saleem MS, Ejaz S, Anjum MA, *et al*. *Aloe vera* gel coating delays softening and maintains quality of stored persimmon (*Diospyros kaki* Thunb.) Fruits. J Food Sci Technol 2022; 59(8): 3296-306.
[http://dx.doi.org/10.1007/s13197-022-05412-5] [PMID: 35876768]

[98] Ajiboye AE, Gboyinde P. Effects of chitosan and *Aloe vera* gel coatings on the preservation characteristics of cucumber samples. Adv J Grad Res 2020; 8(1): 82-90.
[http://dx.doi.org/10.21467/ajgr.8.1.82-90]

[99] Benítez S, Achaerandio I, Sepulcre F, Pujolà M. *Aloe vera* based edible coatings improve the quality of minimally processed 'Hayward' kiwifruit. Postharvest Biol Technol 2013; 81: 29-36.
[http://dx.doi.org/10.1016/j.postharvbio.2013.02.009]

[100] Alkaabi S, Sobti B, Mudgil P, Hasan F, Ali A, Nazir A. Lemongrass essential oil and *aloe vera* gel based antimicrobial coatings for date fruits. Applied Food Research 2022; 2(1): 100127.
[http://dx.doi.org/10.1016/j.afres.2022.100127]

[101] Martínez-Romero D, Alburquerque N, Valverde JM, *et al*. Postharvest sweet cherry quality and safety maintenance by *Aloe vera* treatment: A new edible coating. Postharvest Biol Technol 2006; 39(1): 93-100.
[http://dx.doi.org/10.1016/j.postharvbio.2005.09.006]

[102] Sempere-Ferre F, Giménez-Santamarina S, Roselló J, Santamarina MP. Antifungal *in vitro* potential of *Aloe vera* gel as postharvest treatment to maintain blueberry quality during storage. Lebensm Wiss Technol 2022; 163: 113512.
[http://dx.doi.org/10.1016/j.lwt.2022.113512]

[103] Ullah N, Parveen A, Bano R, *et al*. *In vitro* and *in vivo* protocols of antimicrobial bioassay of medicinal herbal extracts: A review. Asian Pac J Trop Dis 2016; 6(8): 660-7.
[http://dx.doi.org/10.1016/S2222-1808(16)61106-4]

[104] Saks Y, Barkai-Golan R. *Aloe vera* gel activity against plant pathogenic fungi. Postharvest Biol Technol 1995; 6(1-2): 159-65.
[http://dx.doi.org/10.1016/0925-5214(94)00051-S]

[105] Fani M, Kohanteb J. Inhibitory activity of *Aloe vera* gel on some clinically isolated cariogenic and periodontopathic bacteria. J Oral Sci 2012; 54(1): 15-21.
[http://dx.doi.org/10.2334/josnusd.54.15] [PMID: 22466882]

[106] Valverde JM, Valero D, Martínez-Romero D, Guillén F, Castillo S, Serrano M. Novel edible coating based on *aloe vera* gel to maintain table grape quality and safety. J Agric Food Chem 2005; 53(20): 7807-13.
[http://dx.doi.org/10.1021/jf050962v] [PMID: 16190634]

[107] Jhalegar MJ, Sharma RR, Singh D. Antifungal efficacy of botanicals against major postharvest pathogens of Kinnow mandarin and their use to maintain postharvest quality. Fruits 2014; 69(3): 223-37.
[http://dx.doi.org/10.1051/fruits/2014012]

[108] Benítez S, Achaerandio I, Pujolà M, Sepulcre F. *Aloe vera* as an alternative to traditional edible coatings used in fresh-cut fruits: A case of study with kiwifruit slices. Lebensm Wiss Technol 2015; 61(1): 184-93.
[http://dx.doi.org/10.1016/j.lwt.2014.11.036]

[109] Bill M, Sivakumar D, Korsten L, Thompson AK. The efficacy of combined application of edible coatings and thyme oil in inducing resistance components in avocado (*Persea americana* Mill.) against anthracnose during post-harvest storage. Crop Prot 2014; 64: 159-67.
[http://dx.doi.org/10.1016/j.cropro.2014.06.015]

[110] Manoj HG, Sreenivas KN, Shankarappa TH, Krishna HC. Studies on chitosan and *Aloe vera* gel coatings on biochemical parameters and microbial population of bell pepper (Capsicum annuum L.)

under ambient condition. Int J Curr Microbiol Appl Sci 2016; 5(1): 399-405.
[http://dx.doi.org/10.20546/ijcmas.2016.501.039]

[111] Shah S, Hashmi MS. Chitosan–*aloe vera* gel coating delays postharvest decay of mango fruit. Hortic Environ Biotechnol 2020; 61(2): 279-89.
[http://dx.doi.org/10.1007/s13580-019-00224-7]

[112] Ehtesham Nia A, Taghipour S, Siahmansour S. Pre-harvest application of chitosan and postharvest *Aloe vera* gel coating enhances quality of table grape (*Vitis vinifera* L. cv. 'Yaghouti') during postharvest period. Food Chem 2021; 347: 129012.
[http://dx.doi.org/10.1016/j.foodchem.2021.129012] [PMID: 33486359]

[113] Pinzon MI, Garcia OR, Villa CC. The influence of *Aloe vera* gel incorporation on the physicochemical and mechanical properties of banana starch-chitosan edible films. J Sci Food Agric 2018; 98(11): 4042-9.
[http://dx.doi.org/10.1002/jsfa.8915] [PMID: 29377147]

[114] Khoshgozaran-Abras S, Azizi MH, Hamidy Z, Bagheripoor-Fallah N. Mechanical, physicochemical and color properties of chitosan based-films as a function of *Aloe vera* gel incorporation. Carbohydr Polym 2012; 87(3): 2058-62.
[http://dx.doi.org/10.1016/j.carbpol.2011.10.020]

[115] Sari PRP, Darmawati E, Ahmad U. *Aloe vera* and beeswax based coating to maintain shelf life of salak cv. Madu. IOP Conf Ser Earth Environ Sci 2020; 542(1): 012014.
[http://dx.doi.org/10.1088/1755-1315/542/1/012014]

[116] El Fawal GF, Omer AM, Tamer TM. Evaluation of antimicrobial and antioxidant activities for cellulose acetate films incorporated with Rosemary and *Aloe Vera* essential oils. J Food Sci Technol 2019; 56(3): 1510-8.
[http://dx.doi.org/10.1007/s13197-019-03642-8] [PMID: 30956331]

[117] Ozturk B, Karakaya O, Yıldız K, Saracoglu O. Effects of *Aloe vera* gel and MAP on bioactive compounds and quality attributes of cherry laurel fruit during cold storage. Sci Hortic 2019; 249: 31-7.
[http://dx.doi.org/10.1016/j.scienta.2019.01.030]

[118] Khaliq G, Abbas HT, Ali I, Waseem M. *Aloe vera* gel enriched with garlic essential oil effectively controls anthracnose disease and maintains postharvest quality of banana fruit during storage. Hortic Environ Biotechnol 2019; 60(5): 659-69.
[http://dx.doi.org/10.1007/s13580-019-00159-z]

[119] Ortega-Toro R, Collazo-Bigliardi S, Roselló J, Santamarina P, Chiralt A. Antifungal starch-based edible films containing *Aloe vera*. Food Hydrocoll 2017; 72: 1-10.
[http://dx.doi.org/10.1016/j.foodhyd.2017.05.023]

[120] Jiwanit P, Pitakpornpreecha T, Pisuchpen S, Leelasuphakul W. The use of *Aloe vera* gel coating supplemented with Pichia guilliermondii BCC5389 for enhancement of defense-related gene expression and secondary metabolism in mandarins to prevent postharvest losses from green mold rot. Biol Control 2018; 117: 43-51.
[http://dx.doi.org/10.1016/j.biocontrol.2017.08.023]

[121] Díaz-Mula HM, Serrano M, Valero D. Alginate coatings preserve fruit quality and bioactive compounds during storage of sweet cherry fruit. Food Bioprocess Technol 2012; 5(8): 2990-7.
[http://dx.doi.org/10.1007/s11947-011-0599-2]

[122] Ates U, Islam A, Ozturk B, Aglar E, Karakaya O, Gun S. Changes in quality traits and phytochemical components of blueberry (*Vaccinium corymbosum* Cv. bluecrop) fruit in response to postharvest *Aloe vera* treatment. Int J Fruit Sci 2022; 22(1): 303-16.
[http://dx.doi.org/10.1080/15538362.2022.2038341]

[123] Ehtesham Nia A, Taghipour S, Siahmansour S. Effects of salicylic acid preharvest and *Aloe vera* gel postharvest treatments on quality maintenance of table grapes during storage. S Afr J Bot 2022; 147: 1136-45.

[http://dx.doi.org/10.1016/j.sajb.2022.05.010]

[124] Ali S, Anjum MA, Nawaz A, *et al.* Effect of pre-storage ascorbic acid and *Aloe vera* gel coating application on enzymatic browning and quality of lotus root slices. J Food Biochem 2020; 44(3): e13136.
[http://dx.doi.org/10.1111/jfbc.13136] [PMID: 31907949]

[125] Khaliq G, Nisa M, Ramzan M, Koondhar N. Textural properties and enzyme activity of mango (*Mangifera indica* L.) fruit coated with chitosan during storage. J Agric Stud 2017; 5(2): 32-50.
[http://dx.doi.org/10.5296/jas.v5i2.10946]

[126] Hassan B, Chatha SAS, Hussain AI, Zia KM, Akhtar N. Recent advances on polysaccharides, lipids and protein based edible films and coatings: A review. Int J Biol Macromol 2018; 109: 1095-107.
[http://dx.doi.org/10.1016/j.ijbiomac.2017.11.097] [PMID: 29155200]

[127] Ali A, Maqbool M, Ramachandran S, Alderson PG. Gum arabic as a novel edible coating for enhancing shelf-life and improving postharvest quality of tomato (*Solanum lycopersicum* L.) fruit. Postharvest Biol Technol 2010; 58(1): 42-7.
[http://dx.doi.org/10.1016/j.postharvbio.2010.05.005]

[128] Khaliq G, Muda Mohamed MT, Ali A, Ding P, Ghazali HM. Effect of gum arabic coating combined with calcium chloride on physico-chemical and qualitative properties of mango (*Mangifera indica* L.) fruit during low temperature storage. Sci Hortic 2015; 190: 187-94.
[http://dx.doi.org/10.1016/j.scienta.2015.04.020]

[129] Mannozzi C, Cecchini JP, Tylewicz U, *et al.* Study on the efficacy of edible coatings on quality of blueberry fruits during shelf-life. Lebensm Wiss Technol 2017; 85: 440-4.
[http://dx.doi.org/10.1016/j.lwt.2016.12.056]

[130] Rastegar S, Atrash S. Effect of alginate coating incorporated with Spirulina, *Aloe vera* and guar gum on physicochemical, respiration rate and color changes of mango fruits during cold storage. J Food Meas Charact 2020.

[131] Nair MS, Saxena A, Kaur C. Characterization and antifungal activity of pomegranate peel extract and its use in polysaccharide-based edible coatings to etend the shelf-life of capsicum (*Capsicum annuum* L.). Food Bioprocess Technol 2018; 11(7): 1317-27.
[http://dx.doi.org/10.1007/s11947-018-2101-x]

[132] Nair MS, Saxena A, Kaur C. Effect of chitosan and alginate based coatings enriched with pomegranate peel extract to extend the postharvest quality of guava (Psidium guajava L.). Food Chem 2018; 240: 245-52.
[http://dx.doi.org/10.1016/j.foodchem.2017.07.122] [PMID: 28946269]

[133] Moniruzzaman M, Rokeya B, Ahmed S, Bhowmik A, Khalil M, Gan S. *In vitro* antioxidant effects of *Aloe barbadensis* Miller extracts and the potential role of these extracts as antidiabetic and antilipidemic agents on streptozotocin-induced type 2 diabetic model rats. Molecules 2012; 17(11): 12851-67.
[http://dx.doi.org/10.3390/molecules171112851] [PMID: 23117427]

[134] Bhandari AK, Negi JS, Bisht VK, Bharti MK. *In vitro* propagation of *Aloe vera*: A plant with medicinal properties. Nat Sci 2010; 8(8): 174-6.

[135] Abrie AL, Van Staden J, Abrie A áL, *et al.* Micropropagation of the endangered *Aloe polyphylla*. Plant Growth Regul 2001; 33(1): 19-23.
[http://dx.doi.org/10.1023/A:1010725901900]

[136] Murashige T, Skoog F. A revised medium for rapid growth and bio assays with tobacco tissue cultures. Physiol Plant 1962; 15(3): 473-97.
[http://dx.doi.org/10.1111/j.1399-3054.1962.tb08052.x]

[137] Jayakrishna C, Karthik C, Barathi S, Kamalanathan D. *In vitro* propagation of *Aloe barbadensis* Miller, a miracle herb. Res Plant Biol 2011; 1(5): 22-6.

[138] Natali L, Sanchez IC, Cavallini A. *In vitro* culture of *Aloe Barbadensis* Mill.: Micropropagation from vegetative meristems. Plant Cell Tissue Organ Cult 1990; 20(1): 71-4.
[http://dx.doi.org/10.1007/BF00034761]

[139] Shibru S, Olani G, Debebe A. *In vitro* propagation of *Aloe vera* Linn from shoot tip culture. GSC Biol Pharm Sci 2018; 4(2): 001-6.

[140] Heringer AS, Reis RS, Passamani LZ, De Souza-Filho GA, Santa-Catarina C, Silveira V. Comparative proteomics analysis of the effect of combined red and blue lights on sugarcane somatic embryogenesis. Acta Physiol Plant 2017; 39(2): 52.
[http://dx.doi.org/10.1007/s11738-017-2349-1]

[141] Kulus D. Influence of growth regulators on the development, quality, and physiological state of *in vitro*-propagated *Lamprocapnos spectabilis* (L.) Fukuhara. In Vitro Cell Dev Biol Plant 2020; 56(4): 447-57.
[http://dx.doi.org/10.1007/s11627-020-10064-1]

[142] De Oliveira ET, Crocomo OJ, Farinha TB, Gallo LA. Large-scale micropropagation of *Aloe vera*. HortScience 2009; 44(6): 1675-8.
[http://dx.doi.org/10.21273/HORTSCI.44.6.1675]

[143] Sivakumar P, Ashok K, Rajalakshmi K, Sangeetha R, Muthumalai M. A review on *in vitro* propagation of miraculous physician *Aloe vera* (L.). Res J Biotechnol 2019; 14(9): 126-30.

[144] Liao Z, Chen M, Tan F, Sun X, Tang K. Microprogagation of endangered Chinese aloe. Plant Cell Tissue Organ Cult 2004; 76(1): 83-6.
[http://dx.doi.org/10.1023/A:1025868515705]

[145] Gupta S, Sahu PK, Sen DL, Pandey P. *In-vitro* propagation of *Aloe vera* (L.) Burm. f. Br Biotechnol J 2014; 4(7): 806-16.
[http://dx.doi.org/10.9734/BBJ/2014/9747]

[146] Daneshvar MH, Moallemi N, Abdolah Zadeh N. The Effects of different media on shoot proliferation from the shoot tip of *Aloe vera* L. Jundishapur J Nat Pharm Prod 2013; 8(2): 93-7.
[http://dx.doi.org/10.17795/jjnpp-4820] [PMID: 24624195]

[147] Lane WD. *In vitro* propagation of *Spirea bumalda* and *Prunus cistena* from shoot apices. Can J Plant Sci 1979; 59(4): 1025-9.
[http://dx.doi.org/10.4141/cjps79-161]

[148] Al-Rifae'e MAT, Al-Shobaki SA. Twenty one contury techniques for plant improvement by tissue culture. Cairo: Dar Al-Fikr Al-Arabi 2002.

[149] Duhoky MMS. Effect of different concentration of BA and IAA on micropropagation of *Gardenia jasminoides*. Mesopotamia J Agric 2010; 38(2): 16-30.
[http://dx.doi.org/10.33899/magrj.2010.27763]

[150] Danial GH, Ibrahim DA, Yousef AN, Elyas SB. Rapid protocol of *Aloe vera* micropropagation. Iraqi J Agric Sci 2019; 50(5): 1377-82.

[151] Hussey G. *In vitro* propagation of monocotyledonous bulbs and corms. proceedings, 5th International Congress of Plant Tissue and Cell Culture. 677-80.

[152] Jianhua Z, Huoying C, Tianming Z, Xiaoning Z. *In vitro* rapid and efficient propagation of *Aloe vera* L. J Shanghai Agric Coll 2002; 20(2): 122-4.

[153] Neelofar K, Sharma GK. Rapid *in vitro* propagation of *Aloe vera* L. with some growth regulators using lateral shoots as explants. World J Pharm Pharm Sci 2014; 3(3): 2005-18.

[154] Baksha R, Jahan MAA, Khatun R, Munshi JL. Micropropagation of *Aloe barbadensis* Mill. through *in vitro* culture of shoot tip explants. Plant Tissue Cult Biotechnol 2005; 15(2): 121-6.

[155] Hosseini R, Parsa M. Micropropagation of *Aloe vera* L. grown in south Iran. Pak J Biol Sci 2007;

10(7): 1134-7.
[http://dx.doi.org/10.3923/pjbs.2007.1134.1137] [PMID: 19070066]

[156] Aggarwal D, Barna KS. Tissue culture propagation of elite plant of *Aloe vera* Linn. J Plant Biochem Biotechnol 2004; 13(1): 77-9.
[http://dx.doi.org/10.1007/BF03263197]

[157] Molsaghi M, Moieni A, Kahrizi D. Efficient protocol for rapid *Aloe vera* micropropagation. Pharm Biol 2014; 52(6): 735-9.
[http://dx.doi.org/10.3109/13880209.2013.868494] [PMID: 24405115]

[158] Murashige T. Principles of rapid propagation. Propag High plants through tissue Cult A Bridg between Res Appl 1979; 14-24.

[159] Castorena Sanchez I, Natali L, Cavallin A. *In vitro* culture of *Aloe barbadensis* Mill.: Morphogenetic ability and nuclear DNA content. Plant Sci 1988; 55(1): 53-9.
[http://dx.doi.org/10.1016/0168-9452(88)90041-6]

[160] Razib MA, Mamun ANK, Kabir MH, Al Din SMS, Hossain ME, Firoz Alam M. High frequency *in vitro* propagation of *Aloe vera* l. through shoot tip culture. Int J Appl Biol Pharm Technol 2016; 7(4): 28-32.

[161] Nayanakantha NMC, Singh BR, Kumar A. Improved culture medium for micropropagation of *Aloe vera* L. Trop Agric Res Ext 2011; 13(4): 87.
[http://dx.doi.org/10.4038/tare.v13i4.3291]

[162] Nissen SJ, Sutter EG. Stability of IAA and IBA in nutrient medium to several tissue culture procedures. HortScience 1990; 25(7): 800-2.
[http://dx.doi.org/10.21273/HORTSCI.25.7.800]

[163] Davood H, Behzad K. Rapid micro-propagation of *Aloe vera* L. *via* shoot multiplication. Afr J Biotechnol 2008; 7(12): 1899-902.
[http://dx.doi.org/10.5897/AJB2008.000-5038]

[164] Das A, Mukherjee P, Jha TB. High frequency micropropagation of *Aloe vera* L. Burm. f. as a low cost option towards commercialization. Plant Tissue Cult Biotechnol 2010; 20(1): 29-35.
[http://dx.doi.org/10.3329/ptcb.v20i1.5962]

[165] Gantait S, Mandal N, Das PK. *In vitro* accelerated mass propagation and *ex vitro* evaluation of *Aloe vera* L. with aloin content and superoxide dismutase activity. Nat Prod Res 2011; 25(14): 1370-8.
[http://dx.doi.org/10.1080/14786419.2010.541885] [PMID: 21859262]

[166] Lee YS, Yang T jin, Park S, *et al*. Induction and proliferation of adventitious roots from *Aloe vera* leaf tissues for *in vitro* production of aloe-emodin. Plant Omi J 2011; 4(4): 190-4.

[167] Abdi G, Hedayat M, Modarresi M. *In vitro* micropropagation of *Aloe vera* impacts of plant growth regulators, media and type of explants. J Biol Environ Sci 2013; 7(19): 19-24.

[168] Kumari A, Naseem M. An efficient protocol for micropropagation of a medicinal plant *Aloe vera* L. through organ culture. J Indian Bot Soc 2015; 94(1&2): 118-25.

[169] M M, S P, Shekhawat MS. Microstructural and histochemical variations during *in vitro to in vivo* plant developments in *Aloe vera* (L.) Burm.f (Xanthorrhoeaceae). Ind Crops Prod 2021; 160: 113162.
[http://dx.doi.org/10.1016/j.indcrop.2020.113162]

[170] Mani M, Mathiyazhagan CR, Selvam P, Phulwaria M, Shekhawat MS. Foliar micro-morphology: A promising tool to improve survival percentage of tissue culture raised plantlets with special reference to *in vitro* propagation of *Vitex negundo* L. Vegetos 2020; 33(3): 504-15.
[http://dx.doi.org/10.1007/s42535-020-00134-5]

CHAPTER 6

Micropropagation and Phytochemical Studies on *Oroxylum Indicum* (L) Kurz – A Review

Samatha Talari[1] and **Rama Swamy Nanna**[1,*]

[1] *Department of Botany, Kakatiya University, Warangal, Telangana State, India*

Abstract: *Oroxylum indicum* (L) Kurz is a medicinal forest tree with therapeutically active principles owing to its anticancer, anti-inflammatory, antimicrobial, antiulcer, anti-arthiritic, and anti-angiogenic properties and known to be employed in ayurveda, Unani and folk medicine. Due to the possession of biologically active constituents, the tree is uprooted for the isolation of phytoconstituents and preparation of drugs from different parts of a tree and is over-exploited by pharmaceutical industries. Hence the tree is becoming an endangered species. In view of the above, this medicinally important tree species needs conservation and also thorough study on its medicinal properties. *In vitro* culture methodologies have to be employed for large-scale production and to know the importance and the activity of various chemical components of this valuable medicinal tree, as this knowledge plays a vital role in the conservation and synthesis of active principles with specific activity to treat various ailments. The present review focuses on the published data on conservation and also phytochemical studies of *O. indicum* to highlight the traditional usage of this tree species in various health disorders and also to conserve the tree using various *in vitro* culture techniques for its large-scale production.

Keywords: *In vitro* culture studies, Micropropagation, *Oroxylum indicum*, Phytochemical constituents.

INTRODUCTION

Medicinal plants have been used by human beings from time immemorial to cure various ailments and the important bioactive compounds extracted from these were used in ayurvedic, siddha and Unani medicinal systems to treat various ailments. Natural products obtained from medicinal plants are the important ingredients of therapeutics used in traditional medicine as they are easily accessible in the healthcare system of rural people from underdeveloped to developing countries, and these are preferred over western medicine as they are inexpensive and without any negative impact. Most of the world countries depend

---

[*] **Corresponding author Rama Swamy Nanna:** Department of Botany, Kakatiya Universty, Warangal, Telangana State, India; E-mail: swamynr.dr@gmail.com

Mohammad Anis & Mehrun Nisha Khanam (Eds.)
All rights reserved-© 2024 Bentham Science Publishers

on traditional medicine for healthcare needs, and hence many important medicinal plants are disappearing from natural ecosystems at an alarming state and becoming endangered species due to over-exploitation by pharmaceutical industries and have been enlisted in RED DATA BOOK, which need to be conserved not only for their sustainability in the nature but also for their vital therapeutic properties. Among them, *Oroxylum indicum* (L) Kurz is one of the endangered medicinal tree species with a wide range of medicinal values with an urgent need for its conservation [1].

Oroxylum indicum is an important medicinal forest tree known by names **Syonaka** or **Sonpatha** with therapeutic properties such as anticancerous [2], immunomodulatory, antinflammatory [3], analgesic, anti-tussive [4], antidiabetic [5], anti-helminthic, anti-leucodermatic, anti-rheumatic, anti-anorexic, antimicrobial [6, 7], antioxidant [8], and anti-angiogenic activities [9]. Its stem bark is used in the preparation of a **Khakhi** dye, **Agarbathi** and has got an enhancing effect on silk production of *Bombyx mori* and employed in tribal and folk medicines to alleviate a number of diseases [10], and is the main component of **Chyavanaprasha** and **Dasamoolam** [11]. In view of its wide range of therapeutically active properties, the total plant is uprooted and over-exploited from wild populations. Due to its large-scale utilization and indiscriminate collection by pharmaceutical industries, the species is depleted from natural populations and becoming endangered and has been pushed into the **red-listed plants** [12].

In order to multiply and conserve the species, *in vitro* culture technology can be employed for rapid clonal propagation. Conservation of rare and endangered medicinal plants can be achieved by applying *in vitro* multiplication [13] and has been used for the conservation of many medicinally important plants such as *Wrightia tomentosa, Givotia rottleriformis,* and *Oroxylum indicum*, which are endangered through micropropagation technique [14 - 16]. Hence, with the aim of conserving the species and also to evaluate its medicinal properties, we have made an attempt to conserve and propagate the species of *O. indicum*. This review not only discusses efficient micropropagation methods using various explants but also focuses on screening methodologies for analyzing medicinal potentiality in various parts of *O. indicum*.

O. indicum is known as the ***broken bones tree*** or ***mid-night horror tree.*** The tree is deciduous with large bipinnate or tripinnately compound leaves with ovate or elliptical leaflets. The tree is a night bloomer adapted to chiropterophily with many large bell-shaped flowers with five stamens (Bignoniaceae), and produces fruits of 1.0 meter long flat curved capsule enclosing many flat, thin winged seeds (Fig. **1a-d**).

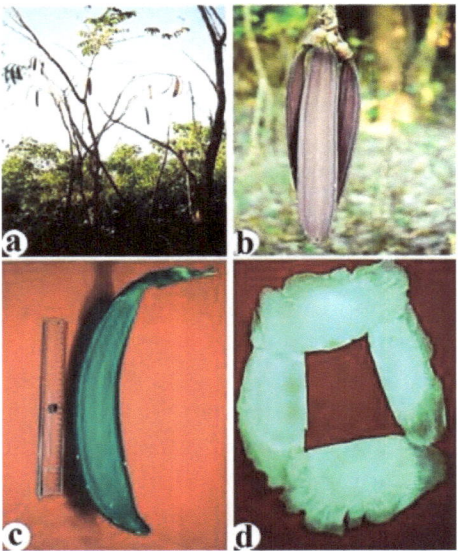

Fig. (1). Showing habit of *Oroxylum indicum*. **a**) Plants growing in Mallur forest, **b**) Bunch of Fruits, **c**) A single pod about 1.0 m long, **d**) Seeds with wings (Samatha *et al.* 2013).

In vitro culture of medicinal plants can be employed for the large-scale production of disease free plants and secondary metabolites for standard plant-based medicines irrespective of the season, which is found to be advantageous over conventional methods. The micropropagation methods increase the multiplication rate and scope for the production of *true-to-type* and pathogen-free plants, and have been reported in many medicinal plants [17].

In vitro micropropagation is advantageous over conventional methods as it paves the way for conservation of rare endangered and threatened plant species with desired characteristics, production of disease-free plants viable for recalcitrant seeds and production of secondary metabolites.

The natural propagation of the *O. indicum* takes place by seeds with 30% of *in-vivo* seed viability, which germinate in the rainy season. This plant has been placed on the list of endangered species of India due to its indiscriminate exploitation for its therapeutic value and problems related to its natural propagation. It has been reported that the estimated demand for the plant material of *O. indicum* in Southern India is 500 Kg/year [1]. Since the demand for *O. indicum* by the pharmaceutical industry is increasing, it is on the verge of extinction. The RAPD analysis of the species has been reported for its conservation and collection strategies; the genetic diversity in different accessions was found to be very low and hence may be the reason for adapting to different environmental conditions [18].

The present review not only focuses on micropropagation but also on various biochemical constituents and phytochemical studies and medicinal properties of *O. indicum*.

In-vitro Seed Germination

A protocol has been developed in *O. indicum* through *in vitro* seed germination using BAP, IAA, and GA_3 to enhance the percentage of seed germination in *O. indicum*, the seeds were also treated with cold and hot water and found significant improvement with cold water treatment at 4°C for 24 h compared to hot water treatment and PGRs with more than 80% seed germination response [19]. This simple economic method for *in vitro* seed germination in *O. indicum* can be used to achieve an increase in the percentage of seed germination and seedling growth [20]. In spite of using various PGRs and different types of media, seed germination was effective (100%) on ½ strength MS and MS media containing different concentrations of sucrose. And the time taken for the germination was also found to be a lesser period (6-7 days) on MS medium (Fig. **2**) and seed germination percentage was found to be lesser on B5 and WPM media.

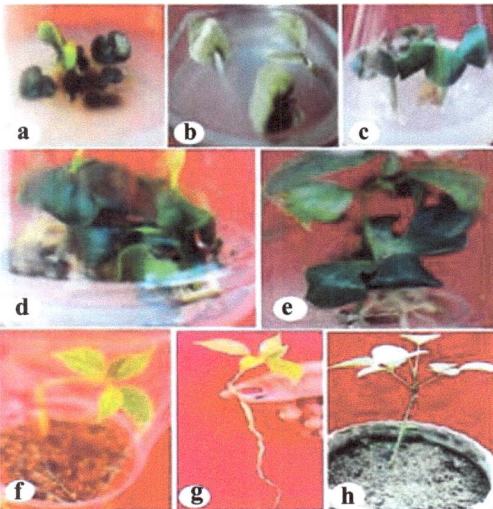

Fig. (2). *In vitro* seed germination in *O. indicum*. **a-c**) Germination of the seeds on WPM+30g/L sucrose and B5+30g/L sucrose on MS medium+30g/L sucrose respectively; **d,e**) Development of seedling on MS medium+30g/L sucrose with fully expanded cotyledonary leaves after 30 days and 45 days of germination respectively; **f**) Acclimatization of plantlet on soilrite in a plastic pot; **g**) *In vitro* raised plantlet showing lengthy root system; **h**) Acclimatized plantlet in an earthenware pot containing garden soil (Samatha 2013 a).

These *in-vitro* raised seedlings were hardened and established successfully with 85% of survival (Fig. 2). These acclimatized plantlets were similar to their mother plant. This method of multiplication can be used for the conservation of the species.

Zygotic Embryo Culture

In vitro zygotic embryo culture was achieved in *O. indicum* [21]. The zygotic embryos were cultured on MS medium+BAP/KIN/TDZ (1.0-7.0 mg/L) individually, also in combination with 0.1 mg/L $AgNO_3$, and the percentage of response recorded was found to be 100% on MS medium fortified with 5.0 mg/L BAP with a maximum number (20±0.82) of shoots/explant in comparison to KIN. The influence of $AgNO_3$ on *in-vitro* zygotic embryo culture in *O. indicum* was found to be effective in inducing multiple shoots/explant, and on MS+0.1 mg/L $AgNO_3$+3.0 mg/L TDZ induced the maximum number (32±0.02) of shoots/explant in comparison to all other PGRs used [22]. *In vitro* raised plantlets were acclimatized and transferred into the field (Fig. 3).

Clonal Propagation

In this technique, shoot tips and nodal segments are used for the mass-scale multiplication and rapid conservation of endangered and threatened medicinally important plants within a short span of time and limited space to produce *true-to-type* plants.

In-vitro micropropagation of *O. indicum* has been reported using cotyledonary node, node, and shoot meristem explants on MS medium supplemented 6-benzyladenine (BA), and with different concentrations of PGRs *viz.*, BAP/KIN/TDZ alone (Fig. 4), respectively, induced multiple shoots, and inclusion of IAA (2.85 µM) with BA augmented resulted in high percentage of response as well as a proliferation of more number of shoots [22, 23]. Among the three types of cytokinins used, the best PGR with a maximum percentage of response and increased number of multiple shoots was found in the BAP as a sole growth regulator, followed by KIN and TDZ [24].

Nodal Culture

A reproducible protocol for micropropagation of *O. indicum* using nodal explants has been reported [25]. The nodal segments were cultured on MS+BAP/KIN (1.0-5.0 mg/L)/TDZ (0.2-1.0 mg/L) as a sole growth regulator. An increase in the concentration of BAP, number of multiple shoots per explant was also found to be enhanced. The maximum percentage of response and more number of multiple shoots (15.0±0.10) formation per explant were found at 5.0 mg/L BAP followed by 3.0-4.0 mg/L BAP. While nodal segments cultured on KIN showed a maximum percentage of response, and the maximum number of multiple shoots was observed at 5.0 mg/L KIN (11±0.10). BAP was found to be more effective than KIN and TDZ in multiple shoot induction in *O. indicum*.

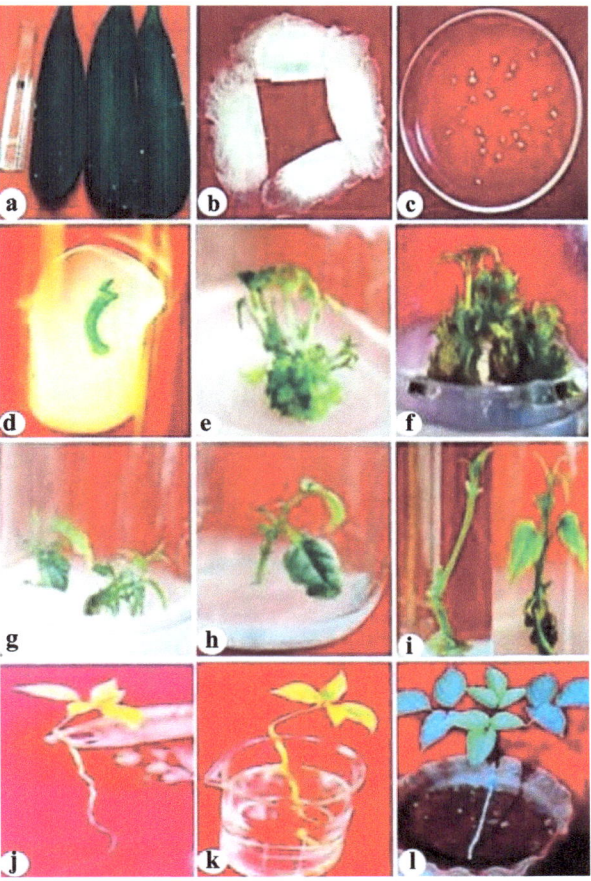

Fig. (3). *In vitro* **zygotic embryo culture in *O. indicum*. a**) Fruits ; **b**) Seeds; **c**) Isolated zygotic embryos; **d**) Germination of zygotic embryo; **e-f**) Multiple shoots formation on MS+3.0 mg/L TDZ after 4 and 6 weeks respectively; **g-h**) Individual micro-shoots; **i**) Elongation of individual shoots on MS+1.0mg/L GA$_3$; **j-k**) *In vitro* raised plant with well developed root system; **l**) Acclimatized plantlet in an earthenware pot containing garden soil (**Samatha *et al.*, 2013 d**).

Mericlone Technology

Shoot apical meristems of *O. indicum* were inoculated on MS medium with 2.0 mg/L BAP for *in vitro* multiple shoots induction. These were cultured on MS+BAP/KIN (1.0-7.0 mg/L)/TDZ (0.2 - 1.0 mg/L) as a sole growth regulator and also on MSO (Fig. 4). At higher concentrations of BAP, low percentages of response as well as the number of shoots per explant was found to be reduced. At higher (5.0 mg/L) levels of KIN, less number of multiple shoots and also callus formation was found. While TDZ showed less percentage of response compared to BAP and KIN. A maximum percentage (60%) of response was observed at 0.4 mg/L TDZ, and high frequency number of shoots development per explant (4.0±0.084) was found at the same concentration of TDZ. At higher concentrations of TDZ, less percentage of response and also a low number of shoots per explant were noted. Shoot tip explants cultured on MS medium without growth regulators (MSO) showed the elongation of only a single shoot.

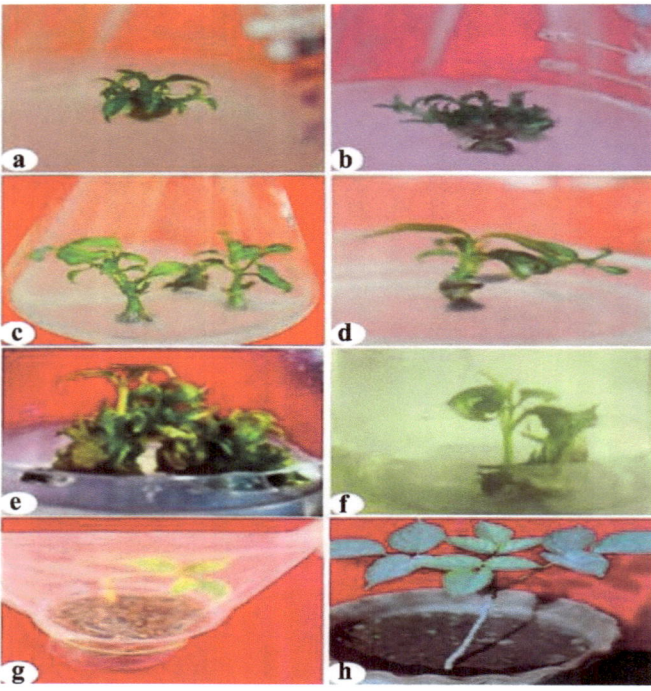

Fig. (4). *In-vitro* micropropagation in *O. indicum*. **a,b,c,d)** Multiple shoots formation from shoot tip, node and cotyledonary nodal segments on MS+3.0 mg/L BAP after 4 weeks of culture respectively; **e)** Further multiplication and elongation of healthy shoots on MS+3.0 mg/L BAP+1.0 mg/L GA3; **f)** *In-vitro* rooting on ½ MS+1.0 mg/L IBA; **g)** Hardening in the culture room; **h)** Healthy and established *in-vitro* raised plantlet in an earthenware pot containing garden soil (Samatha *et al.* 2016 a).

In vitro Rooting and Plantlet Establishment

The developed micro-shoots from direct organogenesis, zygotic embryo culture, nodal and meristem culture of *O. indicum* were elongated and further proliferated on MS+1.0 mg/L GA_3+ 3.0 mg/L BAP (Fig. **4e**) and shifted on to ½ strength MS+0.5- 4.0 mg/L IAA/IBA/NAA for *in vitro* rooting. Profuse rhizogenesis was observed at 1.0 mg/L NAA. *In vitro* rooted plantlets were acclimatized and transferred into the field. The survival percentage was found to be 85% [25]. The elongated micro-shoots developed through the zygotic embryo, nodal and shoot tip cultures were cultured on ½ strength MS medium and ¼ strength WPM+IBA/NAA/IAA for *in vitro* rooting. The root induction is seen on ½ strength MS+1.0 mg/L IBA showed the best hormonal concentration for induction roots lengthy and healthy roots (4f). The well-rooted plantlets were acclimatized and transferred into the field with 80% survival and were found to be normal as donor plants (Fig. **4g-h**) [25].

Callus Induction

In-vitro callus induction was observed using leaf and cotyledonary leaf explants of *O. indicum* on MS+30gm/L sucrose+1.0-5.0 mg/L IAA/IBA/IBA/2,4-D [26 - 28]. Morphogenic response (morphology and texture) of callus was found to be different in leaf and cotyledonary leaf explants cultured on MS with different PGRs (Fig. **5a-f**).

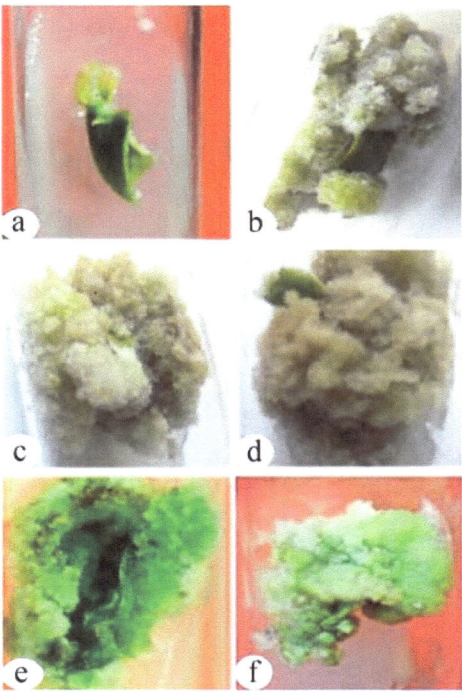

Fig. (5). Callus induction from cotyledonary leaf explants of *O. indicum*.a,b) Callus induction from cotyledonary leaf explant at 2.0 mg/L 2, 4-D and further proliferation of callus on the same respectively; **c)** Formation of white compact callus at 4.0mg/L IBA; **d)** High amount of friable callus at 3.0 mg/L NAA; **e)** Green friable callus at 3.0 mg/L NAA; **f)** Formation of green meristemoids at 2.0 mg/L 2, 4-D/IBA. (Samatha et al. 2016 c).

Proliferation of callus was faster and a very high yield of callus was achieved within 4 weeks from cotyledonary leaf explants on all high and low concentrations of PGRs, particularly at 2.0 and 3.0 mg/L IBA and 2, 4-D followed by IAA/NAA and cotyledonary leaf explants were found be highly efficient for the callus induction in *O. indicum* [28].

Secondary metabolites obtained from callus cultures can be employed in biopharmaceutical production and found to be a promising source at low cost in medicinal plant biotechnology research. The ratio of auxin to cytokinin can be manipulated in the medium for obtaining shoots, roots or somatic embryos from single explants from which whole plants can be produced subsequently. Callus tissue is a good source of genetic or karyotypic variability, so it may be possible to regenerate a plant from genetically variable cells of the callus tissue. Cell suspension culture in a moving liquid medium can be initiated from callus cultures. Thus, the secondary metabolites can be directly extracted from these callus tissues without sacrificing the whole medicinal plant for the preparation of drugs.

Organogenesis

In plant tissue culture, shoots and/or roots are induced directly/indirectly from the explants through organogenesis (development of organs), which starts with stimulation caused by the chemicals of the medium, substances carried over from the original explants and endogenous compounds produced by the culture [29]. Direct *in-vitro* organogenesis, which bypasses the need for a callus phase for large-scale multiplication of the species *O. indicum* has been achieved on MS medium fortified with various concentrations (1.0- 5.0 mg/L) of cytokinins BAP/KIN alone and also in combination with 0.5mg/L 2, 4-D [25].

Somatic Embryogenesis

Somatic embryogenesis has been developed for the conservation of the species *O. indicum* using cotyledonary leaf explants cultured on MS+BAP (0.25-0.75 mg/L) alone and also in combination with 0.5 mg/L IAA/2, 4- D [30]. Induction of somatic embryos was found at 0.5 mg/L 2, 4-D+0.75 mg/L BAP with 95% of response of somatic embryo conversion into bipolar embryo be germinated and converted into plantlets at 0.2 mg/L IAA+0.25 mg/L BAP. These regenerated plantlets were acclimatized in the culture room, which has been successfully achieved in *O. indicum* (Fig. **6a-i**).

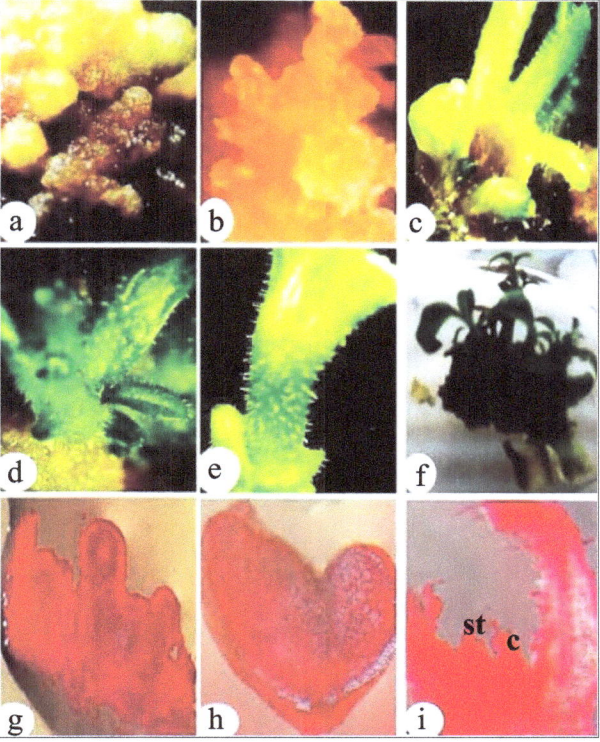

Fig. (6). Somatic embryogenesis and plantlet development in *O. indicum*. a) Induction of somatic embryos on MS+ 0.5 mg/L 2, 4-D+0.75 mg/L BAP; **b)** Development of different types of embryoids; **c)** A group of bipolar embryos; **d-f)** Germination of somatic embryo and complete plantlets respectively (**Samatha and Rama Swamy. 2015**).

This protocol can be used for developing synseeds, conservation and also for rapid multiplication of *true-to-type* species in a short period [31]. In *in-vitro* cuture technology, while dealing with tree species, we encounter problems related to phenolic exudation and callus production. The additives Casein hydrolysate (CH), Activated Charcoal (AC), Coconut milk (CM) and Silver nitrate (AgNO$_3$) in the media enhance the growth and development by promoting metabolic activities [32]. For inducing *in vitro* regeneration of an important endangered medicinal tree species *O. indicum,* among all the additives, AgNO$_3$ acted positively for multiple shoot regeneration, and the enhancement in regeneration efficiency in zygotic embryo culture of *O. indicum* was found [22].

The production, consumption and international trade of medicinal plants and phytomedicine have grown and are expected to grow further in the future. To satisfy growing market demands, surveys are being conducted to unearth new plant sources of herbal remedies and medicines and, at the same time to develop new strategies for better yield and quality, which can be achieved through different methods, including micropropagation. It may help in conserving many medicinally valuable tree species in the process and may open new vistas in medicinal plant biotechnology.

In conclusion, it is mentioned that the effectiveness of plant regeneration depends on the type of PGRs used as a source of explants, and also the stage at which the explant is collected. Thus the potentiality of regeneration is dependent on the type of explants, culture medium and also the conditions at the time of culture. The protocols developed, including our research work for the plant micropropagation, thus open a new perspective that could facilitate the conservation and mass multiplication of vulnerable and valuable medicinal forest tree species like *O. indicum*.

Phytochemical Studies

O. indicum is a multipurpose medicinal forest tree and found to possess bioactive compounds with antiinflammatory, diuretic, antiarthritic, antimicrobial properties and is used in the preparation of many formulations, *i.e.*, *Dashmula* and *Chyawanprasha* [33]. The occurrence of medicinally important chemical constituents such as aldehydes, phenols, fatty acids, flavonoids and alkaloids that are encountered in almost all the parts of *O. indicum* makes it a more valuable medicinal plant and has led to the isolation of phytochemical components such as oroxylin-A, baicalein, chrysin and kaempferol [34 - 38]. *O. indicum* is used in indigenous medicine to cure various ailments such as dysentery and rheumatism in many Asian countries [39]. The bark of the root is used in the treatment of fever, diarrhoea, dysentery, bronchitis, intestinal worms, leucoderma, asthma, inflammation, tuberculosis and nasopharyngeal cancer. The seed extracts exhibit antimicrobial, analgesic, anti-tussive, anti-bronchitic, anti-inflammatory and leucodermatic properties. In folk medicine, root, bark of stem, and leaf of *O. indicum* are being used in the treatment of snake bite and leprosy [40].

Toxicological Activity

The toxicological activities of *O. indicum* have also been reported specifically from root and stem extracts. These are found to be toxic for the growth of brine shrimp nauplii, but the aqueous extract of root bark of *O. indicum* was reported to have a protective effect in DNBS-induced colitis in rat as it did not cause any acute toxic symptoms [41]. The antihelminthic activity of *O. indicum* against Equine strongyle eggs when compared to that of Ivermectin (deworming agent) has been reported [42].

A study on esterase enzyme activity and silk yield in silk worm (*Bombyx mori*) using bark extract of *O. indicum* has been reported (Fig. 7). According to the results, there was a significant increase in the silk yield. Thus, the bark extract of *O. indicum* can be used as a nutrient supplement to enhance the silk yield in *B. mori* [10].

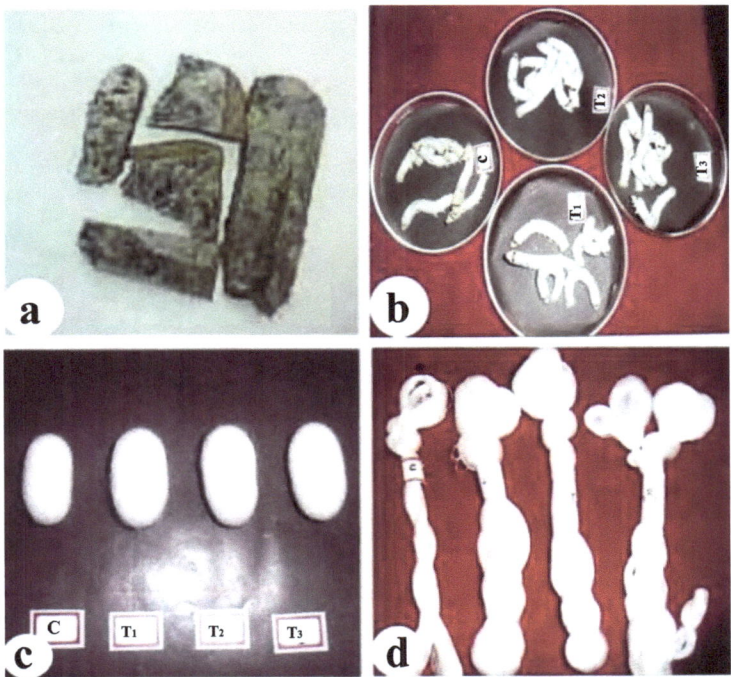

Fig. (7). Effect of stem bark extracts of *O. indicum* on commercial characters of silk worm *B. mori*. a) Stem bark; **b)** Silkworms fed with mulberry leaves coated with stem bark extracts; **c)** Cocoons (C=Control, T1 =1:1, T2=1:2, T3=1:3); **d)** Enhanced quantity of silk production in various treatments (C=Control, T1 =1:1, T2=1:2, T3=1:3 respectively). (**Samatha *et al.* 2014 b**).

The pharmacological studies on the plant extracts have been evaluated pertaining to antibacterial and antifungal activities by many researchers. The antifungal activity of dichloromethane root extract of *O. indicum* against dermatophytes and wood rot fungi was reported [43, 44]. The antibacterial activity of *O. indicum* has also been studied against gram positive and gram negative bacteria [26].

Antimicrobial Activity

Stem bark extract of *O. indicum* was tested for antimicrobial activity against gram-positive and gram-negative bacteria and some fungi *viz.*, *Bacillus cereus*, *B. megaterium*, *B. subtilis*, *B. cereus*, *S. albus*, *S. aureus*, and *S. lutea*, *E. coli*, *P.*

aeruginosa, S. parathyphii, S. typhimurium, S. boydii, S. dysenteriae, Vibrio mimicus, S. cerevisiae, Candida albicans and *Aspergillus niger* [43]. Methanolic extracts were effective against both gram-positive and gram-negative bacteria as well as fungi, and the properties were found to be more effective than the standard antibiotic Ampicillin. *B. subtilis* showed susceptibility towards 1:1 concentration compared with other proportions. The zone of inhibition (ZOI) was the largest (11.7mm) at 1:1 concentration when compared to other microbial strains under study (Fig. **8**).

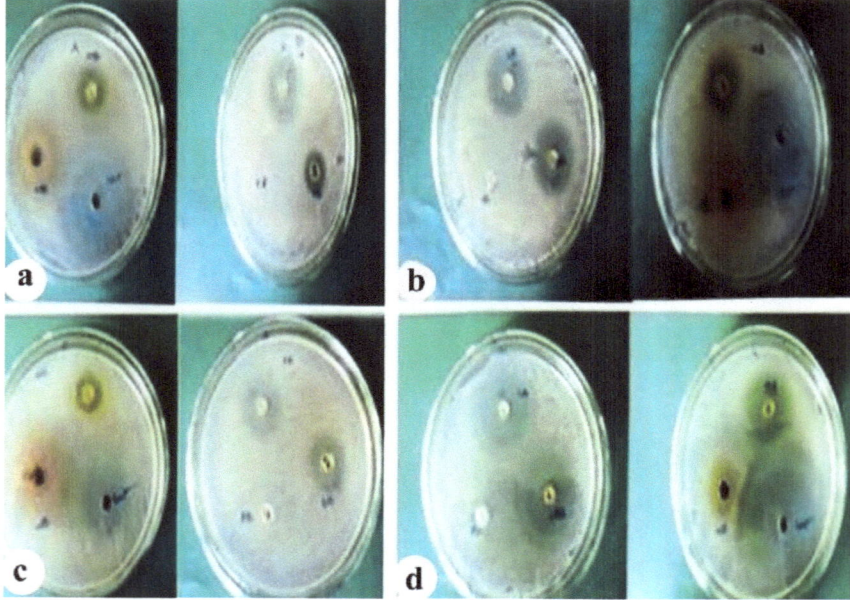

Fig. (8). Anti-bacterial activity of stem bark of *O. indicum* and the zone of inhibition against aqueous extracts. a,b) *Bacillus cereus*; **c,d)** *Staphylococcus aureus*; **c)** *Bacillus subtilis*; **d)** *Staphylococcus albus* (Samatha *et al.* 2013 c).

Antioxidant Activity

Antioxidants of natural origin are gaining enormous importance nowadays in order to prevent several diseases [45]. Modern research is focused on finding antioxidants of natural origin, particularly of plant origin. Generally, medicinal plants are known to be rich in bioactive compounds such as phenolics and flavonoids [36] and have been used for their therapeutic properties. The antioxidant activity of leaves and stem bark of *O. indicum* has also been confirmed by different assays *viz.*, total antioxidant assay and β-carotene bleaching assay. Ethanolic and chloroform extracts exhibited maximum antioxidant potential than other extracts. Methanol and aqueous extracts of stem bark of *O. indicum* have also been found to have free radical (DPPH and OH-)

scavenging activities [46, 47] and the highest value of antioxidant potential (Ic50 of 62.840μg/ml) was exhibited by methanol extract of fruit wall [8].

Anti-inflammatory Activity

Anti-inflammatory activity is being tested with an aqueous extract of leaves of *O. indicum* and found to relieve the pain in rats against carrageenan induced rat paw edema. Analgesic activity has also been studied with the butanol extract of root bark of *O. indicum* in Wistar albino rats and due to the presence of flavonoids present in the roots of *O. indicum*.

Hepatoprotective Activity

Hepatoprotective activity of different extracts of root bark and leaves of *O. indicum* showed a significant protective effect against the liver in Wistar albino rats with alteration of the level of serum enzymes (SGPT, ALP and SGOT) and total bilirubin towards the normal has been found [48].

Nephroprotective Activity

Chrysin at a dose level of 40 mg/Kg BW protected kidney damage from nephrotoxic cisplatin in experimental rats [49]. Antihyperlipidemic activity of root bark extract of *O. indicum* was examined in cholesterol induced hyperlipidemic albino Wistar rat model. The atherogenic index and LDL-C: HDL-C risk ratio was also reduced to significant extent in the group treated with extract. The study scientifically proved the folklore use of *O. indicum* as an ingredient of ayurvedic formulations used in the treatment of cardiovascular diseases [50].

Antidiabetic Activity

Antidiabetic activity of extracts of *O. indicum* in Alloxan induced diabetic Wistar albino rats has been evaluated and found to be effective in reducing the serum glucose, triglyceride, total cholesterol levels and increase in glycogen levels of the liver and muscle when compared with control [5].

Immunomodulatory Activity

The fresh root of *O. indicum* is found to have an immunity-enhancing effect, as rats treated with an n-butanol fraction of root showed a significant rise in antibody titre during secondary antibody responses, indicating potentiating aspects of the humeral response. The immunomodulatory activity of *O. indicum* was observed in broiler chicks which were treated with root and stem bark powder of *O. indicum*. The findings suggested that the root bark of *O. indicum* possessed a significant immunomodulatory activity than its stem bark counterpart [51]. Thus, root bark powder of *O. indicum* may be recommended as a safe and commercially beneficial immunomodulator drug.

The species has been used for ages for the treatment of various gastric disorders. The different fractions of pet ether, chloroform, ethyl acetate and n-butanol of root bark of *O. indicum* were tested against a gastric ulcer in Wistar albino rats and found that n-butanol fraction is more effective with antiulcer and antioxidant activity in gastric mucosal homogenates because of the presence of flavonoids [51].

Anticancer Activity

The anti-cancer property of *O. indicum* has been evaluated in the experimental animals induced by different types of carcinogens, and in human cell lines, which exhibited cytotoxic activity against the Hep 2 cell lines at a concentration of 0.05%. Baicalein, the most abundant flavonoid present in the leaves of *O. indicum* tested for antimutagenic activity and also for induction of apoptosis in the HL-60 cell line. The result of the study indicated that baicalein has inhibited N hydroxylation of Trp-p-2 and anti-tumor effect on human cancer cells [2].

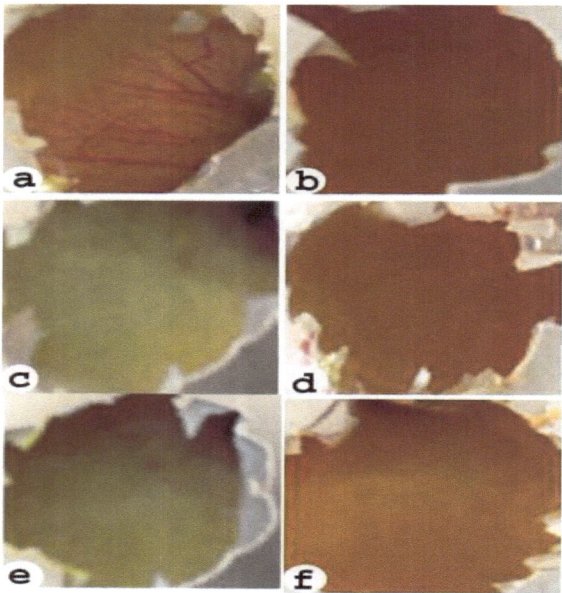

Fig. (9). Showing the anti-angiogenic effect of different extracts of *O. indicum* on chorio allantoic membrane (CAM) of fertilized chicken eggs. **a)** Effect of leaf extract on chick CAM-(No capillary-free area and very weak effect); **b-f)** Effect of stem, petiole, seed, fruit wall and stem bark extracts respectively- Capillary-free area (Strong effect) **(Samatha and Rama Swamy. 2017).**

Anti-angiogenic Activity

The anti-angiogenic activity of different parts *viz.*, petiole, leaf, seed, fruit wall, stem bark and root extracts of *O. indicum* has been evaluated through CAM (chick embryo chorio allantoic membrane) assay (Fig. **9**). The vascular endothelium growth factor (VEGF) has been found to be inhibited, and induction of neovascularization was recorded. According to the above study, all the extracts showed maximum anti-angiogenic effect [9].

Thus, the species has been used as medicine for thousands of years without any known adverse effects in different types of traditional medicinal system for alleviating many disorders. In recent years, there are a number of scientific studies which have been carried out to evaluate the toxic effects of different parts of the plant and proved to be not toxic to experimental animals and humans.

CONCLUSION

Medicinal plants are well known for their enormous usage in traditional medicine for many different purposes in various types of traditional medicine such as ayurveda, herbal, tribal and folk and modern medicine. The preliminary phytochemical screening of crude extracts of seeds, bark of stem and root of *O. indicum* revealed the presence of many bioactive compounds, which are the reason for all the important medicinal properties exhibited by *O. indicum* that are discussed in this review. Hence it can be concluded that the various extracts of this valuable medicinal tree species can be used in preventing many major diseases as the results show therapeutic compositions. We suggest further work can be done on this endangered ethnomedicinally important forest tree species by using an individual bioactive molecule for assessment of its therapeutic value.

LIST OF ABBREVIATIONS

PGR   (Plant growth regulators)

CAM   (Chick embryo chorio allantoic membrane) assay

VEGF   (Vascular endothelium growth factor)

REFERENCES

[1]   Ravi kumar K, Ved DK. 100 Red listed medicinal plants of conservation concern in Southern IndiaFoundation for revitalization of local health traditions. Bangalore India 2000; 5: pp. (1)4-67.

[2]   Roy MK, Nakahara K, Na TV, *et al.* Baicalein, a flavonoid extracted from a methanolic extract of *Oroxylum indicum* inhibits proliferation of a cancer cell line *in vitro via* induction of apoptosis. Pharmazie 2007; 62(2): 149-53.
[PMID: 17341037]

[3]   Upaganlawar B, Tenpe CR. *in vitro* antioxidant activity of leaves of *Oroxylum indicum* (Vent.). Biomed 2007; 2(3): 300-4.

[4] Zaveri M, Jain S. Anti-inflammatory and analgesic activity of root bark of *Oroxylum indicum*. Vent J Global Pharm Technology 2010; 2(4): 79-87.

[5] Ashpak MT, Sujit TK, Sohrab AS, Anil MM, Dattatraya VK. Hypoglycemic activity of extracts of *Oroxylum indicum* (L.) Vent roots in animal models. Pharmacology 2011; 2: 890-9.

[6] Islam MK, Eti ZI, Chowdhury JA. Phytochemical and Antimicrobial analysis on the extract of *Oroxylum indicum* (linn.) stembark. Iranian J Pharma & Therap 2010; 9(1): 25-8.

[7] Samatha T, Sampath A, Sujatha K. Antibacterial activity of stem bark extracts of *Oroxylum indicum* an endangered ethnomedicinal forest tree. J Pharm Biomed Sci 2013; 7(1): 24-8. a

[8] Samatha T, Raju N, Ankaiah M, Rama Swamy N, Venkaiah Y. *In vitro* screening of antioxidant efficiency in different parts of *Oroxylum indicum*. Pharmanest 2014; 5(4): 2210-3.

[9] Samatha T, Shiva Krishna P. Anti angiogenic activity of *Oroxylum indicum* L. Kurz a medicinal tree. Int J Chemtech Res 2017; 10(5): 276-80.

[10] Samatha T, Sampath T, Sujatha K, Nanna RS. Effect of stem bark extracts of *Oroxylum indicum* an ethnomedicinal forest tree on silk production of *Bombyx mori*. Int J Pharm Sci Res 2014; 5(2): 568-71.

[11] Anonymous Uttaranchal Mein Jari Butiyan: Paridrashya wa sambhavanayain, workshop papers,centre for development studies, U.P. Academy of Administration, Nainital 1998; 2: p. (40)132.

[12] Gupta RC, Sharma V, Sharma N, Kumar N, Singh B. *In vitro* antioxidant activity from leaves of *Oroxylum indicum* (L.) (Vent.)-A north Indian highly threatened and vulnerable medicinal plant. J Pharm Res 2008; 1(1): 65-72.

[13] Swamy MK, Paramashivaiah S, Hiremath L, Akhtar MS, Sinniah UR. Micropropagation and conservation of selected endangered anticancer medicinal plants from thewestern ghats of india.Anticancer Plants: Natural Products and Biotechnological Implements. Singapore: Springer 2018.

[14] Purohit SD, Kukda G, Tak K. Micropropagation of *Wrightia tomentosa* (Roxb.) Roem et Schult. J Sustain For 1996; 3(4): 25-35.
[http://dx.doi.org/10.1300/J091v03n04_03]

[15] Rambabu M, Upender M, Ujjwala D, Ugandhar T, Praveen M, Rama Swamy N. *In vitro* zygotic embryo culture of an endangered forest tree *Givotia rottleriformis* and factors affecting its germination and seedling growth. in vitro Cell Dev Biol Plant 2006; 42(5): 418-21.
[http://dx.doi.org/10.1079/IVP2006804]

[16] Samatha T. Micro propagation and Phytochemical studies of an endangered and medicinally important forest tree species *Oroxylum indicum* (L) Kurz. PhD Thesis. India.: Kakatiya University Warangal TS 2013.

[17] Bhatia Saurabh, Kiran Sharma. Modern applications of plant biotechnology in pharmceutical sciences. 116 Applications and merits of micropropagation over conventional plant breeding 2015; 22-4.

[18] Jayaram K, Prasad MNV. Genetic diversity in *Oroxylum indicum* (L.) Vent. (Bignoniaceae), avulnerable medicinal plant by random amplified polymorphic DNA marker. Afr J Biotechnol 2008; 7(3): 254-62.

[19] Singh M, Singh KK, Badola HK. Effect of temperature and plant growth regulators on seed germination response of *Oroxylum indicum*-A 06 High Value Threatened Medicinal Plant of Sikkim Himalaya. J Plant Sci Res 2014; 1(4): 115.

[20] Samatha T, Shyamsundarachary R. *In vitro* micropropagation of *Oroxylum indicum* (L) Kurz an endangered and valuable medicinal forest tree. Indo Amer J Pharm Res 2016; 4(6): 5210-8.

[21] Samatha T, Srinivas P, Rajinikanth M. Embryo culture an efficient tool for conservation of endangered medicinally important forest tree *Oroxylum indicum* L. Kurz. In: Jou of Bio Res 2013; 3(2): 45-52.

[22] Samatha T, Rama Swamy N. Influence of $AgNO_3$ on zygotic embryo culture as an efficient tool for

conservation of a vulnerable medicinal important forest tree *Oroxylum indicum* (L). Kurz Int J Cur Res 2016; 8(2): 26573-9.

[23] Gokhale M, Bansal YK. Direct *in vitro* Regeneration of a medicinal tree *Oroxylum indicum* (L.) Vent. through tissue culture. Afr J Biotechnol 2009; 8: 3777-81.

[24] Tiwari S, Singh K, Shah P. *In vitro* propagation of *Oroxylum indicum*-an endangered medicinal Tree. Biotechnology (Faisalabad) 2007; 6(2): 299-301.
[http://dx.doi.org/10.3923/biotech.2007.299.301]

[25] Talari S, Rudroju S, Nanna R. Conservation of an endangered medicinal forest tree *Oroxylum indicum* (L) Kurz through *in vitro* micropropagation-A review. European J Med Plants 2016; 17(2): 1-13.
[http://dx.doi.org/10.9734/EJMP/2016/30310]

[26] Samatha T. Direct *in vitro* organogenesis and plantlet formation from leaf explants of *Oroxylum indicum* (L) Kurz an endangered medicinal forest tree. Int J Adv Res 2013; 1(8): 431-8.

[27] Madhavi R, Kamdi SR, Wadhai VS. *In vitro* shoot induction and callus induction of a medicinal tree *Oroxylum indicum* (Tattu) through tissue culture. Int J Plant Sci 2010; 6(1): 45-8.

[28] Samatha T. Callus induction in *Oroxylum indicum* (L.). Kurz Int J Herb Med 2016; 4(6): 189-92.

[29] Skoog F, Miller CO. Chemical regulation of growth and organ formation in plant tissues cultured *in vitro*. Symp Soc Exp Biol 1957; 11: 118-30.
[PMID: 13486467]

[30] Samatha T, Rama Swamy N. Conservation of an endangered ethno medicinally important forest tree species *Oroxylum indicum* through somatic embryogenesis. IJBPAS 2015; 4(3): 1064-73.

[31] Ali A, Gull I, Majid A, Saleem A, Naz S, Naveed NH. *In vitro* conservation and production of vigorous and desiccate tolerant synthetic seeds in Ste*via* rebaudiana. Jou of Med Plant Res 2012; 6(7): 1327-33.
[http://dx.doi.org/10.5897/JMPR11.1443]

[32] Bansal YK, Mamatha G. Effect of additives on Micropropagation of an endangered medicinal tree *Oroxylum indicum* (L.) Vent. In: Annarita Leva, Laura M, Rinaldi R, Eds. Recent Advances in Plant in vitro Culture. IntechOpen 2012. https://www.intechopen.com/chapters/40185
[http://dx.doi.org/10.5772/50743]

[33] Vaidya BG. Some controversial drugs of Indian medicine. IX. J Res Indian Med 1975; 10(4): 127.

[34] Zaveri M, Khandhar A, Jain S. Quantification of baicalein, chrysin, biochanin A and ellagic acid in root bark of *Oroxylum indicum* by RP-HPLC with UV detection. Eurasian J Anal Chem 2008; 3: 245-57.

[35] Hari Babu T, Manjulatha K, Suresh Kumar G, *et al.* Gastroprotective flavonoid constituents from *Oroxylum indicum* Vent. Bioorg Med Chem Lett 2010; 20(1): 117-20.
[http://dx.doi.org/10.1016/j.bmcl.2009.11.024] [PMID: 19948405]

[36] Samatha T, Srinivas P, Shyamsundarachary R, Rajinikanth M, Rama Swamy N. Phytochemical Analysis of seeds, stem bark and root of an endangered medicinal forest tree *Oroxylum indicum* (L). Kurz Int J Pharm Bio Sci 2012; 33: 1063-75.

[37] Samatha T, Srinivas P, Shyamsundarachary R, Rama Swamy N. Phytochemical screening and TL C studies of leaves and petioles of *Oroxylum indicum* (L) Kurz. An endangered ethno medicinal tree. International J Pharm & Life Sci 2013; 4(1): 2306-13.

[38] Dinda B, SilSarma I, Dinda M, Rudrapaul P. *Oroxylum indicum* (L.) Kurz, an important Asian traditional medicine: From traditional uses to scientific data for its commercial exploitation. J Ethnopharmacol 2015; 161: 255-78.
[http://dx.doi.org/10.1016/j.jep.2014.12.027] [PMID: 25543018]

[39] Thi Van AT, Stefan S. Dirsch Hermann Stuppner (2015) Screening of Vietnamese medicinal plants for NF-κB signaling inhibitors: assessing the activity of flavonoids from the stem bark of *Oroxylum*

indicum. J Ethnopharmacol. 159: 36-42.

[40] Nadkarni AK. Indian materia medica. Mumbai: Bombay Popular Prakashan 1982; p. 876.

[41] Chowdhury NS, Karim MR, Rana MS. *In vitro* Studies on Toxicological Property of the Root and Stem Bark Extracts of *Oroxylum indicum*. Dhaka University J Pharmaceutical Sciences 2007; 4: 1.

[42] Downing JE. Anthelmintic activity of *Oroxylum indicum* against equine strongyles *in vitro* compared to the anthelmintic activity of Ivermectin. J Biol Res 2000; 1.

[43] Ali RM, Houghton PJ, Hoo TS. Antifungal activity of some Bignoniaceae found in Malaysia. Phytother Res 1998; 12(5): 331-4.
[http://dx.doi.org/10.1002/(SICI)1099-1573(199808)12:5<331::AID-PTR305>3.0.CO;2-W]

[44] Kawsar Uddin, Abu Sayeed, Anwarul Islam, Seatara Khatun, Ali A, Khan AM. Biological activities of extracts and two flavanoids from *Oroxylum indicum* Vent. J Biol Sci 2003; 3(3): 371-5.
[http://dx.doi.org/10.3923/jbs.2003.371.375]

[45] Abdalla AE, Roozen JP. Effect of plant extracts on the oxidative stability of sunflower oil and emulsion. Food Chem 1999; 64(3): 323-9.
[http://dx.doi.org/10.1016/S0308-8146(98)00112-5]

[46] Kumar V, Chaurasia AK, Naglot A, *et al*. Antioxidant and antimicrobial activities of stem bark extracts of *Oroxylum indicum* Vent. (Bignoniaceae) Ã¢â‚¬Â A medicinal plant of northeastern India. South Asian J Exp Biol 2011; 1(3): 152-7.
[http://dx.doi.org/10.38150/sajeb.1(3).p152-157]

[47] Zaveri M, Jain S. Hepatoprotective effect of root bark of *Oroxylum indicum* on carbon tetrachloride (CCl_4)-induced hepatotoxicity in experimental animals. Biol 2009; (2): 1-15.

[48] Adikay S, Usha U, Koganti B. Effect of chrysin isolated from *Oroxylum indicum* against cisplatin induced acute renal failure. Recent Res Modern Med 2011; 302-7.

[49] Shetgiri PP, Dargi KK, Mello PM. Evaluation of antioxidant and antihyperlipidemic activity of extracts rich in polyphenols. Int J Phytomed 2010; 2(3): 267-76.
[http://dx.doi.org/10.5138/ijpm.2010.0975.0185.02038]

[50] Zaveri M, Gohil P, Jain S. Immunostimulant activity of n-butanol fraction of root bark of *Oroxylum indicum*, vent. J Immunotoxicol 2006; 3(2): 83-99.
[http://dx.doi.org/10.1080/15476910600725942] [PMID: 18958688]

[51] Khandhar M, Shah M, Santani D, Jain S. Antiulcer activity of the root bark of *Oroxylum indicum* against experimental gastric ulcer. Pharm Biol 2006; 44(5): 363-70.
[http://dx.doi.org/10.1080/13880200600748234]

CHAPTER 7

Exploring Plant Tissue Culture in *Ocimum basilicum* L.

Priyanka Chaudhary[1], Shivika Sharma[2] and Vikas Sharma[2,*]

[1] Department of Botany, DPG Degree College, Gurgaon, India
[2] Biochemical Conversion Unit, SSSNIBE, Kapurthala, India

Abstract: *Ocimum basilicum* is a well-known, economically important therapeutic plant that belongs to the family Lamiaceae. Basil is marvelous in the environment as the complete plant has been used as a conventional remedy for domestic therapy against numerous illnesses since ancient times. *O. basilicum* exhibited interesting biological effects due to the presence of several bioactives such as eugenol, methyl eugenol, cineone and anthocyanins. *O. basilicum* possesses antimicrobial, anti-inflammatory, hepatoprotective, hypoglycemic, immunomodulator, antiulcerogenic, antioxidant, chemomodulatory and larvicidal activities. The oil of this plant has been found to be valuable for the cure of wasp stings, snakebites, mental fatigue, and cold. The demand of this multipurpose medicinal plant is growing day by day due to its economic importance, pharmacological properties and its numerous uses in cooking and folk medicine. Thus seeing the exciting biological activities of *O. basilicum*, micropropagation could be a fascinating substitute for the production of this medicinal plant because numerous plantlets can be achieved in fewer times with the assurance of genetic stability. An overview of the current study showed the use of the plant tissue culture technique for micropropagation, which is very beneficial for duplicating and moderating the species, which are problematic to regenerate by conventional methods and save them from extinction.

Keywords: Bioactives, Eugenol, Lamiaceae, Micropropagation, Medicinal, Pharmacological, Therapeutic.

INTRODUCTION

Ocimum species are outstanding commercially significant healing plants on the globe [1]. It belongs to the family Lamiaceae. The genus *Ocimum* L. comprises almost 150 species, having an abundant dissimilarity in plant morphology, essential oil and chemical composition [2]. The term basil is supposed to be originated from the Greek word "Basileus", meaning "Royal or King". It is

---

* **Corresponding author Vikas Sharma:** Biochemical Conversion Unit, SSSNIBE, Kapurthala, India; E-mail: biotech_vikas@rediffmail.com

Mohammad Anis & Mehrun Nisha Khanam (Eds.)
All rights reserved-© 2024 Bentham Science Publishers

frequently referred to as King of the Herbs [3]. *Ocimum basilicum* is universally recognized as "Sweet basil" and is used in both Ayurvedic and Unani systems of medication [4]. *Ocimum basilicum* in English is known as Basil or Sweet Basil [5], whereas in Hindi and Bengali [6], it is named Babui Tulsi. This plant is extensively developed as a decorative as well as a field crop all over Burma, India and numerous Mediterranean nations, including Turkey. It is a tropically dispersed genus with two-thirds of the 160 species reported from Africa and the remaining one-third from Asia and America. Around nine species are reported from India distributed in tropical and peninsular regions. In India, 300 hectares of basil is cultivated, and the total yield is approximately 300 tons of oil in the states of Uttar Pradesh, Haryana and Punjab [7]. This herbaceous plant is also extensively cultivated in France, U.K., U.S.A and Egypt [8].

Ocimum basilicum is a widespread herb rich in perfumed imperative oil and is cherished for its spicy, mildly peppery flavour with a trace of mint and clove. This plant is utilized as a cookery herb [9] and has been used as a food ingredient for flavouring baked foods and meat products [10]. The plant is diaphoretic, antihelminthic, carminative, antipyretic and stimulant [11]. *O. basilicum* is used to cure many ailments, such as migraines, cold, diabetes, tension, fevers, feminine spasms, cardiovascular infections, nerve torment, and abdominal pain reliever [12, 13]. The *O. basilicum* is utilized as a characteristic coagulant for the management of material wastewater [14] and as a biosorbent for copper and chromium uptake with its high biosorption limit of the seeds [15, 16].

O. basilicum majorly comprises a number of bioactives such as methyl eugenol, α-linalool, β-linalool, estragole 1, 8-cineole, linalool, estragole, Camphor, limonene and thymol. Methyl eugenol is the foremost compound of *O. basilicum*. It has been observed that 1,8-cineol (5.61%), methyl eugenol (18.74%) and Linalool (52.42%) are the chief phytoconstituents, whereas myrcene, neral and borneol are the minor compounds present at 5%, 8%, and 9% w/w respectively [17, 18]. *O. basilicum* possess several pharmacological properties such as anti-thrombotic [19], anti-hyperlipidemic, antiplatelet property [20], anticonvulsant [21], anti-aging, antiviral, anticancer and anti-microbial [22], immunomodulatory [23] and cytotoxicity effect [24], anti-inflammatory [25] and also antioxidant [26, 27].

The actual difficulty during the utilization of Lamiaceae species for pharmacological desires lies in the genomic and proteomic variability [28]. Propagating the Lamiaceae family through conventional modes is *via* seed, but poor seed viability and low rate of germinating seeds limit their proliferation to a huge extent [29]. Therefore *in vitro* micropropagation is the best alternative technique for the rapid multiplication of species to obtain a high offspring uniformity. *In vitro* micropropagation can guarantee a large-scale production of several true-to-type plants in precise conditions in a small period of time without adverse effects on habitats [30].

The vanishing of this medicinal important plant in some areas is growing step by step; therefore, to generate attentiveness to the therapeutic significance of this plant to stop its disappearance is important. Moreover, speedy economic development and urbanization have led to overexploitation and damage of valued natural resources, together with numerous therapeutically imperative herbaceous plants [31]. Hence enthusiasm for utilizing *in vitro* culture procedures for quick and expansive scale proliferation of therapeutic plants has definitely improved. The miscellaneous species of genus *Ocimum* were exposed to *in vitro* studies by means of diverse explants, such as nodal segments [32] *via* leaf segments [33]. *In vitro* flowering has also been reported [34].

EVALUATION OF HEREDITARY STABILITY OF TISSUE CULTURE RAISED PLANTLETS *VIA* MOLECULAR MARKERS

Genetic uniformity is the preservation of the hereditary structure of a specific copy through its lifespan era [35] and is a vital pre-necessity in the multiplication of plant species, and is affirmed through molecular investigation [36]. *Ocimum* species exhibited hereditary along with biochemical differences because of interspecific hybridization. Therefore, it is essential to establish the best micropropagation procedure for the establishment of hereditarily identical plants before it is ready for profitable purposes. Moreover, there is a need to frequently check the clonal fidelity of micro propagated plantlets to confirm their true-to-t-e-type nature in order to avoid variations, which, if introduced, can proliferate very rapidly and lead to damage to the desirable characters of the parental genotypes. A lot of factors may possibly affect the firmness of the *in vitro* raised plantlets, such as genotype, time of culture period and nature of explants.

The biochemical stability of micropropagated plantlets has been established using various valuable tools such as GC profiling, molecular markers and flow cytometry [37]. To examine the hereditary uniformity and unsteadiness of *in vitro* culture-derived plantlet, random amplified polymorphic DNA (RAPD) and inter-simple sequence repeats (ISSR) have been commonly utilized because of easiness, rapid execution and requirement of a minute quantity of DNA devoid of little prior information regarding the genome [38]. The utilization of more than one marker has been more significant for the examination of the hereditary strength of plants, as they target various regions of the genome [39].

As assessed, fingerprinting sketches of *in vitro* cultured and donor plants of *O. basilicum* by utilizing ISSR and RAPD markers check the true-to-type nature of the plants [40]. It has been observed that all banding profiles generated during

ISSR and RAPD analysis from micropropagated plantlets were monomorphic and like those of the mother plantlet, which showed that there was no inherited variety in the regenerated plantlet populace and thus confirmed the hereditary stability of the *in vitro*-raised clone. The absence of hereditary variation using RAPD has been described in numerous therapeutic plants such as *Ocimum gratissimum* [41] and *Celastrus paniculatus* [42] and the Indian basil *variety* CIM-Saumya [43]. Henceforth it is recommended that the molecular marker technique is a significant approach in the assessment of hereditary constancy of *in vitro*-propagated plantlets. The nodal explants can be effectively utilized for the industrial increase of *Ocimum* without much hazard of genetic instability.

SHOOT INDUCTION

The cell and tissue culture practises have unbelievable benefits in horticulture, agriculture, forestry and other commercially imperative fragrant harvest plants, especially those which are propagated vegetatively. The accomplishment of tissue culture practices finally depends on the explant selection, size of the explants, age and the way in which it is cultured [44]. Several studies revealed that kernels and young tissues were used for *in vitro* multiplication more than the tissues from matured plants [45]. Moreover, the choice of an appropriate nutrient medium is also necessary for the accomplishment of tissue culture practices [46].

The type and concentration of cytokinin are significant variables for effective *in vitro* duplication. It has been observed that BAP and kinetin are incredibly efficient in stimulating propagation [47]. Cytokinin participates in numerous processes that initiate the division of cells in the callus along with auxin and subsequently prompt shoot or root advancement straightforwardly on the explant or from the calli [48]. Therefore, the presence of cytokinin in the basal medium may have been necessary to rupture the apical dominance and for the initiation of shoots in Basil. The shoot cultures are generally placed on basal media supplemented with cytokinins such as BAP, TDZ, kinetin and zeatin. Various researchers have observed a strong correlation between the auxin/cytokinin ratio in the basal media and shoot establishment by using different explant sources and genotypes [49, 50].

When cotyledon leaves of *Ocimum basilicum* were used as explant, the maximum number of the shoot (3.46) having 66.7% shoot regeneration per explant on MS medium fortified with BAP (5mg/L) and NAA (0.2 mg/L) was obtained. Further maximum number of shoots (5.88%) having 90% shoot formation was observed on medium containing 0.2mg/L BAP from shoot tip explants [51]. Later, the third subculturing of *O. basilicum* maximum shoot elongation and development rate was achieved on a medium devoid of hormones [52]. The age of the explant is considered to be an imperative characteristic; therefore, the maximum number of shoots (5.1) with 85% multiplication rate was attained when young seedlings were used as explant.

A maximum number of multiple shoots (6.2) were produced when nodal segments were cultured on basal media along with BAP (0.5mg/L). It has been noticed that in this medium, shoots reached a maximum height of 3.7 cm after 4 weeks. It has also been observed that a concentration of BAP of more than 0.5 mg/L inhibits the length as well as the number of multiple shoots. The hormone-free medium did not show any response as the explants swelled and turned necrotic after 2 weeks of culture. The MS medium fortified with BAP (11µM) yielded 5.6 number of shoots with a maximum percentage of shoot multiplication (96%) [53]. The supplementation of MS medium with 8.88 µM BAP was the most excellent medium for the initiation and multiplication of shoots from the apical bud. This agrees with earlier findings [54] where *in vitro* propagation of *O. basilicum* by axillary bud proliferation on MS medium supplied with 4.44 µM BAP.

MS medium fortified with 2.67 µM BAP showed maximum shoot regeneration and shoot elongation (9.5 shoots/explants and 14.0cm, respectively) with no hyperhydricity. Several researchers reported the importance of BAP and IAA for shoot revival in *O. basilicum* [55], but it also has been reported that after 3 to 4 subculturing of *in* vitro plantlets, they are unable to stay alive due to a high rate of hyperhydricity. It has been observed that treatment of BAP at higher concentrations and for longer duration showed lethal effects that lead to retardation in the growth of shoots. The effect of BAP was more as compared to Kinetin in direct shoot regeneration of *Ocimum gratissimum* [56]. BAP at 1mg/l produced a maximum number of shoots (31.2 + 0.37) per culture with (7.84 + 0.06 cm) average length after eight weeks of inoculation followed by 1.5mg/l Kinetin producing (25.8 + 0.37) multiple shoots with (6.5 + 0.03cm) average length. Numerous other workers also observed the stimulatory effect of BAP on multiple shoot induction in several medicinal plants, including *Eclipta alba* [57], *Ocimum gratissimum*, *Ulmus parvifolia* [58] and *Ocimum sanctum*.

In Vitro Rooting

The formation of roots is a challenging process in several medicinally important plants [59]. The induction and development of adventitious roots require some metabolic substrates, primarily the carbohydrates next to the phytohormones [60]. The shoot is the site of producing auxin, and translocation of auxin to the stem base stimulates *in vitro* rooting [61].

It has been reported that *O. basilicum* showed the maximum rate of *in vitro* rooting when MS medium was fortified with 1.0µM IBA. In *O. basilicum* the *in vitro* root formation was not attained on ½ strength MS media devoid of phytohormones after four weeks from *in vitro* culturing. Moreover, researchers also observed that IBA was generally more efficient as compared to the other auxins. Similarly, The *in vitro* rooted plantlets were acclimatized and moved to the normal environmental conditions with 90% survival. The capability of *in vitro* plantlets to stay alive in a natural environment is of ultimate significance since it governs the achievement of tissue culture practices [62].

The addition of NAA along with different concentrations of auxins prevents root formation in *O. basilicum*. Whereas, he MS medium when fortified with 0.27 µM NAA, maximum root induction frequency (75.09%), root number and length (4.6/explant and 6.7cm), respectively, was achieved after the third week of subculturing. This research agrees with the previous reports where ½ strength MS media containing auxins appeared to be best for the initiation of roots.

ACCLIMATIZATION

The *in vitro* raised plantlets are dissimilar from field-grown plantlets. High demise is detected upon transferring micro shoots to *ex vitro* environments. *In vitro* plantlets are maintained under high humidity and have poorly developed cuticle and non-functional stomata. Therefore *in vitro* plants require acclimatization before transferring to natural climatic conditions [63]. After *in vitro* root induction, rooted shoots were washed gradually with water so as to eradicate the nutrient medium and successively shifted to plastic mugs having sterilized soilrite. These plants were protected with transparent polythene so as to maintain high humidity and then watered the plants after 3 days with ½ strength MS solution for 14 days. After 28 days, acclimatized plantlets were shifted to containers that contain soil and kept in a greenhouse under normal environmental conditions. It has been found that about 90% of the plantlets of *O.basilicum* stay alive and developed normally. These plantlets did not show any phenotypic dissimilarity and resemble the mother plantlets.

SOMATIC EMBRYOGENESIS

Amongst the tissue culture techniques, somatic embryogenesis is a proficient procedure for the regeneration of *in vitro* raised plantlets [64]. In somatic embryogenesis, somatic cells under appropriate conditions go through numerous biochemical and morphological changes to separate into somatic embryos which can develop into a completely new plant which looks like the zygotic stages of growth [65]. In direct somatic embryogenesis, a somatic embryo might be formed either directly from a cell or tissue without the callus phase [66] whereas in indirect somatic embryogenesis, embryos can differentiate indirectly from undifferentiated cells [callus] or cell suspension culture [67]. The Development of somatic embryos occurs in two steps. In the first step, the callus is cultured onto auxins-rich medium, and embryogenic clumps are made. In the second step these clumps are transferred into the fresh medium without auxins and mature embryos are formed [68]. The accomplishment of somatic embryogenesis relies upon several aspects such as genotype, phytohormones, and type of explant used. Among the various phytohormones, auxins [2,4-D] are considered to be best for the induction of somatic embryogenesis. The development of embryonic callus in *O.basilicum* was obtained in MS medium containing 1.0mg/L 2,4-D and 0.5mg/L BAP was best for the development of embryonic callus in *O.basilicum* [69] whereas medium containing 0.5mg/L 2,4-D and 0.5mg/L BAP was appropriate for *O. gratissimum* when cotyledonary leaves were used as an explant.

The maximum percentage of differentiated embryos with 80% survival was noticed when the germinated embryos were shifted from a medium containing BAP and NAA (1.0mg/L) and KIN (0.5mg/L) to the *ex vitro* environment [70]. The regeneration potential of *O.basilicum via* somatic embryogenesis is an efficient protocol in nanobiotechnology [71]. It has been found that the percentage of somatic embryos and an average number of regenerated plantlets/explants (18.7) were enhanced by the use of 5μM Cu-NPs as compared to 5μM $CuSO_4·5H_2O$.

PRODUCTION OF SECONDARY METABOLITES IN *O.BASILICUM*

There are many biotechnological methods for increased production of secondary metabolites from numerous plant species such as utilization of precursor feeding method, elicitation, screening of high-yielding cell line, media adjustment, immobilization of plant cells, hairy root culture and biotransformation [72]. The plant cells adjust themselves to the fresh atmosphere in the lag stage whereas plant cells store the highest biomass and secondary metabolites production in the exponential phase [73]. Thus secondary metabolites' productivity was completely reliant upon the basal medium fortified with phytohormones, nitrogen and carbon sources and the exponential stage of cell lines [74]. The improved formation of bioactives *via* these strategies unfolded an innovative vicinity of studies that can have critical financial advantages for the pharmacological industry.

ORGAN AND CALLUS CULTURE

Organs of *in vitro* regenerated plantlets are a very noble resource of several bioactives. The unorganized proliferative mass of cells and tissues once grown aseptically on the artificial nutrient medium is called a callus. The callus was treated with different phytohormones or elicitors at varying concentrations to increase the enhancement of bioactives. When the MS medium was supplemented with 2,4-D and KIN (0.1 mg/L), maximum biomass development (17.8 g/l) and rosmarinic acid content (104 mg/l) were observed [75]. Better production of rosmarinic acid (ranged from 9.42 to 38.25g/mg DW) was observed when shoot and callus cultures of *O. basilicum* were cultured on a medium containing 5mg/L BAP and 1mg/L NAA [76].

The use of LEDs is an encouraging approach for improving the enhancement of phytoconstituents in the callus cultures of *O. basilicum*. In the callus culture of *O. basilicum*, the most favourable levels of total phenolic and flavonoid value were estimated that were developed under blue and red light, respectively [77]. It has been revealed by HPLC analysis that the highest concentrations of rosmarinic acid (96.0 mg/g DW) and (0.273 mg/g DW) were accumulated under blue light which was 2.46 and 2.45 higher than the control.

The callus culture of *O. basilicum* was established by using different concentrations of phytohormones such as TDZ, NAA, and BAP either alone or in combination with 1.0mg/L NAA by using leaves. It has been evaluated that 2.5 mg/L NAA showed maximum biomass accumulation (23.2 g/L DW) as well as total phenolic mg/L) and flavonoid (210.7mg/L, 196.4 mg/L) production, respectively [78]. The callus extracts of *O. basilicum.* showed greater production of chcoric acid and rosmarinic acid (35.77mg/g DW and 7.35 mg/g DW) respectively associated with higher antioxidant capacity when compared to commercial leaves [79].

CELL SUSPENSION CULTURE

In suspension culture, single cells or minute duplicate cell aggregates are suspended in liquid media. This suspension is a standout amongst the most essential approach for the production and enhancement of bioactives. The increased production of rosmarinic acid and anthocyanins in the cell suspension cultures of *O. basilicum* and found diverse rosmarinic acid related molecules such as coumaric acid, caffeic acid and flavones [80, 81].

The impact of different concentrations of cadmium chloride ($CdCl_2$), silver nitrate ($AgNO_3$) and yeast extract (YE) in the suspension culture of *Ocimum bacilicum was also evaluated* [82]. The investigation (RP-HPLC) revealed that the highest increase in chicoric acid and rosmarinic acid was achieved in yeast extract treatment. The optimal biosynthesis of rutin and isoquercetin was attained with 50 mg/L yeast extract with an increase of 1.91 times and 1.86 times respectively. The highest values of linalool and estragole were obtained from $AgNO_3$ as 4.37 µg/g DW (25 µM treatment) and 3.30 µg/g DW (5 µM treatment) respectively. Therefore, it has been concluded that $CdCl_2$, $AgNO_3$ and YE might be a good approach for enhancing the active ingredients in the cell suspension cultures of *O. basilicum*. The suspension culture of Basil was common for the enhancement of rosmarinic acid, phenolics, anthocyanins as well as oleanolic and ursolic acid [83]. The cambial meristematic cell (CMC) is an outstanding type of plant cell culture derived from vascular cambium. CMCs are developed at a quicker speed, aggregate less and accumulate more product than dedifferentiated cells of the same plant [84]. CMCs were termed for *Taxus cuspidata* [85], *Catharanthus roseus* [86] and *Tripterygium wilfordii* [87].

IN VITRO PLANT CELL ELICITATION

A current method to improve the bioactive formation in plant tissue culture is recognized as elicitation [88]. The utilization of an elicitor is to elicit the pressure or defense associated responses in the plant cells [89]. An elicitor will enhance the production of particular bioactives when introduced in minute concentration into a living cell system [90]. The classification of elicitors is on the basis of their 'nature' such as abiotic (Jasmonic acid) and biotic elicitors (*e.g.*, Yeast extract and enzymes) and 'origin' like exogenous elicitors (Glucans and glycoproteins) and endogenous elicitors (Alginate oligomers and hepta--glucosides) [91]. The exogenous application of several elicitors such as jasmonates [92], gibberlic acid, chitosan and salicylates [93] stimulates the enhanced biosynthesis of secondary metabolites. The various factors like elicitor specificity, time of exposure, concentration, culture conditions and growth stage of the cultured cells greatly affect the elicitation process [94].

The chitosan was found to enhance the rosmarinic acid (RA) and eugenol concentrations to 2.5 and 2 folds respectively in sweet basil [95]. Similarly, elicitation of hairy root culture of *O. basilicum* with fungal cell wall elicitors derived from *Phytophthora cinnamon*, enhanced the rosmarinic acid accumulation (2.67fold) [96]. Methyl jasmonate (MeJa) is a signal molecule involved in plant defence when applied to plant cell culture. The production of secondary metabolites can be increased, for example, rosmarinic acid in *O. basilicum* [97] or flavonoids in *Pueraria candollei* [98]. Subsequently, the elicitor can acceptably activate the phytochemicals in plants which might be an alternate option as opposed to inherited adjustment.

CONCLUSION AND FUTURE PERSPECTIVES

The main objective of this review is to explore the plant tissue culture research and opportunities for the improvement of the therapeutic potential of *Ocimum basilicum*. In addition to highlighting the phytochemical profile, the pharmacological properties such as antibacterial, antifungal and antioxidant properties have also been compiled to recall its potential in the pharmaceutical arena. The bioactives present in this medicinal plant are used for the preparation of therapeutic drugs for human consumption so as to treat numerous health problems. Different *in vitro* plant proliferation protocols through somatic embryogenesis, meristem culture and organogenesis have been developed for *O. basilicum* that could effectively be utilized for large-scale clonal propagation. The regenerated plantlets will be helpful for a steady supply of crude materials for secondary metabolite isolation for commercial purposes. The molecular data clearly established the true character of the *in vitro* raised clones because the monomorphic pattern showed no variability among regenerants and thus maintained genetic integrity. Therefore, it has been concluded that *Ocimum basilicum* can be effectively explored for commercial/industrial utilization without much risk of genetic instability. The wide range of studies on this medicinal plant shows that it is very beneficial for the betterment of drugs and more work can be done to take advantage of the potential remedial qualities of it. More emphasis should be given to metabolic engineering procedures for better development of secondary metabolites which incorporate recognition and overexpression of genes associated with the metabolic pathway. For future perspective, the novel methods of plant-microbe interactions can be utilized using this plant as a model system to screen beneficial plant-microbe interactions and further application in biofertilizer production.

REFERENCES

[1] Saha S, Ghosh PD, Sengupta C. An efficient method for micropropagation of *Ocimum basilicum* L. Indian J Plant Physiol 2010; 15(2): 168-72.

[2] Danesi F, Elementi S, Neri R, Maranesi M, D'Antuono LF, Bordoni A. Effect of cultivar on the protection of cardiomyocytes from oxidative stress by essential oils and aqueous extracts of basil (*Ocimum basilicum* L.). J Agric Food Chem 2008; 56(21): 9911-7.
[http://dx.doi.org/10.1021/jf8018547] [PMID: 18928294]

[3] Neelam LD, Nilofer SN. Preliminary immunomodulatory activity of aqueous and ethanolic leaves extracts of *Ocimum basilicum* Linn in mice. Int J Pharm Tech Res 2010; 2(2): 1342-9.

[4] Muralidharan A, Dhananjayan R. Cardiac stimulant activity of *Ocimum basilicum* Linn. extracts. Int J Pharmacol 2004; 36: 163-6.

[5] Jayaweera DMA. Medicinal Plants, (Indigenous and Exotic) Used in Ceylon. Part III. Colombo. National Sci Foundation of Sri Lanka 1981; 1981: 101-3.

[6] Dymock W, Warden CJH, Hooper D. Pharmacographica Indica.A history of the principal drugs of vegetable origin. New Delhi: Shrishti book distributors 2005; 3: pp. 82-5.

[7] Varshney SC. Trends in essential oil production in india. Symposium on future trends in essential oil industry in india, organized by essential oil association of. Jammu. India, New Delhi: At Regional Res Laboratory 1997.

[8] Leelavathi D, Kuppan N, Yashoda A. calibrated protocol for direct regeneration of multiple shoots from *in vitro* apical bud of *Ocimum basilicum* : An Important Aromatic Med Plant. J Pharm Res 2014; 8(6): 733-5.

[9] Vieira RF, Simon JE. Chemical characterization of basil (ocimum spp.) found in the markets and used in traditional medicine in brazil. Econ Bot 2000; 54(2): 207-16.
[http://dx.doi.org/10.1007/BF02907824]

[10] Chang X, Alderson PG, Wright CJ. Variation in the essential oils in different leaves of Basil (*O.*

basilicum L.) at day time. Open Horticul J 2009; 2(1): 13-6.
[http://dx.doi.org/10.2174/1874840600902010013]

[11] Phippen WB, Simon JE. Shoot regeneration of young leaf explants from basil (*Ocimum basilicum* L.). In Vitro Cell Dev Biol Plant 2000; 36(4): 250-4.
[http://dx.doi.org/10.1007/s11627-000-0046-y]

[12] Kirtikar KR, Basu BD. Indian Medicinal Plants with Illustrations. 2nd. Uttaranchal: Oriental Enterprises 2003; pp. 2701-5.

[13] Ch M, Naz S, Sharif A, Akram M, Saeed M. Biological and pharmacological properties of the sweet basil (*ocimum basilicum*). Br J Pharm Res 2015; 7(5): 330-9.
[http://dx.doi.org/10.9734/BJPR/2015/16505]

[14] Sarwan B, Pare B, Acharya AD, Jonnalagadda SB. Mineralization and toxicity reduction of textile dye neutral red in aqueous phase using BiOCl photocatalysis. J Photochem Photobiol B 2012; 116: 48-55.
[http://dx.doi.org/10.1016/j.jphotobiol.2012.07.006] [PMID: 22964463]

[15] Melo JS, D'Souza SF. Removal of chromium by mucilaginous seeds of *Ocimum basilicum*. Bioresour Technol 2004; 92(2): 151-5.
[http://dx.doi.org/10.1016/j.biortech.2003.08.015] [PMID: 14693447]

[16] Sahay R, Patra DD. Identification and performance of sodicity tolerant phosphate solubilizing bacterial isolates on *Ocimum basilicum* in sodic soil. Ecol Eng 2014; 71: 639-43.
[http://dx.doi.org/10.1016/j.ecoleng.2014.08.007]

[17] Poonkodi K. Chemical composition of essential oil of *osimum basilicum* l(basil) and its biological activities : An Overview. J Critic Rev 2016; 3(3): 56-62.

[18] Radulović NS, Blagojević PD, Miltojević AB. α-Linalool : A marker compound of forged/synthetic sweet basil (*Ocimum basilicum* L.) essential oils. J Sci Food Agric 2013; 93(13): 3292-303.
[http://dx.doi.org/10.1002/jsfa.6175] [PMID: 23584979]

[19] Tohti I, Tursun M, Umar A, Turdi S, Imin H, Moore N. Aqueous extracts of *Ocimum basilicum* L. (sweet basil) decrease platelet aggregation induced by ADP and thrombin *in vitro* and rats arterio–venous shunt thrombosis *in vivo*. Thromb Res 2006; 118(6): 733-9.
[http://dx.doi.org/10.1016/j.thromres.2005.12.011] [PMID: 16469363]

[20] Amrani S, Harnafi H, Gadi D, *et al.* Vasorelaxant and anti-platelet aggregation effects of aqueous *Ocimum basilicum* extract. J Ethnopharmacol 2009; 125(1): 157-62.
[http://dx.doi.org/10.1016/j.jep.2009.05.043] [PMID: 19505553]

[21] Nguyen PM, Kwee EM, Niemeyer ED. Potassium rate alters the antioxidant capacity and phenolic concentration of basil (*Ocimum basilicum* L.) leaves. Food Chem 2010; 123(4): 1235-41.
[http://dx.doi.org/10.1016/j.foodchem.2010.05.092]

[22] Sakr SA, Al-Amoudi WM. Effect of leave extract of *Ocimum basilicum* on deltamethrin induced nephrotoxicity and oxidative stress in albino rats. J Appl Pharm Sci 2012; 2(5): 22-7.
[http://dx.doi.org/10.7324/JAPS.2012.2507]

[23] Okazaki K, Nakayama S, Kawazoe K, Takaishi Y. Antiaggregant effects on human platelets of culinary herbs. Phytother Res 1998; 12(8): 603-5.
[http://dx.doi.org/10.1002/(SICI)1099-1573(199812)12:8<603::AID-PTR372>3.0.CO;2-G]

[24] Aarthi N, Murugan K. Larvicidal and repellent activity of Vetiveri azizanioides L, *Ocimum basilicum* L. and the microbial pesticide spinosad against malarial vector, Anopheles 76 stephensi Liston (Insecta: Diptera: Culicidae). J Biopesticides 2010; 3: 199-204.

[25] Raina P, Deepak M, Chandrasekaran CV, Agarwal A, Wagh N, Kaul-Ghanekar R. Comparative analysis of anti-inflammatory activity of aqueous and methanolic extracts of *Ocimum basilicumin* RAW 264.7, SW1353 and human primary chondrocytes. J Herb Med 2016; 6(1): 28-36.
[http://dx.doi.org/10.1016/j.hermed.2016.01.002]

[26] Pandey V, Patel A, Patra DD. Integrated nutrient regimes ameliorate crop productivity, nutritive value, antioxidant activity and volatiles in basil (*Ocimum basilicum* L.). Ind Crops Prod 2016; 87: 124-31.
[http://dx.doi.org/10.1016/j.indcrop.2016.04.035]

[27] Guo Q, He J, Zhang H. Oleanolic acid alleviates oxidative stress in Alzheimer's disease by regulating stanniocalcin-1 and uncoupling protein-2 signalling. Clin Exp Pharmacol Physiol 2020; 1440–1681: 13292.
[http://dx.doi.org/10.1111/1440-1681.13292]

[28] Dode LC, Bobrowski VL, Braga EJB, Seixas FK, Schunch W. *In vitro* propagation of *Ocimum basilicum* L. Maringa 2003; 25: 435-7.

[29] Heywood VH. Flowering Plants of the World. UK: Oxford Univ. Press 1978.

[30] Canhoto J. Biotecnologia Vegetal – Da clonagem de plantas à transformação genética. Coimbra: Imprensa da Universidade de Coimbra 2010.
[http://dx.doi.org/10.14195/978-989-26-0404-6]

[31] Arora R, Bhojwani SS. *In vitro* propagation and low temperature storage of*Saussurea lappa* C.B. Clarke? An endangered, medicinal plant. Plant Cell Rep 1989; 8(1): 44-7.
[http://dx.doi.org/10.1007/BF00735776] [PMID: 24232594]

[32] Begum F, Amin N, Azad MAK. *In vitro* rapid clonal propagation of *Ocimum basilicum* L. Plant Tissue Cult 2002; 27-35.

[33] Gopi C, Sekhar NY, Ponmurugan P. *In vitro* multiplication of *Ocimum gratissimum* L. through direct regeneration. Afr J Biotechnol 2006; 5: 723-6.

[34] Sudhakaran S, Sivashankari V. *In vitro* flowering response of *Ocimum basilicum* L. J Plant Biotechnol 2002; 4: 181-3.

[35] Chaterjee G, Prakash J. Genetic stability in commercial tiss cultPlant Biotechnology: commercial prospects and problems. New Delhi, India: Oxford IBH Publishing Co 1996; pp. 11-121.

[36] Alizadeh M, Singh SK. Molecular assessment of clonal fidelity in micropropagated grape (Vitis spp.) rootstock genotypes using RAPD and ISSR markers. Iran J Biotechnol 2009; 7(1): 37-44.

[37] Prasad A, Shukla SP, Mathur A, Chanotiya CS, Mathur AK. Genetic fidelity of long-term micropropagated Lavandula officinalis Chaix.: An important aromatic medicinal plant. Plant Cell Tissue Organ Cult 2015; 120(2): 803-11.
[http://dx.doi.org/10.1007/s11240-014-0637-7]

[38] Williams JGK, Kubelik AR, Livak KJ, Rafalski JA, Tingey SV. DNA polymorphisms amplified by arbitrary primers are useful as genetic markers. Nucleic Acids Res 1990; 18(22): 6531-5.
[http://dx.doi.org/10.1093/nar/18.22.6531] [PMID: 1979162]

[39] Lakshmanan V, Venkataramareddy SR, Neelwarne B. Molecular analysis of genetic stability in long-term micropropagated shoots of banana using RAPD and ISSR markers. Electron J Biotechnol 2007; 10(1): 0.
[http://dx.doi.org/10.2225/vol10-issue1-fulltext-12]

[40] Saha S, Sengupta C, Ghosh P. Evaluation of the genetic fidelity of *in vitro* propagated *Ocimum basilicum* L. using RAPD and ISSR markers. J Crop Sci Biotechnol 2014; 17(4): 281-7.
[http://dx.doi.org/10.1007/s12892-014-0050-0]

[41] Saha S, Kader A, Sengupta C, Ghosh P. *In vitro* propagation of *Ocimum gratissimum* L. (Lamiaceae) and its evaluation of genetic fidelity using RAPD marker. Am J Plant Sci 2012; 3(1): 64-74.
[http://dx.doi.org/10.4236/ajps.2012.31006]

[42] Phulwaria M, Rai MK, Patel AK, Kataria V, Shekhawat NS. A genetically stable rooting protocol for propagating a threatened medicinal plant--*Celastrus paniculatus*. AoB Plants 2013; 5(0): pls054.
[http://dx.doi.org/10.1093/aobpla/pls054]

[43] Kumari M, Agnihotri D, Chanotiya CS, Mathur AK, Lal RK, Mathur A. Chemical and genetic stability of methyl chavicol-rich Indian basil (*Ocimum basilicum* var. CIM-Saumya) micropropagated *in vitro*. S Afr J Bot 2017; 113: 186-91.
[http://dx.doi.org/10.1016/j.sajb.2017.08.018]

[44] George EF, Sherrington PD. Plant propagation by tiss cult handbook and dictionary of commercial laboratories exergetics Ltd. England 1984.

[45] Bonga JM. Clonal propagation of mature trees: Problems and possible solutions.Cell And Tiss Cult In Forestry. Dordrecht: Martins Nighoff Publishers 1987; pp. 249-71.
[http://dx.doi.org/10.1007/978-94-017-0994-1_15]

[46] Bhojwani SS, Razdan MK. Tiss cult media.Plant Tiss Cult: Theory And Practice. Amsterdam: Elsevier Sci Publishers 1983; pp. 25-41.

[47] Grattapaglia D, Machado MA. Micropropagação.Cultura de tecidos e transformação genetic de plantas. Brasília: EmbrapaSPI/Embrapa-CNPH 1998; 1: pp. 183-260.

[48] Taiz L, Zeiger E. Auxin: The growth hormone. Plant Physiol 2002.

[49] Singh NK, Sehgal CB. Micropropagation of "Holy basil" (*Ocimum sanctum* L.) from young inflorescens of mature plants. Plant Growth Regul 1999; 29(3): 161-6.
[http://dx.doi.org/10.1023/A:1006201631824]

[50] Sharzad A. SA *In vitro* organogenesis in *Ocimum sanctum* L. – A multipurpose heb. Phytomorphology, Delhi 2000; 50(1): 27-35.

[51] Banu LA, Bari MA. Protocol establishment for multiplication and regeneration of *Ocimum sanctum* Linn. An important medicinal plant with high religious value in Bangladesh. J Plant Sci 2007; 2(5): 530-7.
[http://dx.doi.org/10.3923/jps.2007.530.537]

[52] Siddique I, Anis M. An improved plant regeneration system and *ex vitro* acclimatization of *Ocimum basilicum* L. Acta Physiol Plant 2008; 30(4): 493-9.
[http://dx.doi.org/10.1007/s11738-008-0146-6]

[53] Asghari F, Hossieni B, Hassani A, Shizad H. Effect of explant source and different hormonal combinations on direct regeneration of basil plants (*Ocimum basilicum* L.). Aus J Agri Eng 2012; 3: 12-7.

[54] Sahoo Y, Pattnaik SK, Chand PK. *In vitro* clonal propagation of an aromatic medicinal herb *Ocimum basilicum* L. (sweet basil) by axillary shoot proliferation. *In vitro*. Cell Dev Biol Plant 1997; 33(4): 293-6.
[http://dx.doi.org/10.1007/s11627-997-0053-3]

[55] Daniel A, Kalidass C, Mohan VR. *In vitro* multiple shoot induction through axillary bud of *Ocimum basilicum* L. an important medicinal plant. Int J Biotechnol 2010; 24-8.

[56] Chaudhary P, Sharma V. *In vitro* mass multiplication and molecular validation of *Ocimum gratissimum* using DNA based markers. Res J Biotechnol 2019; 14(2): 32-41.

[57] Dhaka N, Kothari SL. Micropropagation of *eclipta alba* (L.) hassk—An important medicinal plant. *In Vitro*. Cell Dev Biol Plant 2005; 41(5): 658-61.
[http://dx.doi.org/10.1079/IVP2005684]

[58] Thakur RC, Karnosky DF. Micropropagation and germplasm conservation of Central Park Splendor Chinese elm (Ulmus parvifolia Jacq. 'A/Ross Central Park') trees. Plant Cell Rep 2007; 26(8): 1171-7.
[http://dx.doi.org/10.1007/s00299-007-0334-7] [PMID: 17431632]

[59] Custódio L, Martins-Loução MA, Romano A. Influence of Sugars on *in vitro* rooting and acclimatization of Carob Tree. Biol Plant 2004; 48(3): 469-72.
[http://dx.doi.org/10.1023/B:BIOP.0000041107.23191.8c]

[60] Thorpe TA. Callus organization and de novo formation of shoots, roots and embryos in vitro.Application of Plant Cell and Tiss Cult to Agriculture & Industry. Ontario, Canada: Plant Cell Cult Centre, Univ. of Guelph 1982; pp. 115-38.

[61] Barcelo CJ, Nicolas RG, Sabater GB, Sánchez TR. Fisiologia Vegetal Pirámide. Madri 1988.

[62] Mathur A, Mathur AK, Verma P, *et al.* Biological hardening and genetic fidelity testing of micro-cloned progeny of *Chlorophytum borivilianum.* Afr J Biotechnol 2008; 7: 1046-53.

[63] The background. 3rd ed. George EF, Hall MA, Klerk GJD. Plant propagation by Tiss cultDordrecht, London: Springer Publisher 2008; 1.

[64] Xie H, Hu X, Zhang CR, Chen YF, Huang X, Huang XL. Molecular characterization of a stress related gene MsTPP in relation to somatic embryogenesis of alfalfa. Pak J Bot 2013; 45(4): 1285-91.

[65] Komamine A, Murata N, Nomura K. 2004 SIVB Congress Symposium Proceeding: Mechanisms of somatic embryogenesis in carrot suspension cultures—Morphology, physiology, biochemistry, and molecular biology. In Vitro Cell Dev Biol Plant 2005; 41(1): 6-10.
[http://dx.doi.org/10.1079/IVP2004593]

[66] Pierik RLM. *In vitro* culture of higher plants. Dordrecht: Martinus Nijhoff 1987; pp. 183-230.
[http://dx.doi.org/10.1007/978-94-009-3621-8_20]

[67] Srivastava LM. Plant growth and development. New York: Hormones and environment academic press 2002; pp. 140-3.

[68] Razdan MK. Introduction to Plant Tissue Culture. 2nd. New Hampshire: Science Publishers, Inc. 2003; p. 132.

[69] Mathew R, Sankar PD. Comparison of somatic embryo formation in *Ocimum basilicum* L., *Ocimum sanctum* L. & *Ocimum gratissimum* L. Int J Pharma Bio Sci 2011; 2(1): 356-67.

[70] Gopi C, Ponmurugan P. Somatic embryogenesis and plant regeneration from leaf callus of *Ocimum basilicum* L. J Biotechnol 2006; 126(2): 260-4.
[http://dx.doi.org/10.1016/j.jbiotec.2006.04.033] [PMID: 16759731]

[71] Ibrahim AS, Fahmy AH, Ahmed SS. Copper nanoparticles elevate regeneration capacity of (*Ocimum basilicum* L.) plant *via* somatic embryogenesis. Plant Cell Tissue Organ Cult 2019; 136(1): 41-50.
[http://dx.doi.org/10.1007/s11240-018-1489-3]

[72] Vanishree M, Lee CY, Lo SF, Nalawade SM, Lic CY, Tsay HS. Studies in the production of some important metabolites from medicinal plants by plant tiss cult. Bot Bull Acad Sin 2004; 45: 1-22.

[73] Liu JY, Guo ZG, Zeng ZL. Improved accumulation of phenylethanoid glycosides by precursor feeding to suspension culture of Cistanche salsa. Biochem Eng J 2007; 33(1): 88-93.
[http://dx.doi.org/10.1016/j.bej.2006.09.002]

[74] Sivanandhan G, Kapil Dev G, Jeyaraj M, *et al.* A promising approach on biomass accumulation and withanolides production in cell suspension culture of Withania somnifera (L.) Dunal. Protoplasma 2013; 250(4): 885-98.
[http://dx.doi.org/10.1007/s00709-012-0471-x] [PMID: 23247920]

[75] Hakkim FL. Somatic embryogenesis, embyogenic cell suspension from *Ocimum sanctum* (L.) leaf callus cult and their rosmarinic acid accumulation. Int J Biol Med Res 2011; 2(4): 1064-9.

[76] Rahman A, El-Wakil H, Abdelsalam NR, Elsaadany RMA. *In vitro* Production of rosmarinic acid from basil (*Ocimum basilicum* L.) and lemon balm (Melissa officinalis L.). Middle East. J Appl Sci 2015; 5(1): 47-51.

[77] Nadeem M, Abbasi BH, Younas M, Ahmad W, Zahir A, Hano C. LED-enhanced biosynthesis of biologically active ingredients in callus cultures of *Ocimum basilicum.* J Photochem Photobiol 2018; 2018.

[78] Nazir M, Tungmunnithum D, Bose S, *et al.* Differential production of phenylpropanoid metabolites in

callus cultures of *Ocimum basilicum* L. with distinct *in vitro* antioxidant activities and *in vivo* protective effects against UV stress. J Agric Food Chem 2019; 67(7): 1847-59.
[http://dx.doi.org/10.1021/acs.jafc.8b05647] [PMID: 30681331]

[79] Nazir M, Asad Ullah M, Mumtaz S, *et al.* Interactive effect of melatonin and uv-c on phenylpropanoid metabolite production and antioxidant potential in callus cultures of purple basil (*ocimum basilicum* l. var purpurascens). Molecules 2020; 25(5): 1072.
[http://dx.doi.org/10.3390/molecules25051072] [PMID: 32121015]

[80] Kintzios S, Makri O, Panagiotopoulos E, Scapeti M. *In vitro* rosmarinic acid accumulation in sweet basil (*Ocimum basilicum* L.). Biotechnol Lett 2003; 25(5): 405-8.
[http://dx.doi.org/10.1023/A:1022402515263] [PMID: 12882562]

[81] Strazzer P, Guzzo F, Levi M. Correlated accumulation of anthocyanins and rosmarinic acid in mechanically stressed red cell suspensions of basil (*Ocimum basilicum*). J Plant Physiol 2011; 168(3): 288-93.
[http://dx.doi.org/10.1016/j.jplph.2010.07.020] [PMID: 20943285]

[82] Açıkgöz MA. Establishment of cell suspension cultures of *Ocimum basilicum* L. and enhanced production of pharmaceutical active ingredients. Ind Crops Prod 2020; 148: 112278.
[http://dx.doi.org/10.1016/j.indcrop.2020.112278]

[83] Pandey P, Singh S, Banerjee S. *Ocimum basilicum* suspension culture as resource for bioactive triterpenoids: Yield enrichment by elicitation and bioreactor cultivation. Plant Cell Tissue Organ Cult 2019; 137(1): 65-75.
[http://dx.doi.org/10.1007/s11240-018-01552-9]

[84] Ochoa-Villarreal M, Howat S, Jang MO, *et al.* Cambial meristematic cells: A platform for the production of plant natural products. N Biotechnol 2015; 32(6): 581-7.
[http://dx.doi.org/10.1016/j.nbt.2015.02.003] [PMID: 25686717]

[85] Lee EK, Jin YW, Park JH, *et al.* Cultured cambial meristematic cells as a source of plant natural products. Nat Biotechnol 2010; 28(11): 1213-7.
[http://dx.doi.org/10.1038/nbt.1693] [PMID: 20972422]

[86] Zhu J, He S, Zhou P, *et al.* Eliciting effect of *catharanthine* on the biosynthesis of vallesia chotamine and isovallesiachotamine in Catharanthus roseus cambial meristematic cells. Nat Prod Commun 2018; 13(5): 1934578X1801300.
[http://dx.doi.org/10.1177/1934578X1801300508]

[87] Song Y, Chen S, Wang X, *et al.* A novel strategy to enhance terpenoids production using cambial meristematic cells of *Tripterygium wilfordii* Hook. f. Plant Methods 2019; 15(1): 129.
[http://dx.doi.org/10.1186/s13007-019-0513-x] [PMID: 31719835]

[88] Thakur M, Bhattacharya S, Khosla PK, Puri S. Improving production of plant secondary metabolites through biotic and abiotic elicitation. J Appl Res Med Aromat Plants 2019; 12: 1-12.
[http://dx.doi.org/10.1016/j.jarmap.2018.11.004]

[89] Narayani M, Srivastava S. Elicitation: A stimulation of stress in *in vitro* plant cell/tissue cultures for enhancement of secondary metabolite production. Phytochem Rev 2017; 16(6): 1227-52.
[http://dx.doi.org/10.1007/s11101-017-9534-0]

[90] Radman R, Saez T, Bucke C, Keshavarz T. Elicitation of plants and microbial cell systems. Biotechnol Appl Biochem 2003; 37(1): 91-102.
[http://dx.doi.org/10.1042/BA20020118] [PMID: 12578556]

[91] Namdeo AG. Plant cell elicitation for production of secondary metabolites: A review. Pharmacogn Rev 2007; 1(1): 69-77.

[92] Walker TS, Bais HP, Vivanco JM. Jasmonic Acid induced hypercin production in *Hypericum perforatum* L. (St. John wort). Phytochemistry 2002; 60: 289-93.
[http://dx.doi.org/10.1016/S0031-9422(02)00074-2] [PMID: 12031448]

[93] Woerdenbag HJ, Lüers JFJ, Van Uden W, Pras N, Malingré TM, Alfermann AW. Production of the new antimalarial drug artemisinin in shoot cultures of Artemisia annua L. Plant Cell Tissue Organ Cult 1993; 32(2): 247-57.
[http://dx.doi.org/10.1007/BF00029850]

[94] Krzyzanowska J, Czubacka A, Pecio L, *et al.* The effects of jasmonic acid and methyl jasmonate on rosmarinic acid production in Mentha × piperita cell suspension cultures. Plant Cell Tissue Organ Cult 2012; 108(1): 73-81.
[http://dx.doi.org/10.1007/s11240-011-0014-8]

[95] Kim HJ, Chen F, Wang X, Rajapakse NC. Effect of chitosan on the biological properties of sweet basil (*Ocimum basilicum* L.). J Agric Food Chem 2005; 53(9): 3696-701.
[http://dx.doi.org/10.1021/jf0480804] [PMID: 15853422]

[96] Bais HP, Walker TS, Schweizer HP, Vivanco JM. Root specific elicitation and antimicrobial activity of rosmarinic acid in hairy root cultures of *Ocimum basilicum*. Plant Physiol Biochem 2002; 40(11): 983-95.
[http://dx.doi.org/10.1016/S0981-9428(02)01460-2]

[97] Pandey H, Pandey P, Singh S, Gupta R, Banerjee S. Production of anti-cancer triterpene (betulinic acid) from callus cultures of different Ocimum species and its elicitation. Protoplasma 2015; 252(2): 647-55.
[http://dx.doi.org/10.1007/s00709-014-0711-3] [PMID: 25308098]

[98] Udomsin O, Yusakul G, Kitisripanya T, Juengwatanatrakul T, Putalun W. The Deoxymiroestrol and Isoflavonoid production and their elicitation of cell suspension cult of *Pueraria candollei* var. mirifica: From Shake Flask to Bioreactor. Appl Biochem Biotechnol 2020; 190(1): 57-72.
[http://dx.doi.org/10.1007/s12010-019-03094-y] [PMID: 31301012]

CHAPTER 8

Plant Tissue Culture: A Perpetual Source for the Production of Therapeutic Compounds from Rhubarb

Shahzad A. Pandith[1,*] and **Mohd. Ishfaq Khan**[1,2]

[1] *Department of Botany, University of Kashmir, Srinagar, Jammu and Kashmir, India*

[2] *Plant Biotechnology Section, Department of Botany, Aligarh Muslim University, Aligarh, Uttar Pradesh, India*

Abstract: Plants are interesting natural resources that have had a close association with mankind since their existence. Their utility ranges from simple food, fodder, varied commercial and industrial products, and above all, as efficacious medical agents to cure various human health ailments. Amongst this vast reservoir of natural economical wealth, Rhubarb (*Rheum* Linn; Family: Polygonaceae), a perennial herb represented by about 60 extant species occurring across Asian (mostly restricted to China) and European countries, is one of the oldest and best-known medicinal plant species which finds extensive use in different traditional medical systems. Over the past several decades, and owing to the pharmacological efficacy of Rhubarb, the plant species has been subjected to different natural and anthropogenic pressures in the regions of its occurrence, rendering it threatened. In this context, the present chapter provides the basic account of Rhubarb while giving a gist of its therapeutic potential vis-à-vis major bio-active secondary chemical constituents. Additionally, the focus has been given to the *in vitro* production system of this wondrous drug for its sustainable conservation and meticulous utilization while highlighting various attributes of the technique of tissue culture such as somatic embryogenesis, cell suspension cultures, hairy roots, *etc*. , as projected potential approaches for desirable benefits from the genus *Rheum*.

Keywords: Conservation, Pharmacological efficacy, Phytochemicals, Polygonaceae, Rhubarb, Threatened, Tissue culture.

INTRODUCTION

RHUBARB: A General Account

Throughout the ages, nature has been the home for basic human needs. In particu-

---

[*] **Corresponding author Shahzad A. Pandith:** Department of Botany, University of Kashmir, Srinagar, Jammu and Kashmir, India; E-mail: drshahzad@uok.edu.in

Mohammad Anis & Mehrun Nisha Khanam (Eds.)
All rights reserved-© 2024 Bentham Science Publishers

lar, man has been using plants as food and fodder besides benefitting from various services plants could afford. Owing to the close association with plants and the experiences gained thereby, humans have identified a rich treasury of medicinal wealth among the diverse flora inhabiting the earth alongside other creatures. This knowledge traversed through generations while getting enriched with every passing generation with the addition of new plants and improvement in the ways this medically rich repository could be utilized. Certainly, the same information ultimately gave rise to the widely practiced and majorly accepted (in developing and underdeveloped countries) traditional medical systems, which include Ayurveda, Chinese, Homeopathic, Naturopathy, Siddha, Tibetan, Unani systems, *etc*. The WHO has estimated that about 75% of Asians and Africans still believe and use this ethnomedicinal knowledge to cure varied ailments [1, 2]. They employ conventional measures in using single or conjugate cheap herbal extracts with higher safety and minor adverse effects. Importantly, previous decades have seen the use of plants and plant products as efficacious chemotherapeutic and/or chemopreventive agents for the effective treatment of various diseases. Pertinently, sumptuous investigations have shown that the medicinal properties of this flora are due to the presence of special metabolites literally known as secondary chemical constituents, that have a very wide diversity in their form and function. These natural products (NPs) have indeed provided a platform for renewed attention from both practical and scientific viewpoints for the evidence-based development of novel phototherapeutics and nutraceuticals.

Earth holds a vast reservoir of angiosperm taxa, among which Rhubarb is one of the important plant species with a complex history of use and trade over centuries and across borders. This perennial species has remained as a source of fascination vis-à-vis its role in cathartic therapy and as a tonic without considerable warnings in eighteenth- and nineteenth- century America and Europe. Indeed, in therapeutic history, and in all probability, there hasn't been a contemporary to the medicinal Rhubarb (*R. officinale*) owing to its efficacy and wide use among immense number of people [3]. Rhubarb, besides being an important object of botanical, horticultural and commercial interest, has seized considerable attention of both theoretical and clinical physicians by vigorously helping with the major medical requirements in the eighteenth century [4]. Nevertheless, there were certain misperceptions associated with the existence of true medicinal Rhubarb. Pertinently, it was clear in the second half of the nineteenth century that high valued Rhubarb roots were native to west China highlands, northern Tibet and southern Mongolia. Moreover, other extant species of Rhubarb with comparatively lesser medical efficacy are known to occur in countries like Bhutan, India, and Nepal, besides South East Europe and South West Asia [3, 4]. The perennial *Rheum,* commonly known as Rhubarb, finds wide use as a medicinal herb for ages in the traditional medical system of China. This

Polygonaceous member has limited most of its species to China vis-à-vis its distribution centers in north-western and western China. It is called "Chun-tza" and "Ta-huang" in traditional Tibetan and Chinese medical systems, respectively [5]. Indeed, owing to its wide distribution in China, the common names for them are readily found in Chinese and/or Tibetan languages; suffix 'huang' is used with different species of this wondrous drug. On the other hand, Rhubarb species also receive their names based on their utility (such as 'ornamental Rhubarb', *etc.*) or the country of their occurrence (such as 'Turkish Rhubarb', 'Chinese Rhubarb', 'Indian/Himalayan Rhubarb' or 'Russian Rhubarb').

Rhubarb as a genus includes familiar ethnomedically important plant species which are mostly confined to the mountainous regions of the Qinghai-Tibetan Plateau (QTP) and its adjacent areas [6]. The literature reports 60 extant congeneric species of this perennial herb, which exhibit wide distribution from temperate to alpine regions growing at an elevation of 500 to 5400 m asl. Owing to wide habitat tastes, the genus *Rheum* is found across Asian and some European countries with 19 species of them reportedly known to be endemic to China [7]. Moreover, earlier investigations have reported the existence of 10 species [8, 9] in the Indian subcontinent, which were later restricted to a mere 8 (*R. acuminatum* Hook. f. & Thomson, *R. australe* D. Don, and *R. webbianum* Royle., *R. globulosum* Gage, *R. moorcroftianum* Royle, *Rheum nobile* Hook. f. & Thomson, *R. spiciforme* Royle, and *R. tibeticum* Maxim. ex Hook. f.) [10]. Pertinently, we have reported *R. moorcroftianum* for the first time from Kashmir Himalaya [11]. The widespread diversity in habitat characteristics, which in turn has shaped and thereby resulted in varied morphological attributes of this dynamic vegetable and medicinal herb, has been employed by various investigators to classify Rhubarb into different sections. The pioneering work on such distinguishing traits of *Rheum* was done in 1936 [12], the authors divided this perennial herb into nine different sections based on the morphology, pollen exine structure and trnL-F region (of cpDNA). Nonetheless, around four decades later, this sectioning faced modifications by researchers [13] who firstly accredited only 5 sections from the above classification while adding two more sections *viz.* sect. Acuminata (based on the morphology of leaf) and sect. Globulosa (based on the inflorescence). Furthermore, aerobiology (study of pollen grains) came to the rescue, and, as of now, 8 sections are recognized, accepted and acknowledged within the genus *Rheum* that are based on six different types of pollen grains the species produce [14]. The pollen grain types include: Verrucate-rugulate (sect. Nobilia); verrucate-perforate (sect. Globulosa); rugulate (sect. Spiciformia); microechinate-perforate (sects: Rheum, Palmata, Deserticola and Spiciformia); microechinate-foveolate (sects; Rheum, Palmata, Acuminata, Deserticola, Orbicularia, Nobilia and Spiciformia); and finely-reticulate (sects; Rheum and Palmata). The miniature pollen grains present a great taxonomic significance in families like

Polygonaceae, which are highly multipalynous (also called eurypalynous) wherein the pollen types show considerable differences in size, aperture, and stratification of exine, *etc.* in contrast to the stenopalynous or unipalynous cases where the pollen type is usually constant and characteristic of the family (such as Asclepiadaceae, Brassicaceae, and Poaceae, *etc.*). Indeed, it is also believed that the inter-generic/specific differences within the external ornamentations of these pollen grains indicate possible genetic erraticism within a group and might explain the speciation events. Pertinently, in order to comprehend Rhubarb on phylogenetic and evolutionary levels vis-à-vis its geographical/ecological distribution in NW China, a couple of investigations in the recent past have described the morphology of pollen grains in forty different species of *Rheum* distributed in all the eight sects mentioned above [15, 16]. The authors presented Rhubarb as a natural group while proposing an arbitrary evolutionary trend for its different pollen types. The microechinate -foveolate/perforate type pollens were regarded as the most primitive type (appears in species occurring at lower altitudes) due to their common prevalence within the genus. This was followed by the finely-reticulate type, rugulate type, verrucate-perforate type and the verrucate-rugulate type, which were considered the most advanced ones (appear in species occurring at lower altitudes). Moreover, they found multiple pollen types within the same sect, which opened the gates for the anticipatory existence of parallel evolution of pollens within the genus *Rheum*. Additionally, and even after the reports of the presence of 60 extant species [17] of this wondrous drug belonging to the monophyletic tribe Rumiceae [18], 'Plant List' (http://www.theplantlist.org/1.1/browse/A/Polygonaceae/Rheum/)— accessed on 6th June 2021) recognizes only 44 names as the accepted species names making their percentage to mere 38.6% from a total of 114 names available; others include 32 names as synonyms and 38 names marked unassessed amounting to a total of 28.1% and 33.3%, respectively.

Rheum L., the extensively diversified and radiated genus mainly present in the QTP and its adjoining areas, which reportedly form its centers of both origin and diversification, is known to prefer habitats like dry and cold meadows of alpine regions, dry slopes and steppe deserts [6]. The genus is believed to have faced rapid radiations, possibly due to enormous uplifts of the QTP [6, 17]. Certainly, a notable phenotypic variation is evident in the Rhubarb as a measure of its adaptation to unlike habitat changes [19]. For example, various congeneric species of *Rheum* have attained dwarf stature and have either drooping bracts or leathery basal leaves. The evolutionary compensation for decumbent (horizontal) species is thought to be the shield against harsh winds, and for the limp bracts in the maintenance of inflorescence temperature and protection against UV radiations, thereby ensuring their dispersion along the snowline up to an altitude of 5000 m asl [20]. Different morphologies of this intricate genus have been reported so far, which best fits their respective environments. Amongst various habitat factors, the temperature is known to have a significant impact on different phenophases of the plant. Throughout the sub-zero temperatures of winter months, the rhizomes remain dormant under the soil (November to March/April). In fact, the evidences from altered phenological episodes within the genus vis--vis their habitats, altitudinal gradient and pollination requirements would help in developing accurate plans for proper cultivation and conservation practices of this important medicinal herb. In the perspective of a brief and general morphological account, the genus *Rheum* comprises of herbaceous perennial plants, which may be as long as 3 meters (*R. australe*) or as short as 2 cm (*R. globulosum*) in height. The species bear an erect and glabrous but hollow stem. Nevertheless, it is solid in some, like *R. ribes*. Moreover, the stem is also absent in some of the Rhubarb species including *R. globulosum*, *R. moorcroftianum*, *R. nanum etc.* [21]. The leaves are mostly basal, cauline, simple, palmate or sinuate dentate. Ocrea is generally large, with leaves petiolated. The length as well as the color of the petioles vary from 5 cm (*R. ribes*) to 45 cm (*R. australe*), and yellow to purple red, respectively. Leaves range in shape and may be (among others) orbicular (*R. nanum*), rhombic (*R. rhomboideum*), ovate-elliptic (*R. spiciforme*), triangular-ovate (*R. reticulatum*) or ovate (*R. moorcroftianum*) with papery appearance in some species like *R. tataricum* and *R. uninerve, etc.* The inflorescence is simple or branched (sparsely or profusely), generally paniculate or spikelike with much color variations that aid in easy and primary recognition of different congeneric species of this medical herb. Pedicel is articulate and bears bisexual or polygamo-monoecious flowers, which are small, dense or 1-5 fascicled. The characteristic feature of Rhubarb is the absence of petals; instead, a dogged perianth composed of 6 elliptic tepals is present. The androecium usually consists of 9 (6+3) or less (7 or 8) yellow to purple-red stamens with variations in anther colour. On the other hand, the female part, the gynoecium, is composed of an inflated large stigma, three short horizontal styles, and ovoid to rhomboid shaped ovary. The fruit is trigonous winged achene [7]. Fig (**1**) provides some representative images of the species of *Rheum* reported in India.

ETHNOBOTANY AND PHARMACOLOGY OF RHUBARB

Science in the current era is growing at an untraceable pace with the proportionate advancements in related technologies. In this backdrop, present researchers do not show much and anticipated interest in herbal medicines, which demands extra effort, scientific knowledge and time for complete and successful exploration and subsequent laboratory investigations. However, Traditional Chinese Medicine (TCM), among others, has served as an eye opener while portraying the interesting and promising opportunities in the field of Herbal Medicine (HM). In fact, ethnomedicine plays a significant role in existing drug discovery programs and has become a motivating and potential source for novel drugs being generated against old and evolving diseases. Pertinently, the production of such drug agents/candidates or lead molecules appears reasonable if the contemporary scientific community is inspired to look back at the elapsed momentous medical works and the accompanying remedies.

Fig. (1). Representative species of genus Rheum in natural habitats of Indian Himalaya.

Rhubarb is a natural treasure with age-old potential utility against varied human ailments. The evidences for promising, safe and efficacious use of this wondrous drug by humans since the eighteenth century are available in different traditional medical systems, which are still in vogue, particularly in the underdeveloped and developing countries. In fact, it has remained a common part of kitchen recipes with the help of various herbalists who presented it as a vegetable; so, it got space in kitchen gardens and greenhouses also as a cultivated medicinal/vegetable herb (petiole is the preferably consumed part). Besides proteins, carbohydrates, and dietary fiber, Rhubarb is considered a rich source of vital mineral elements (calcium and potassium) and vitamins (C and K) with no cholesterol content as per the USDA Nutrient Database [22]. The dietary fiber in Rhubarb is believed to be similar in amount to that found in apples, oranges or celery. Owing to the presence of malic and oxalic acid [23], Rhubarb is probably amongst the most sour-tasting vegetables in usage, which has inspired its use with some sweetener.

Importantly, the fleshy stalks (petioles) of Rhubarb are consumed (with added sugar) in varied forms *viz.* sauces, sweet soups, jams, cocktails, crumbles, pies or tarts, and as Rhubarb wine or an ingredient in baked goods. Interestingly, a traditional dessert made of Rhubarb pies is available in North America and the United Kingdom, giving the plant a new name as the 'pie plant.' Nonetheless, this high-altitude medicinal herb is a good source of the non-dietary/anti-nutrient form of calcium, calcium oxalate (CaOx), which our body is unable to absorb proficiently [24], instead becomes a major health issue at times. The CaOx is a common form of oxalic acid which appears in dynamic forms (as crystals) and varied concentrations in the Rhubarb; the diversity exists at the species level, across geographies and spatially (within the same plant). The harvesting of Rhubarb is advocated in early spring by folk tradition as the oxalate content is believed to rise from spring towards summer. Further, the enhanced CaOx levels are known to cause hyperoxaluria (deposition of CaOx crystals within body organs) and may even lead to kidney failures following the formation of kidney stones [25]. Consequently, herbalists and dieticians suggest its consumption (usually not by people predisposed to kidney stones) in a balanced form, preferably cooked, as later is believed to reduce CaOx concentration by 30-87% [26 - 28].

The main utility of rhubarb as an herbal medicine is essential because of its productive and harmonizing effect on the digestive system, which has certainly made it an efficacious laxative. This property of the wondrous drug has made it an effective agent to cure various basic health problems, *viz.* constipation, diarrhea, gastroenteritis, heartburn, hemorrhoids, stomach pain, *etc.* , and to reduce the strain during bowel movements. Indeed, HMPC (Herbal Medicinal Product Committee) has documented the administration of rhubarb roots to treat erratic constipation as a "well-established medicinal use". The HMPC has the special authority at European level for autonomous valuation of the registration of herbal medicines. Further, the peripheral application of the alcoholic root extracts of Rhubarb is known to be effective in treating gum inflammation [29 - 32]. The TCM serves as good evidence for the varied and profound utility of Rhubarb against a spectrum of human ailments that, too, without any documented adverse effects. During Han Dynasty, the Divine Husbandman's Classic of Materia Medica (compiled in ~ 200 AD) has summed up the wide and efficacious uses of Rhubarb. Pertinently, Chinese Rhubarb (Da huang) is considered among the best-known ancient herbal medicines. The chemical constituents with which the medical worth of Rhubarb is associated are believed to accumulate in this perennial herb at the age of about 6 to 7 years. So, traditionally, the mature plants are harvested, preferably for rhizomes (leaves, along with petioles, are consumed by some people as vegetables) at the age of 6 or more years in fall and then shade dried. The dried rhizomes are used in various forms like powder, paste, decoction, *etc.* , by the respective herbalists to treat a range of human health issues: antiseptic, antispasmodic, aperient, demulcent, diuretic, purgative, and stomachic. The herbalists from Europe had recommended an auspicious use of Rhubarb as a laxative and diuretic agent in addition to its ability to treat urolithiasis, jaundice, and diarrhea in a concentration and dose-dependent manner. Studies from growing economies like China, Japan, and India also present Rhubarb as well-anticipated and effective treatment agent against some of the serious health issues such as kidney failure, lung infections, and sepsis in human and animal models [33]. It has also been advocated against the disorders in the upper body, including eye contagions, hemorrhages, respiratory infections, nose bleeds, *etc.* , besides some menstrual issues and endometriosis [34]. Rhubarb is also believed to boost the appetite, treat burns, relieve pain in inflammation/injury, endorse blood circulation, minimize autoimmune reactions, and prevent bowel infections [35 - 39]. However, all such proposed effects need to be further assessed and properly examined in light of the current field-related state-of-the-art technologies using different *in vitro* assays and suitable *in vivo* models. In general, the ethnopharmacological effects of some of the well-known species from the genus *Rheum* is more or less similar and without any adversative effects. Further, owing to the projected efficaciousness which Rhubarb is delivering in light of the available traditional medical systems, this herbal drug would surely be an instrumental supplementation to any human- and eco-friendly herbal program.

As discussed above, Rhubarb finds mention among one of the ancient, good and widely used medicinal herbs in different traditional medical systems *viz*. Unani, Tibetan, Ayurvedic, Unani and Chinese, *etc.*. This dynamic (with a wide range of occurrence across varied habitats) herb has to its credit a long history of ethnic and pharmaceutical use against diverse human health complications ranging from simple laxatives to major global threats like diabetes and even cancer. Rhubarb formulations, either alone or as an adjunct with some other herbal phytomedicine, see profound use in HM, TCM, in particular against different human ailments. Virtuous and in-depth scrutiny of the literature presents a good number of studies which have been carried out on the herbal extracts of different congeneric species of *Rheum* or on the isolated/extracted individual bio-active compounds from them. The frequently occurring key bio-active phytoconstituents of Rhubarb are purportedly known for various activities, including cytotoxic, anti-tumor, -proliferative, -oxidant, -microbial, -diabetic, and nephro-/hepato- protective actions [2, 4, 40]. Nonetheless, further and progressive investigation vis-à-vis contemporary state-of-the-art technological interventions are needed to actually ascribe the functions of this medical herb to some of its compounds which could be potential lead molecules for future drugs against different diseases.

Starting with the antimicrobial benefits, anthraquinone derivatives are presumably known as vital and effective antimicrobial agents in Rhubarbs [22, 41]. In fact, the major anthraquinone constituents *viz.* chrysophanol, rhein, physcion and aloe-emodin have proven very successful against common-pathogenic bacteria such as methicillin-resistant *Staphylococcus aureus* [42] and *Escherichia coli* K12 [43], *etc.* In addition to anthraquinones, the phytoceutical flavonoids *viz.* myricetin, rutin and quercetin also possess antibacterial activity. The mechanism of antibacterial effects of Rhubarb has been attributed to the inhibition of the electron transport system in bacteria *via* multiple enzyme inhibitions in mitochondria [44] by Rhubarb constituents. Other than bacteria, various studies have shown the active role of many *R. australe* extracts and the isolated bio-active anthraquinones against various fungal species such as *Aspergillus fumigatus*, *A. niger*, *Rhizopus oryzae* and *Candida albicans*, *etc.* [45, 46]. And also, a few recently undertaken investigations have shown up important antibacterial [47] and antiviral [48] activities in anthraquinone derivatives from rhizomes of *R. australe*. Most of the extracts from rhizomes of Rhubarb species have indicated virucidal effects against HSV-I, polio, influenza and the measles virus *in vitro* [49 - 52]. Also, extracts from *R. palmatum* serve as a potential drug against SARS coronaviruses because of their 3CL protease inhibitory activity [53]. Regarding human respiratory and entero-viruses, emodin is a novel and secure antiviral agent among all anthraquinone derivatives. Additionally, by inhibiting the viral replication and protein expression, another phenolic agent, kaempferol, has also been noted to be efficacious against the Japanese encephalitis virus [54]. Hence, from the above-mentioned studies, it is clear that Rhubarb extracts show synergistic and a multitude of actions against different pathogens and may find use in the treatment of various pathogen-induced diseases in crop plants in the future.

On the other side, Rhubarb is considered a good source of phytoconstituents with promising antioxidant activities. In a study based on two Polish varieties of Rhubarb, namely, Victoria and Red Malinowy, later was found to exhibit the highest and former the lowest total antioxidant activity. The antioxidant activities in these varieties were determined on the basis of the percentage of flavonols in them [55]. Further, *R. australe* has been explored much for evaluation of the antioxidant potential of its bio-active phytoconstituents. It was found that key constituents like phenolics, stilbenoids and anthraquinones serve as potential antioxidative agents and find their application in the food and pharmaceutical industries [56, 57]. Further investigations on the antioxidant effects of anthraquinone constituents from Rhubarb have been carried out, and based on the inhibition of peroxidation of linoleic acid, the activities of these compounds were found to follow the trend; aloe-emodin > rhein > emodin [58]. Nevertheless, additional and advanced studies are required for the clear elaboration of the mechanism of action of these rhubarb extracts to recommend them as prospective antioxidative agents.

The anticancer effects of Rhubarb are also immense. In a study for the determination of cytotoxic potential against breast (MDA-MB-435S), liver (Hep3B) and human prostrate (PC-3) cancer cell lines, the methanolic and aqueous extracts of the rhizomes of *R. australe* were examined *in vitro* to check their potency in combating the tumor. It was found that both the extracts, especially the methanolic extract, exhibited anti-cancer activities in a concentration dependent manner [59] and also showed the ability to induce and push the cells toward apoptosis [60]. Another report reveals the apoptotic effects of anthraquinone constituents of Rhubarb in different human and animal cancer cell lines [61]. In fact, among the key anthraquinone constituents, most of the anti-tumor studies, both *in vivo* and *in vitro*, have been focused on maesopsin. It has also been reported that emodin, an integral constituent of Rhubarb, by targeting androgen receptors, suppresses prostate cancer cell growth [62]. In addition to these, aloe-emodin present in the rhizomes of Rhubarb induces apoptosis in various human cancer cell lines [63]. Besides anthraquinones, studies have shown that most of the flavonoids present in Rhubarb, such as quercetin, myricetin, kaempferol and daidzein, also possess anticancer activity by inhibiting cell cycle progression and down-regulating various anti-apoptotic compounds. Apart from anthraquinones and flavonoids, various stilbenoid compounds (stilbenes) from Rhubarbs *viz.*, resveratrol, rhapontigenin and piceatannol have also been reported to possess anticancer properties.

Interestingly, Arvindekar *et al* . [64] reported that five foremost anthraquinone constituents isolated from *R. australe* possess the anti-hyperglycemic activity and inhibitory effects of α-glucosidase, with aloe-emodin exhibiting the maximum action to reduce the blood glucose levels. Furthermore, the hydroalcoholic extract of *R. turkestanicum,* by inhibiting lipid peroxidation, and reducing the serum levels of lipids and glucose, was reported to inhibit myocardial damage, liver injury and nephropathy in diabetes [65]. Rhein has a wide scope in the treatment of diabetes in clinical trials and animals [66]. This anthraquinone is also used in experimental diabetic nephropathy. Moreover, different flavonoids *viz.* epigallocatechin, naringenin, epigallocatechin gallate and kaempferol are proposed to be potential antidiabetic drugs.

Besides anti-diabetic activities, Rhubarb also possesses anti-inflammatory effects. From the crude extracts of the four tested Rhubarb species *viz.R. palmatum, R. officinale, R. nobile,* and *R. franzenbachii*, it has been reported that *R. franzenbachii* possesses the greatest anti-inflammatory activity [67]. Based on another study, it was found that *R. undulatum* L. exhibits a definite anti-inflammatory and vasorelaxant activity by suppressing the vascular inflammatory process [68].

The nephroprotective activity of Rhubarb is also well-known. Gentamicin from Rhubarb is an aminoglycoside antibiotic with investigated nephrotoxic effects [69]. In a study based on methanolic extracts of *R. australe* rhizome, the nephroprotective activity has been determined by monitoring the levels of urea nitrogen and creatinine in serum against chemical-induced (CCl_4, gentamicin, mercuric chloride and potassium dichromate) kidney damage in Wistar albino rats [70]. From *R. Palmatum,* a Chinese herbal preparation WH30+, was administered to rats at the dose of 50 mg/kg/day for 10 days, which was seen to reverse the adenine-induced renal failure in them [71]. Another study on the treatment of chronic renal failure by Rhubarb extracts was administered in dogs. Extracts of *R. officinale,* under the trade name Rubenal®, were found to induce tremendous improvements in serum blood urea nitrogen and creatinine levels, level of proteinuria, serum phosphorus concentration and clinical score (by vet and clients) in dogs after 6 months of treatment with no side-effects [72]. Additionally, various individual constituents of Rhubarb are believed to have a nephroprotective role against various renal toxins/ailments. Due to the antioxidant and anticancer properties of one of the constituents of Rhubarb *viz.* emodin, it serves as an important nephroprotective agent. Another anthraquinone, rhein, is also a powerful nephroprotective agent and anti-hyper-uricemic agent for clinical applications.

In addition to the above activities, Rhubarb is also known to have a protective role against chronic liver injury, especially the development of liver fibrosis leading to hepatocellular liver cirrhosis and carcinoma. Ibrahim *et al* . [73] investigated the hepatoprotective effects of *R. australe* ethanolic extracts by prompting hepatotoxicity in rat hepatocyte primary cultures *via* carbon tetrachloride (CCl_4). In another study, chloroform and methanolic extracts of *R. australe* rhizomes were subjected to hepatoprotective activity in male Wistar rats *in vivo* against toxicity induced by paracetamol (acetaminophen). With the help of these extracts, the liver retained normal function by reducing the concentration of bilirubin, AST, ALT and ALP [74]. In a recent finding, Rhubarb extract (0.3%) has been shown to reduce liver tissue injury, oxidative stress, hepatic lipid accumulation and various inflammatory disorders induced by acute alcohol administration in male C57Bl6J mice [75, 76]. Besides, emodin is also a remarkable hepatoprotective agent. It has been shown that emodin from Rhubarb possesses restoring activity on cholestatic hepatitis by improving hepatic microcirculation, antagonizing pro-inflammatory mediators and cytokines, controlling neutrophil infiltration, inhibiting oxidative damage, and reducing impairment signals [77].

In this backdrop, the need of the hour is to investigate the exact mechanism of action of individual constituents from crude extracts of Rhubarb and different herbal preparations from the same. Moreover, their bioavailability, pharmacokinetics and pharmacodynamics and dosage/toxicity levels need to be well evaluated. This would somehow help in upgrading the health care system for the overall betterment of society.

PHYTOCONSTITUENTS FROM RHUBARB WITH THERAPEUTIC SIGNIFICANCE

As literature reports, plants are a treasure of a spectrum of secondary chemical constituents, many of which have stood at the forefront to treat various human ailments for ages owing to their bio-active nature and harmless pharmacological efficacy to which well-known herbal/traditional pharmacopeia is the witness. Importantly, these low molecular weight chemical constituents are actually produced by their bearer (plant) as natural protection agents against various biotic and abiotic stressing agents in their habitats. Now, over the centuries, humans have learned to use such medicinal plants for their adverse health issues, and the knowledge has traversed immense generations to lead us to explore and investigate this natural wealth for the overall benefit of mankind. In fact, using these medically important plants in varied formulations, and against a spectrum of health issues, has remained a common and effective practice nearly in all cultures across centuries. Further, and as a substitute to evade the side effects of synthetic medicines, it is necessary to explore and comprehensively investigate the herbal (medical) flora, which has anticipated promising benefits to achieve high effectiveness at low cost and improve patient compliance. Pertinently, this thing has somehow encouraged part of the scientific community and thereby endorsed the prospective generation of the spectrum of lead molecules as potent and safe drug agents/molecules [78].

Rhubarb, like other medicinal plants, is recognized to synthesize a suit of secondary compounds primarily as naturally defensive agents against herbivory and microbial attack. The remedying properties of this wondrous drug with historical ethnic use in various traditional herbal systems are ascribed to different classes of phytoconstituents, including phenolics, anthraquinones and stilbenoids. Certainly, the pharmacological use of these metabolites and/or their derivatives is being scientifically examined as lead molecules for the treatment of varied human illnesses [22]. Topical studies examining the chemical composition of various species of *Rheum* have found them to be a rich repository of different compounds, including anthracene-, catechin-, and naphthalene- derivatives besides stilbene glycosides and some tannin-related compounds [79]. The occurrence and amounts of such major bio-active chemical constituents are not equivalent in different species of Rhubarb, and within different samples (and even tissues) of the same species vis-à-vis their eco-physiological attributes, altitude and habitat characteristics, in particular. Indeed, altitude is known to exhibit a profound impact on the yield of secondary chemical constituents; secondary compounds are found in higher concentrations in plants growing at higher altitudes as compared to the ones which grow in low lying areas. The number of compounds and their concentration has also been found to vary with the age of the plant and the time of harvest (Malik *et al*., 2010) [2, 22, 80]. Further, and for obvious reasons, such disparities have also been found in cultivated and *in vitro* plants when compared to the naturally growing species [81]. Additionally, studies have shown the occurrence of metabolite channeling between defense mechanisms (the main aim to produce secondary metabolites) and growth (normal activity of the plant) of the plant species based on the nature and level of risk involved. Investigators have found the existence of trade-offs between such processes of growth and defense wherein shuffling between the production of secondary metabolites and the introduction of an enzymatic anti-oxidative defense system is known to occur [82]. So, to understand the dynamic behavior of these metabolites and their shifts, vis-à-vis varied environmental attributes, it is important to investigate and assess the production of these low molecular weight chemical compounds under natural and laboratory-controlled conditions. Rhubarb is a rich repository of major bio-active chemical constituents; the structures of many of them have been elucidated and even taken to the level of assessing SAR (structure activity relationship) activities. These chemical constituents belong to different classes of compounds which chiefly include anthraquinones, stilbenes and flavonoids [83]; Pandith *et al*., 2018).

Flavonoids

Flavonoids are a large and diverse group of polyphenolic secondary compounds exhibiting ubiquitous occurrence in the plant kingdom (Fig. **2**). With the basic structure of two benzene rings (A and B) linked by a heterocyclic pyrane/pyrone ring (C), flavonoids have attained diversity in the chemistry by various modifications including prenylation, methoxylation, hydroxylation and glycosylation that ultimately resulted in their categorization into different classes *viz*. flavonols, flavones, isoflavones, flavanones and anthocyanin pigments [84]. Generally, among all flavonoid classes, flavonols are regarded as the most abundant group that typically amass as glycosides of quercetin, myricetin or kaempferol in the vacuoles of plant cells. Besides performing diverse roles in plants, flavonoids are often regarded as potent phytoceuticals for human health, and have been reported to act as vital agents against some prominent illnesses [84 - 86]. They are known to exhibit anti-cancer, -oxidative, -hypertensive, -cholinesterase activities besides acting as antihypertensive (during pregnancy) agents. Owing to their promising effects on health, flavonoids (exhibiting their abundance in dietary plants/herbs) have received much attention as potential bioactive therapeutic phytoconstituents. Unexpectedly, we (and animals) need these chemical constituents as external supplements in the form of fruits, vegetables, tea and wine, *etc.* , as we are unable to synthesize flavonoids *in situ*. In fact, their regular consumption is believed to exert cardio-vasculoprotective effects while diminishing the inception or advancement of many cardiovascular ailments, hypertension, in particular [87 - 89]. Moreover, due to the considerable antioxidant potential in both *in vitro* and *in vivo* systems, flavonoids are strongly believed to offer health-endorsing properties as dietary constituent elements. Nevertheless, like some other natural product classes, there isn't required and considerable scientific evidence which freely and fairly addresses the issues of absorption, bioavailability, metabolism and possible (dose-dependent) toxicity of these polyphenolic compounds in humans. This stresses the need to properly assess and evaluate the proposed biological efficacies of these flavonoids to make them human friendly with no or lesser side effects. So far, about 10 flavonoids have been characterized from the genus *Rheum* and include quercetin, kaempferol and rutin as major constituents, among others [83]. Further, Rhubarb is also known to contain 17 different polyphenol types, which mainly consist of gallic acid that is present as glucogallin and catechin with slight quantities of tannin. The principle compound glucogallin (and also catechin) are supposed to possess antitumor and antidiabetic potential [90] (Anonymous. 2000). Additionally, the tannins are thought to play a role in minimizing the risk of heart disease (Corder *et al* . 2001).

Anthraquinones

Anthraquinones (9,10-dioxoanthracene or 9,10-anthracenedione; $C_{14}H_8O_2$; (Fig. 3) are one of the robust and diverse classes of organic natural products which exhibit their occurrence across varied life forms ranging from simple unicellular prokaryotes (bacteria) to the highly evolved and complex multicellular systems like plants. Their occurrence has also been reported in marine environments [91]. So far, about 700 natural anthraquinones have been reported from lichens, fungi (including endophytes), sponges, marine invertebrates and a few voluminous families of angiospermic plants, including Polygonaceae (Rhubarb is a member), Rhamnaceae, Rubiaceae and Xanthorrhoeaceae [92 - 96]. Indeed, the advanced techniques of mass spectrometry have decrypted the identity of about 100 phenolics from the genus *Rheum*, which majorly include the bio-active anthraquinones and stilbenoids, besides catechins, sennosides, naphthalenes and glucose gallates [83, 97]. The anthraquinones are synthesized through two different biosynthetic routes which actually present structural differences in them; succinyl-benzoate pathway, wherein ring C of the basic skeleton is hydroxylated, and acetate/malonate pathway (polyketide pathway) where both the rings A and C get hydroxylated. These phenolic constituents can be isolated from their sources both in glycosidic form (basic anthraquinone structure attached with a sugar residue) or the free aglyconic (without sugar moiety) form. Later is usually the pharmacologically active form of anthraquinones [95]. The anthraquinones and/or their derivatives have established roles in varied pharmacological activities, as discussed in the previous section. Indeed, they are a promising class of secondary chemical constituents which offer wide pharmacological efficacies against a range of health ailments ranging from simple laxatives to anticancer drugs [91]. Further, anthracyclines (belonging to the family of anthraquinoids) consist of an anthraquinone tricyclic ring system which is linked to an amino-sugar moiety, and are regarded as effective antitumor agents. Interestingly, FDA (Food and Drug Administration, Maryland, USA) has approved six members of the anthracycline group (both natural and semi-synthetic) as anticancer drugs. These lucky chemical compounds include adriamycin, daunorubicin, epirubicin, idarubicin, mitoxantrone and valrubicin [98]. Rhubarb is a rich source of some of the major and therapeutically active anthraquinones *viz.* emodin, chrysophanol, aloe emodin, rhein and physcion, *etc.* A thorough investigation of the therapeutic potential of various phenolic (and other) constituents from Rhubarb with prospective anticipations as potential drugs/drug-agents is available in our recent voluminous communications [22] Pandith *et al.* 2020).

Despite all positivity's, there is a dark side associated with these potential therapeutic agents, emodin, in particular. Emodin, the major anthraquinone from Rhubarb, has been projected to have considerable anticancer potential against various cancers and with varied mechanisms of action [99]. However, there exists a loophole between preclinical observations and their translation to the level of clinical studies/trials. For example, the selectivity toward targeted cancerous cells has not been addressed to a satisfactory level. Genotoxicity and the carcinogenic potential attributed to anthraquinones are the key issues of imminent concern. There is an absence of any 'green line' from EFSA (European Food and Safety Authority) advocating required or harmless intake doses of anthraquinones. In fact, they have warned of possible adverse health issues following long-term intake of anthraquinones in non-specific doses [100]. As with other therapeutic agents, the proposed toxicological profile of anthraquinones needs to be properly measured to ensure its appropriate risk/benefit assessment. Pertinently, it would be better to thoroughly examine the pharmacokinetic and genotoxic profiles of anthraquinones and their derivatives through proper *in vitro* and *in vivo* studies while employing different innovative approaches, as done with emodin [101]. It is pertinent to mention here that the European Commission recently (April 2021) published a regulation which states that emodin, aloe-emodin and danthron, and all the extracts in which any of these substances are present, are prohibited for use in food [91]. In a nutshell, further and advanced studies are needed which would cover the gap from preclinical to clinical trials besides defining the actual toxicological profile of anthraquinones to ensure their utility as safe and efficacious anticancer agents.

Fig. (2). Representative phenolic (flavonoid) constituents from Rhubarb.

Stilbenoids

Stilbenoids (C_6–C_2–C_6) are the hydroxylated derivatives of stilbenes, a small class of phenolic compounds which are structurally characterized by a 1,2-diphenylethylene nucleus (Fig. **4**). They occur in monomeric (*e.g.*, resveratrol, piceatannol, and combretastatin A-4, *etc.*), dimeric as well as complex oligomeric (*e.g.*, α-viniferin, hopeaphenol A, miyabenol C and kobophenol B, *etc.*) forms. Stilbenes are synthesized by phenylpropanoid pathway and show their presence in several varied and phylogenetically disparate plant families, such as Vitaceae, Polygonaceae, Leguminosae, Gnetaceae, Dipterocarpaceae and Cyperaceae *etc.* These low molecular weight (~200-300 g/mol) naturally occurring organic compounds are constitutively synthesized in some plant tissues, including leaves, roots, bark and fruits. Nevertheless, their production can also be encouraged by biotic (*e.g.*, herbivore or pathogen attack) as well as abiotic (*e.g.*, wounding, ozone and UV irradiation) stressors [102]. The stilbenes have wide roles in the plant kingdom. They guard the host plant against disproportionate ultraviolet exposure and viral and microbial attacks [103]. Moreover, they have attained growing interest from different researchers due to their varied biological activities and potential pharmacological utilities. In addition to the anthraquinones, the genus *Rheum* is also a rich repository of stilbene compounds, including resveratrol, piceatannol, rhaponticin, deoxyrhaponticin, rhapontigenin, rhaponticin-β-D-glucoside, and deoxyrhapontigenin, *etc*. [83]. Resveratrol (trans-3,5,4'-trihydroxystilbene) is the major stilbene found in Rhubarb, whose concentration equals to that in peanuts and wines that too even under cultivated conditions [81]. It is probably the most widely studied compound among stilbenoids with wide and effective therapeutic attributes. Indeed, the satisfactory cardiovascular effects of red wine are mainly attributed to this favorable compound which has also been shown to exhibit various other biological activities such as anti-cancer, -oxidant, -inflammatory, -diabetic, and -aging, besides being a recognized phytoestrogen [104]. Nonetheless, like that of anthraquinones, there are certain reservations associated with the use of resveratrol as well. It is linked to some toxic side effects, *viz*. renal toxicity, ulcerogenicity, and detrimental cardiovascular effects, *etc*. , which are believed to be due to its biphasic dose-dependent effects. In general, the low doses of this drug have been found to be stimulating, whereas the higher doses are reportedly known to present detrimental effects *in vitro* and *in vivo*. Some studies have also shown resveratrol to interact with other drugs and enzymes associated with their metabolism and transport, thereby affecting their bioavailability and overall effect/action. A recent review by David *et al* . [105] gives an account of stilbenes and their pharmacological efficacy, controversies and toxicological effects. Therefore, stilbenes, resveratrol, in particular, also need further attention to reinvestigate their proposed therapeutic potential with valid support from preclinical and clinical studies to get the nod for their usage as safe and harmless drugs among common masses. Moreover, like resveratrol, piceatannol (a naturally occurring hydroxylated analogue of resveratrol) is also known to display a suit of biological activities [106].

Fig. (3). Representative anthraquinone constituents from Rhubarb.

Fig. (4). Representative stilbenoid constituents from Rhubarb.

IN VITRO PROPAGATION OF RHUBARB: CHALLENGES AND PROSPECTS

As discussed in previous sections, different congeneric species of Rhubarb have varied household and phototherapeutic utility with proven non-adversative effects against several health issues of mankind. Due to this food and herbal importance, many species of *Rheum* face varied kinds of anthropogenic pressures, which have made them figure various positions in the IUCN recognized threat categories [107]. The local pressures include the construction of roads and residential buildings through/around the areas occupied by the natural vegetation, deforestation, industrialization, landslides, unwarranted tourist flow surpassing the normal carrying capacity of a particular health resort, overgrazing, particular unlawful extraction, and over-exploitation for local use among others [22, 108]. Adding to this, nature has also chosen the highly grazed alpine (and adjoining) regions as the normal growing habitat for this medical herb which further supplements in diminishing their natural populations [109]. As a representative case, and owing to the natural as well as anthropogenic pressures, various researchers (including us) have observed a shift in the habitat of this one of the most sought after herbs from lower to higher altitudes in Kashmir Himalayas (Pandith *et al* . 2020). This alarming statistics of different species of Rhubarb in the natural habitat is a matter of concern for biodiversity loss. This worrying situation of *Rheum* (and other threatened species belonging to different angiospermic families) has incited the related scientific community to ensure sustainable utilization vis-à-vis regional socio-economic concerns. The scientific communities, conservationists, in particular, have also felt the need to plan, design and implement different conservation strategies wherein the method of *in vitro* propagation holds prime importance among others.

The techniques of *in vitro* propagation and multiplication system provide exciting benefits when compared to the naturally grown or cultivated individual plants as it ensures a regular stock with even production, quality, yield, and above all, the maintenance of the genetic composition of the cultured species. Besides extensive micropropagation, such approaches are usually employed for reproduction purposes, which include ploidy manipulations, embryo rescue, somatic embryogenesis, and protoplast fusions, which are proper tools for short- and mid-term storage of plant genetic resources. Pertinently, while realizing the importance of axenic culture production, the task of investigating the crucial aspects of *in vitro* plant conservation was assigned to a working group of specialists established by the International Board for Plant Genetic Resources (IBPGR) in 1980s [110]. In comparison to the laborious and (sometimes) ineffective traditional multiplication approaches, it is rather considered a vibrant approach for both propagation and preservation of the chosen genetic stock of aromatic/medicinal plants to the desired levels. Roggemans and Claes made openings in this field through their efforts to develop an efficient and reproducible *in vitro* regeneration system of the medicinal herb *R. rhaponticum* L. They succeeded in observing the axillary bud initiation from explants in addition to the rooting when grown on MS medium supplemented with 1 mg/l of 3-indolebutyric acid (IBA) and 1 mg/l of 6-benzylaminopurine (BAP) [111]. Since then, various researchers have put in their efforts to propagate different species of Rhubarb through *in vitro* multiplication system with the aim of conservation and/or production of bio-active secondary chemical constituents. We have also successfully carried out the *in vitro* micropropagation of two important and threatened (from North West Indian Himalayas) species of Rhubarb *viz*. *R. spiciforme* (Khan *et al* . 2021) and *R. australe* [80]. Two of our recent communications provide a detailed account of such studies on *Rheum* (Khan *et al* . 2021; Pandith *et al* . 2020). So, we don't deem it necessary to repeat the contents here; instead, focus on the challenges associated with this laboratory-based exercise and the possible future perspectives will be highlighted.

The *in vitro* culture of plants is actually a biotechnological approach to re-grow *in vivo* plants or their parts under precisely controlled *in vitro* laboratory conditions, wherein we can easily observe and manipulate plant growth in a relatively small space. Though the protocol for tissue culturing of plants is common, each species is different and may accordingly behave differently throughout various stages of *in vitro* micropropagation under aseptically controlled conditions. Moreover, the artificial environment created for culturing of plants (explants) is known to be stressful for plant cells, posing a challenge when you intend to maintain the generated germplasm of the specific plant species for the extended time period. This stressed environment has been projected to be the reason for non-recombinational changes in DNA sequences literally called somaclonal variations [112], which might have epigenetic [113] or possible exogenous hormone-based [112] origins, among other factors hitherto unreported. Pertinently, the process of successful propagation (and its maintenance) through various stages (of callusing, embryo-/organo- genesis, shoot multiplication and rooting) until final acclimatization and hardening is a difficult task to accomplish *per se*. Additionally, if you intend to cryopreserve the *in vitro* raised tissues, later, you must be able to survive the method usually employed for cryopreservation, which involves exposure to liquid nitrogen. While certain species are able to survive and move on with all these steps easily, others face difficulty at one or the other level, thereby seeking relative and result-oriented modifications. Some of the common issues which the subject-specific experts face during more time-/resource-consuming *in vitro* propagation of (usually) aromatic/medicinal plant species include: failure of the explant to grow in culture medium, contamination (small arthropods like mites, thrips *etc.* can cause severe fungal contaminations in tissue cultures which are tough to eliminate) or browning of the primary explant used, failure of somatic embryos to germinate or of shoots to root, yellowing and/or necrosis of cultures, hyperhydricity, etiolation (observed by us in *R. tibeticum* and to some extent in *R. spiciforme* also), leaf-drop in shoots, moulding or wilting of plants during acclimatization, failure of somatic embryos or shoot tips to survive cryoprotection, and occurrence of cellular ageing, senescence during prolonged cultivation, *etc.* and grave issues in maintaining different genetic lines of a particular species or preserving its genetic fidelity, besides the funding issues, human errors and technical malfunctions. The strategies to combat some (if not all) of the bottlenecks associated with this *in vitro* propagation system may include the implementation of management plans for quality checks, cleaning, and regular monitoring of the stored materials/cultures. To ensure higher-level safety, a convenient approach is to duplicate the collection in *in vitro* form or cryopreserved, preferably at a far-off location. Certainly, the wider applications of the technique of *in vitro* propagation are possible with fundamental-researc--based comprehension of the associated factors which otherwise create hindrances in the easy-going of this conservative and economical approach to preserve and benefit from the desired flora. In particular, the genotypes which differ in *in vitro* responses would be useful in such studies. Further, sustained morphological, physiological, developmental and molecular research focused at the responses required for *in vitro* propagation and preservation can only advance the utility of these methods.

SOMATIC EMBRYOGENESIS AS AN ALTERNATIVE TO *IN VITRO* MICROPROPAGATION OF RHUBARB

Somatic embryogenesis (SE) is a developmental process that involves a series of morphological and biochemical changes in asexual/somatic cells to produce embryogenic cells under suitable induction conditions. These cells result in the formation of a somatic embryo and eventually the generation of new plants, which are believed to be true-to-type [114, 115]. Somatic embryos grow from a diverse set of tissues, although the developmental stages remain the same as that of zygotic embryos [116]. Their rapid capacity of regeneration has offered studies on the morphological, biochemical, physiological, and molecular aspects occurring during the course of embryogenesis in higher plants, besides ensuring the propagation of plants *in vitro* within short periods of time. SE was first observed in carrot cell suspension cultures by [117, 118], and later on, it was demonstrated as characteristic of a wide variety of both dicotyledonous and monocotyledonous species of plants [119, 120]. Afterwards, different factors and mechanisms regulating SE were evaluated [121, 122] by virtue of which the technique was used in genetic manipulation experiments which, among others, included the production of a large number of elite and/or transgenic plants [123, 124]. SE involves the acquisition of embryogenic competence and proliferating capacity in differentiated somatic cells with the potential to redifferentiate into embryogenic cells [125, 126]. This redifferentiation pathway involves the activation of the embryogenic gene expression program replacing its existing counterpart in the explant [119]. This reprogramming in the gene expression system is believed to be mediated by plant growth regulators (PGRs), the physiological state of the explant used, and stress conditions [127, 128]. Among different PGRs, auxins are known to play the primary role in inducing metabolic, genetic, and physiological reprogramming leading to the embryogenic competence of somatic cells [129]. Further, cytokinins [130], abscisic acid [131], jasmonates, oligosaccharides, polyamines, and brassinosteroids have been demonstrated to play important roles in the induction and progression of SE [123]. In addition, nitrogen source has also been reported as an important factor in the induction and development of somatic embryos. For instance, in *Dactylis glomerata*, an additional nitrogen source was shown to enhance dedifferentiation and proliferation of epidermal cells to induce the formation of secondary embryos [132].

Somatic embryos are formed either directly from the explant (direct SE, DSE) or from the callus (indirect SE, ISE) [133]. It is proposed that DSE does require a little genetic and physiological reprogramming or minimal proliferation to proceed with embryo formation. However, major genetic, biochemical and physiological reprogramming is required to induce ISE in callus [121, 134]. DSE has been reported in wide variety of explants, including seedlings, microspores, ovules, and zygotic embryos [135, 136]. Moreover, ISE has been used in extensive studies [137 - 140], especially somatic hybridization and transformation experiments [123, 124, 141]. Additionally, during the process of ISE, both embryogenic (regenerative) and non-embryogenic (non-regenerative) calluses are formed. The non-embryogenic callus is usually friable, rough, and translucent, whereas embryogenic calluses have a smooth surface, and cells are generally small and isodiametric in shape [142]. The cellular origin of somatic embryos depends on the synchronized behavior of neighboring cells and may be unicellular [143] or multicellular [144], wherein later depicts a lack of synchronized cells [119].

In Rhubarbs, the technique of SE has been utilized as a strategy to conserve this precious medicinal herb and to elicit different metabolites of therapeutic significance and study their synthesis pathways (Pandith *et al* . 2020). Not all species of Rhubarb have been evaluated for their response under *in vitro* conditions, but certainly, most of the known species have been studied and include *R. palmatum* (Ishimaru *et al* . 1990; Kasparova and Siatka 2001a, b; Cui et *al. 2008), R. moorcroftianum* [145], *R. officinale* [146], *R. rhabarbarum* [147], *R. ribes* [148], *R. coreanum* [149] and *R. tanguticum* [150]. Although ISE has not been described in all Rhubarb species studied, however, some authors have reported its occurrence within a few species of the genus *Rheum* [151] demonstrated DSE in *R. emodi* on the leaf lamina and intact petiolar bases grown on MS medium containing 1.0 - 5.0 mg/l BAP. According to them, SE was influenced by the auxin in the medium, and its absence suppressed the formation of somatic embryos. Further, IBA at 1.0 mg/l or IAA at 0.25 mg/l in combination with 2.0 mg/l BAP was shown to exhibit DSE that developed into normal shoots within 6 weeks. In addition, leaf explants cultured on MS medium fortified with 1.0 mg/l BAP + 1.0 mg/l IBA showed induction of somatic embryos on the adaxial surface, and formation of roots on the abaxial surface. In one of our recent communications, we reported DSE in *R. spiciforme* (Khan *et al*. 2021). When leaf explants were grown on MS medium fortified with 25 µM BA, 1.0 µM NAA, 2.0 µM GA_3, 50 µM glutamine and adenine sulphate each, emergence of mean 7.83±0.98 direct somatic embryos were observed on leaf lamina and petiole. It was further observed that the addition 4 µM 2,4-D and 8 µM GA_3 improved the rate of germination of these somatic embryos up to 76.08±2.31%. In a similar study on *R. webbianum* callus derived from the rhizome was found to form somatic embryos when supplemented with BAP and 2,4-D [152]. Nonetheless, inadequate or very old (even if performed) studies are available that have focussed on the *in vitro* regeneration of different congeneric species of this wondrous drug which would otherwise have served as a valuable source to understand the intricacies of specific pathways typically linked with production of bio-active secondary chemical constituents. This would in turn had led us to enhance the concentration of desired metabolites by taking the cultures (cell suspension cultures, for instance) to the bioreactor levels. Indeed, owing to the utilities and accompanying threats from natural habitats of different (most sought after) congeneric species of this medical herb, proper and planned measures should be taken to *in vitro* propagate it for both medical/pharmaceutical/commercial benefits and for its effective conservation.

SE relates to the process of induced cellular totipotency wherein embryos typically develop from the vegetative/somatic cells while bypassing the natural phenomenon of fertilization. Owing to its better potential for multiplication as compared to organogenesis, SE is often considered as the more favored pathway of *in vitro* proliferation. It is well-known to offer a model *in vitro* system for studies which relate to the basic biology of plant cells, and embryo development, in particular [153]. Indeed, SE is an ideal platform to investigate the process of differentiation of cells/tissue and to comprehend the mechanisms of cellular totipotency. Importantly, SE has many advantages for mass propagation [154] that include ease of use of the liquid medium, the high multiplication rate, the possible use of bioreactors, the formation of an enormous number of embryos at one time and ease in their handling [155]. Further, the somatic embryo regenerated plantlets bear the tap root system in contrast to the adventitious one. Therefore, the technique of inducement of SE can play a vital role in *in vitro* regeneration and multiplication of plants which either do not possess the tap root system or have its formation in low percentage under *in vitro* conditions [156]. Moreover, some of the recent studies focus on the relations in inducing SE by hormones (like auxin) or by the TFs (transcription factors) [157]. Nevertheless, further studies are needed to actually assess the mechanism of SE and ascertain the potential of cellular totipotency in *in vitro* systems while employing advanced genetic techniques, which may include an inducible gene silencing approach (to unearth the relationship between hormone- and TF- induced SE) or fluorescence-activated cell sorting (where cells marked by an embryo reporter can be specifically selected), *etc.*

INITIATION AND ESTABLISHMENT OF CELL SUSPENSION CULTURES: A PROSPECTIVE APPROACH FOR RHUBARB

Cultivation of plants under controlled *in vitro* conditions has led to the assessment of the minimal concentration of macro- and micro-nutrients required for healthy plant growth and development. This application has paved the way to grow plants as liquid plant cell cultures (literally called cell suspension cultures, CSCs) in a sterile environment consisting of water, sucrose, inorganic salts, vitamins and plant hormones [158]. The CSCs allow long-term maintenance of plant cell cultures under constant shaking of media in which cells grow. They can also be cryopreserved. Mostly used plant CSC media include Murashige and Skoog medium (MS) [159] and Gamborg B5 medium [160]. These easy-going liquid cultures have anticipated potential for industrial applications in comparison to plant tissue or organ cultures. However, CSCs are yet to attain the prospect of long-term commercial success, as evident in several studies concerning its optimization [161 - 166]. Not bearing the projected and prospective utilities on commercial/industrial platforms, CSCs have seen meager introduction vis-à-vis industrial applications. Nevertheless, they have been considered ideal for bioreactor experiments, and the choice of plant CSC bioreactors is essential. The conventional stirred tank reactor with bubble aeration is still considered a better choice for CSCs, but the most emphasis has been placed on its modifications [167 - 170]. The CSCs have also been implicated in SE, which involves the transition from undifferentiated somatic cells into embryogenic cells that eventually form entire plants [171]. The process begins with the proliferation of an embryogenic culture which is achieved in a liquid medium with continuous shaking as an embryogenic CSC [172, 173]. Later are preferred as friable embryogenic calluses need less force to achieve cell separation [174]. Moreover, it needs large quantities of high-quality embryogenic calluses and transparent proembryos to initiate the culture, while meristematic globules, compact calluses and cotyledonary stage embryos are supposed to be removed with care [175].

Interestingly, the CSC is often considered superior to the callus culture as it offers high explant-to-medium contact, which then promotes a rapid tissue multiplication system [176]. This, in turn, helps in the development of a robust and consistent culture due to the homogenous transfer of nutrients, and the easy access (of cells) to medium and other supplements [173]. Another advantage of using CSC is that the nutrient gradients are not formed around the explant, and the concentration of additives is kept uniform in an agitated system. In addition, the inhibition of secretion of metabolites by the plant cells during *in vitro* growth is minimized by the fast and frequent dilution and dispersion of these substances in the liquid medium in contrast to agar-gelled media where toxins remain in close proximity to the explant [177]. Further, minimal efforts and equipmentation are required to maintain the CSC system with a high cell population density in a single vessel. Studies have shown that the plants grown in suspension cultures possess large leaves as the explant has more surface area in contact with the medium for the uptake of nutrients and PGRs [178]. Moreover, there is a better exchange of O_2 and CO_2 in liquid shake cultures (compared to the gelled ones) due to the constant agitation of the medium [179]. These things make CSCs a convenient platform to study cell responses which correspond to the processes of cellular growth, their differentiation and regeneration [180]. On the commercial scale, Vitis, Lilium and Begonia, and many other plant species have been propagated using CSCs [181 - 183], however, the *in vitro* propagation of Rhubarb has majorly been evaluated using agar-solidified medium, with very few reports of the CSC production [184]. Importantly, we require economical procedure in terms of time, media and space for the commercial propagation and conservation of rare and endangered species like that of Rhubarb. These might be some of the possible reasons for finding scanty *in vitro* (*senso lato*) and CSC (*senso stricto*) studies in different species of Rhubarb. Pertinently, in *R. emodi* [151], CSCs were used and shoot proliferation was observed within 7 days with a drop in the medium requirements by 50%, a decrease in time by 37.5%, and interestingly with a 1.5 - 2.2 fold increase in growth and multiplication rate. The authors observed the generation of healthy shoots with yellowish-green axillary shoot buds within 9 days. With other observable benefits during sub-culturing, the CSC-raised plantlets simplified transplantation with 90-95% survival after transfer to potting mix in glasshouse. Owing to these observations, the authors strongly advocated the use of CSCs for *in vitro* propagation of Rhubarb species in contrast to the laborious, expensive, time-consuming and comparatively less efficient (as part of the nutrients in the medium remain unused in case of gelled media) agar-solidified culture media. In another study on the same species of Rhubarb (the most sought after from Indian Himalayas) [184], it was detected that shoots grown in suspension cultures with constant shaking showed a maximum growth, followed by those in liquid static medium with least growth observed in agar gelled media. Therefore, it is alarming to realize the need (for conservation and commercial/industrial utilization of aromatic/medicinal plants) and potential for *in vitro* production of medicinally important plant species, threatened ones, in particular, using both approaches of agar-solidified and CSC (preferably) approaches.

IN VITRO SYSTEMS FOR THE PRODUCTION OF SECONDARY METABOLITES IN RHUBARB

The chemical compounds which plants produce have been broadly categorized into two classes *viz.* primary metabolites and secondary metabolites; the former are involved in the normal growth and development of the plants, whereas the latter include a vast diversity of chemical constituents with varied roles both in plants and for the other creatures, humans, in particular. Though produced as defensive constituents by plants, most of these compounds have been shown to exhibit biological activities against various human ailments, as discussed in previous sections. Besides, these chemical moieties have also been projected as insecticides and flavoring and dying agents, *etc.* Importantly, with the advancement in the technique of tissue culture, it became possible to cultivate plant cells in large quantities and hence production of important plant metabolites (as the parent plant produces and accumulates in nature) inside laboratories. Pertinently, and owing to certain discrete advantages [185] of secondary metabolite production in plant cell culture rather than *in vivo* in the whole plant, the sturdy and mounting demand in today's marketplace for various natural products has refocused the attention of field-specific experts on *in vitro* plant materials as potential factories for secondary phytochemical products [186, 187]. Importantly, the phytochemical compounds produced under *in vitro* conditions have supplementary benefits of uniformity in terms of both yield and quality. Moreover, the otherwise toxic issues associated with some compounds under certain circumstances can be bypassed by the engineered production of promising phytochemicals. In fact, a recent review by Khan *et al.* has highlighted the usefulness of *in vitro* culture technology as a promising factory for the production of efficacious secondary chemical constituents against the recent COVID-19 pandemic [188].

The biosynthesis of secondary metabolites appears to be related to the differentiation of the plant cells/organs [189, 190]. In higher plants, the secondary metabolism is generally active in the differentiated cells, whereas in the dedifferentiated callus cells (under *in vitro* conditions), the primary metabolism is activated to achieve proliferation, and conversely, the secondary metabolite production is decreased. Therefore, for the production of plant products by callus cultivation, it is necessary to find an optimal condition promoting both primary and secondary metabolisms. The secondary metabolites in Rhubarb have been produced from callus cultures besides seedling proliferation and suspension cultures. Certainly, more attention has been paid to the production of sennosides among other chemical constituents like anthraquinones and their derivatives [191]. The greater production of secondary metabolites proportionately requires higher production of callus, hence the effective concentration of growth regulators. In Rhubarb, 2,4-D has been found more effective than IAA in the proliferation of callus on solid media. However, for the extensive cultivation of Rhubarb callus, suspension culture has been attempted in the medium containing IAA (1 ppm) and N-(2-chloro-4-pyridyl)-N'-phenylurea (4-PU-30) (1 ppm) for 4 weeks. Further, for chemical profiling, the authors cultivated these calluses on the MS medium containing 2,4-D (1 ppm) and K (0.1 ppm) in the dark at 21 °C for 4 weeks, which followed their extraction using different chemical solvents that yielded different compounds in varying concentrations *viz.* physcion (0.4%), chrysophanol (0.47%), emodin (0.7%), aloe-emodin, rhein-8-glucoside (0.09%), *etc.* The presence of sennoside-A was determined in oxalic acid-THF extract of Rhubarb callus and sennoside-B in 70% methanolic extract. Further, the highest quantity of sennosides was formed in callus cultures of a stable hybrid of *R. coreanum* and *R. palmatum* [191].

The tissue culture of Rhubarb species was first reported by researchers [192], who also showed the production of emodin and chrysophanol from *in vitro* raised plantlets. Thereafter, several studies were published on *in vitro* culture of *R. rhaponticum* L. and *R. palmatum*, showing the production of anthraquinone glycosides but not sennosides [193]. Production of sennosides and anthraquinone derivatives in *in vitro* raised seedlings of a stable hybrid of *R. palmatum* and *R. coreanum* (Shinshu-Daiwo) had also been achieved [194]. Researchers [195] cultured Rhubarb tissue, and it was observed that ethylene-generating reagent 2-chloroethylphosphonic acid (2- CEPA) enhances the production of chrysophanol and emodin under *in vitro* conditions [196]. cultured different parts of sterile seedlings on MS medium to study the generation of calli. The explants were cultured with different combinations of PGRs and the best combination for the generation of callus was IBA (1 mg l^{-1}) and BA (1mg l^{-1}). They observed the growth of callus declined with increase in rate of secondary metabolite production. Further, myoinositol (100 mg l^{-1}) improved the anthraquinone production in medium to the maximum with NO3/NH4; 1/1.

In a nutshell, the commercial value and pharmaceutical implications of plant secondary metabolites have remained the key stimulatory factors for the overwhelming research efforts which have been invested in understanding and further manipulating their biosynthetic pathways. In fact, the cell and/or tissue culture of medicinally important plant species (which embody bio-active chemical constituents) for the production of selective metabolites holds vital significance on commercial platforms. Over the past several years, the increased use of plant cell culture systems is possibly due to the enhanced comprehension of the secondary metabolite pathways responsible for the synthesis of desired compounds in medically and/or economically important plants. Moreover, the advances in plant cell culture systems could also act as a prelude to the cost-effective and commercial production of rare/exotic plants, their cells, and the desired chemical compounds that they will produce. Though substantial progress has been made so far in improving the production of these desired chemical constituents, however, a holistic approach is needed to further understand and then modulate the pathways to produce and also enhance the yield of required and valuable chemicals.

PRECURSOR ADDITION AS A MEANS TO IMPROVE THE PRODUCTION OF BIO-ACTIVE CONSTITUENTS IN RHUBARB

Higher plants are major sources of useful secondary metabolites used in agrochemical, pharmaceutical, and flavor/aroma industries. Priority should be to search for new plant-derived chemicals for current and future use as a part of a sustainable conservation approach and rational utilization of biodiversity [197]. In this regard, the technique of plant tissue culture plays a vital role in the search for alternatives to the synthesis of desirable medicinal compounds from plants [198]. In today's marketplace, growing demand for more natural, organic and renewable products has enforced attention on *in vitro* plant tissue culture as potential factories for secondary metabolite production and has paved the way for new research exploring avenues for their enhancement. Besides its commercial significance, *in vitro* tissue culture technique provides an excellent forum for in-depth investigation of metabolic and biochemical pathways under highly controlled micro-environmental conditions [186]. On a global scale, plants are used as a major source of crude drug extracts, including some of the potent compounds such as codeine (antitussive), morphine (pain killer), papaverine (phosphodiesterase inhibitor), ajmaline (antiarrhythmic), ephedrine (stimulant), quinine (antimalarial), scopolamine (travel sickness), reserpine (antihypertensive), berberine (psoriasis), galantamine (acetylcholine esterase inhibitor), caffeine (stimulant), capsaicin (rheumatic pains), yohimbine (aphrodisiac), pilocarpine (glaucoma), colchicine's (gout) and various cardiac glycosides (heart insufficiency) [199]. Production of such and similar compounds under *in vitro* conditions would act as sustainable and ready-to-use systems for desired outputs. Certainly, efforts have been made to obtain high yields suitable for commercial exploitation, focused on optimizing the culture conditions, transformation and elicitation methods, isolation and characterization of high-producing strains, immobilization techniques and employing precursor feeding [200]. Moreover, for secondary metabolite production, the technique of tissue culture has many advantages over conventional plant cultivation such as production under controlled conditions, microbe and insect free cultures, range of explants that can be used, and automated control of cell growth requirements [201]. Importantly, the advancement in tissue culture techniques to produce specific medicinal compounds at a rate comparable to or higher than that of intact plants is accelerating at a good pace. As a result of these advancements, research in the field of tissue culture technology for the production of plant metabolites has flourished beyond expectations [187]. Indeed, unorganized callus or suspension cultures have been widely used for the production of numerous valuable secondary metabolites, however, certain cultures require more differentiated micro-plant or organ cultures [202]. These include cultures where metabolite of interest is produced in specialized plant cells, tissues or glands in the explant. A prime example is provided by saponin production in ginseng (*Panax ginseng*) roots, and hypericin's and hyperforin's production in foliar glands of *Hypericum perforatum* [203], *etc.* Additionally, the plant secondary metabolite accumulation in response to pathogen attack triggered and activated by the signal compounds of plant defense responses [204] has led to the concept of elicitation in plants under *in vitro* conditions.

On the basis of knowledge of biosynthetic pathways, certain organic compounds may be added to the culture medium in order to enhance the synthesis of secondary metabolites [205]. This process is known as "precursor feeding". It is based on the idea that the external introduction of a compound, serving as a key or primary intermediate of a secondary metabolite biosynthetic route, helps in increasing the yield of the final product [198]. Precursor feeding has shown very promising results in some plant cell culture systems. For example, the amino acid precursors of alkaloid biosynthesis added to cell suspension cultures enhanced the production of tropane and indole alkaloids, *etc.* Similarly, ferulic acid feeding to cultures of *Vanilla planifolia* resulted in an increase in vanillin accumulation [206]. Further, a recent study [207] on *Larrea divaricate* plant cell cultures reported a 2.2-fold increase in nordihydroguaiaretic acid upon feeding of 0.5 mM L-phenylalanine. Additionally, the use of different biotic and/or abiotic elicitors is a useful strategy to enrich secondary metabolite production in plant cell cultures [186]. Explants grown *in vitro* show morphological as well as physiological responses to chemical, microbial or physical factors known as "elicitors". The elicitation as a process involves stimulating, inducing or enhancing the production of secondary metabolites by the plants under *in vitro* conditions to ensure their survival, persistence, and competitiveness [186, 208]. Some of the most frequently used elicitors include amino acids, yeast extract, fungal carbohydrates, methyl jasmonate and chitosan. Methyl Jasmonate is demonstrated as the most effective elicitor of ginsenoside production in *Panax ginseng* and taxol production in *Taxus chinensis* Roxb [199]. Amino acids valine and isoleucine have been used in the biosynthesis of adhyperforin and hyperforin's in *H. perforatum* shoot cultures [209]. Similarly, leucine enriches triterpene asiaticoside accumulation in leaf-derived callus and CSCs of *Centella asiatica* [210]. In addition, chitosan has been used as a biotic elicitor in the production of anthraquinones in *Rubia akane* cell culture [211]. Besides using elicitors in leaf, node, root and petiole cultures, the elicitation technique has also been employed to improve secondary metabolite production in hairy root cultures. However, in most cases, elicitor treatments were found to inhibit the growth of the hairy roots. Benzo(1,2,3)-thiadiazole-7-carbothionic acid S-methyl ester and lysate of cell suspension of bacteria-*Enterobacter sakazaki* has been demonstrated to improve the production of secondary metabolites in callus, and hairy root cultures of *Ammi majus* [212]. Rhubarbs are a repository of many valuable and medicinally important compounds including flavonoids, anthraquinones and their glycosides, besides terpenes and phenolics, *etc.* Although many tissue culture studies has been carried out on different Rhubarb species, very few had focused on the elicitation and precursor feeding in this group of plants. Nevertheless, some earlier studies used biotic elicitor *Pseudomonas aeruginosa* in the form of a homogenate and an aqueous suspension of dead cells and salicylic acid for the production of anthracene derivatives in *R. palmatum* L [213, 214]. Moreover [215], has studied the elicitation of catechin from callus cultures of *R. ribes* using different concentrations of PGRs. The need is to attempt precursor feeding and elicitation techniques in other Rhubarb species for the production of important secondary metabolites like emodin, aloe-emodin, physcion, chrysophanol, *etc.* , to meet their increasing market demand. Further advancements in the elicitation process include metabolic engineering that involves the targeted and persistent alteration of metabolic pathways for energy transduction, chemical transformation, and supramolecular assembly.

HAIRY ROOTS: A POTENTIAL SOURCE FOR SECONDARY CHEMICAL CONSTITUENTS

Hairy roots were studied majorly between the 1930s and the 1960s, primarily as a sign of pathogen invasion in plants [216]. During the 1970s and 1980s, *Agrobacterium rhizogenes* was identified as the bacterium that induces hairy root syndrome through the gene transfer of the bacterial Ri (root-inducing) plasmid [217, 218]. Thereafter several studies were carried out to develop the hairy root culture technology. Hairy roots arise from the plantlets when they are infected by *A. rhizogenes*, currently taxonomically renamed *Rhizobium rhizogenes*, causing "hairy root syndrome" leading to non-geotropic branching root overgrowth at the site of infection [219]. In order to develop a better hairy root system, plant intrinsic properties, the safety of the plant species and the complexity of the extracting molecules to be produced need to be considered very early in the process [220]. Further, the two most important criteria used globally to select the most appropriate plant species are its biomass production capacity and the ability to secrete molecule(s) of interest in higher amounts. Ultimately hairy root system is used to produce plant-expressed compounds of interest in large-scale bioreactors, which enable growth in a sterile environment that contains only liquid nutrients [216]. The hairy root culture has become prevalent in the last two decades as a method of producing secondary metabolites in roots [186, 221]. Indeed, they produce the same metabolites as that of plants but with higher yield [222]. Higher yield, together with rapid growth and genetic stability in simple media lacking phytohormones, makes them suitable for biochemical studies not that easily possible in root cultures of an intact plant [201]. During the process of infection, *A. rhizogenes* transfers a part of its DNA (T-DNA) of the root-inducing plasmid (Ri) to plant cells. The genes transferred are expressed in the same way as the endogenous genes of the plant cells. The hairy roots are induced wounded and aseptic parts of plant by inoculating them with *A. rhizogenes* [186]. Any other foreign gene of interest can be transferred *via* the *A. rhizogenes*-mediated transformation as well [223].

The soil bacterium *A. rhizogenes* engineers dicotyledonous plant genetically into its opine producing source. This genetic transformation leads to the production of the by-product 'hairy roots' in plants. These hairy roots have genetic and biochemical stability, fast growth and the capability to grow in hormone-free media. Hairy root cultures, besides serving as model systems for plant metabolism and physiology, also act as a good source of a wide range of therapeutics and specialty chemicals depending on plant source [224]. Moreover, plants harbor a wide range of therapeutic products as lead compounds for drug development. The same compounds are produced in hairy roots as well. However, novel compounds can be generated by transformation and by the introduction of heterologous genes into the hairy roots. Certainly, one of the advantages of hairy root cultures is that certain metabolites extracted from the whole plant or shoot cultures may be produced in hairy roots. An example is provided by the accumulation of red-orange dye lawsone (2-hydroxy-1,4-naphthoquinone), which is restricted to the aerial parts of *Lawsonia inermis* (henna plant) and not in the roots of wild type henna; however, lawsone has been found in substantial quantities in hairy root cultures [225]. The hairy roots are also known to accumulate higher levels of secondary metabolites than undifferentiated cultures [226]. In hairy root cultures, elicitation, cell permeabilization, precursor feeding, and trapping of the molecules have been recognized to increase productivity. Additionally, various elicitors and permeabilizing agents (*e.g.*, calcium chelators, detergents, sonication, pH, temperature and oxygen stresses) have been shown to stimulate the secretion of secondary metabolites from hairy roots [227, 228]. Besides production of secondary metabolites, hairy root systems are used to produce recombinant proteins like monoclonal antibodies [229], green fluorescent protein [230], murine interleukins [231], human acetylcholinesterase [232], recombinant human EPO [233], thaumatin sweetener [234], humaninterferonalpha-2b [235], and recombinant alpha-L-iduronidase [236]. These recombinant proteins can be either secreted into the culture medium or produced internally [236].

In Rhubarbs, only a few species have been employed for the production of secondary metabolites using the hairy root culture system. In *R. palmatum* L. hairy root culture about 28% of the total free anthraquinones were observed to be released into the liquid medium [237]. In *R. wittrockii* Lundstr *A. rhizogenes* 9402 strain has been used to induce hairy roots and production of free anthraquinones such as aloe emodin, emodin, rhein, chrysophanol, physcion, and 8-O-methyl chrysophanol. It was found that the quantity of these free anthraquinones was much higher than the control plants [238]. As mentioned in previous sections, literature is witness to the inadequate efforts made to cultivate different medically high-valued and even threatened species of Rhubarb under *in vitro* conditions. On the other hand, the same Rhubarb species field various kinds of anthropogenic pressures due to their pharmacological efficaciousness. This thing surely serves as an alarm for sustainable conservation of important bio-active congeneric species of *Rheum*, wherein hairy root cultures could be one of the important platforms to work on. Indeed, the technological induction of hairy roots could be potentially implemented in such species with different and high-end advantages over the CSCs and open-field grown plants. These can serve as optimal model systems to study the biosynthesis of plant-derived compounds, for recombinant protein production and for discovering the complex interactions involved in phytoremediation. Indeed, owing to the multiple applications of the hairy root system, Rhubarb could act as a good source for its broader commercialization from the production of dyes to therapeutic compounds, *etc.* , wherein novel and contemporary molecular biology tools (like CRISPR/Cas9 technology) can be employed for desired products as done in transgenic tomato [239], *Glycine max* and *Medicago truncatula* [240, 241] and more recently in *Brassica carinata* [242].

CONCLUSION AND FUTURE PROSPECTS

Plants are well-known to form the basis of human existence since the old stone age, and with the passage of time, those with therapeutic significance found their place in different traditional medical systems as cheap, safe and efficacious chemotherapeutic/chemo-protective agents for the effective treatment of a vast array of human ailments. One such medical herb with pharmacological utility is Rhubarb (*Rheum* L.) which is an extensively diversified and radiated genus bearing sixty congeneric extant species reportedly growing across the continents of Asia and Europe in eighteen countries at an elevation of 500 to 5400 m asl [6]. Undoubtedly, the literature available on Rhubarb defends its substantial medical implications when used alone or as an adjunct in dissimilar herbal formulations, as mentioned in different traditional medical systems. It is pertinent to mention that the medical effectiveness of this wondrous drug has been ascribed to some of the foremost secondary chemical constituents *viz.* anthraquinones, stilbenoids, chromones, *etc.* , with recognized pharmacological efficacy toward an array of human illnesses. A series of new and effective derivatives of these bio-active phytochemical constituents have been produced and evaluated for several biological activities while employing contemporary synthetic and systems biology approaches. Nonetheless, additional and advanced studies are needed to further adventure on this valuable wondrous drug which may include extraction and isolation of bio-active compounds, including the production of highly effective novel derivatives to well screen and authenticate the traditional knowledge of Rhubarb. Similarly, the actual mechanism (both *in vitro* and *in vivo*) of the action of the different herbal formulations using Rhubarb in various forms needs to be screened and well understood. Moreover, serious examination of the toxicity levels, pharmacodynamic and pharmacokinetic mechanisms, and bioavailability are the issues which need prime attention to modulate the crucial bio-active constituents and/or their potent derivatives from *Rheum* as central frameworks for upcoming drugs. Additionally, due to the wide and well-known pharmacological efficacy of the low molecular weight secondary metabolites (as bio-actives) so far isolated from different species of *Rheum*, their improved production in homo- and/or hetero- logous host systems has a vibrant significance in the field of secondary metabolite pathway engineering as attempted in various related studies across the globe.

Loss of endemic and endangered taxa which have fundamental ecological and economic implications has remained a crucial issue with many medicinal plants across the globe, including the most sought-after (among others) and selected species of the genus *Rheum*. For that matter, the need of the hour is to combat and reverse this losing trend by looking for alternate and advanced approaches for the sustainable utilization of such natural and economic wealth. Indeed, in order to fill the huge appetite of the herbal industry, privatization and commercialization should be the priority. Pertinently, *in vitro* culture of such (and related) species would serve as a model system for both conservations as well as industrial benefits vis-à-vis their therapeutic utilities. Unfortunately, and so far, not much has been done in the case of the wondrous drug Rhubarb at *in vitro* levels, while on the other hand, its populations are dwindling from natural existence. So, subjecting this plant species to tissue culture and its related applicatory parts (as discussed in previous sections) *viz.* SEs, CSCs and hairy root systems would serve as important and yield-based efforts in conserving and utilizing Rhubarb in a desired and sustainable way.

REFERENCES

[1] Wani BA, Ramamoorthy D, Rather MA, *et al.* Induction of apoptosis in human pancreatic MiaPaCa-2 cells through the loss of mitochondrial membrane potential ($\Delta\Psi m$) by *Gentiana kurroo* root extract and LC-ESI-MS analysis of its principal constituents. Phytomedicine 2013; 20(8-9): 723-33.
[http://dx.doi.org/10.1016/j.phymed.2013.01.011] [PMID: 23453831]

[2] Pandith SA, Hussain A, Bhat WW, *et al.* Evaluation of anthraquinones from Himalayan rhubarb (*Rheum emodi* Wall. ex Meissn.) as antiproliferative agents. S Afr J Bot 2014; 95: 1-8.
[http://dx.doi.org/10.1016/j.sajb.2014.07.012]

[3] Clifford M. Rhubarb: The wondrous drug. Princeton, NJ, USA: Princeton University Press 1992; pp. 9-155.

[4] Foust CM. Rhubarb: The wondrous drug. Princeton, NJ, USA: Princeton University Press 2014; Vol. 191: pp. 6-63.

[5] Xiao P, He L, Wang L. Ethnopharmacologic study of chinese rhubarb. J Ethnopharmacol 1984; 10(3): 275-93.
[http://dx.doi.org/10.1016/0378-8741(84)90016-3] [PMID: 6748707]

[6] Wan D, Wang A, Zhang X, Wang Z, Li Z. Gene duplication and adaptive evolution of the CHS-like genes within the genus Rheum (Polygonaceae). Biochem Syst Ecol 2011; 39(4-6): 651-9.
[http://dx.doi.org/10.1016/j.bse.2011.05.016]

[7] Wu Z, Raven PH, Hong D. Flora of china. Ulmaceae through Basellaceae: Science Press. Bijing: China Science Publishing Group 2003; p. 145.

[8] Hooker JD. The Flora of British India: L. Reeve. London.: Covent garden 1897; pp. 209-464.

[9] Santapau H, Henry AN. A dictionary of the flowering plants in India. New Delhi: CSIR 1973; 7: p. 95.

[10] Srivastava R. Family polygonaceae in India. Indian Journal of Plant Sciences 2014; 3: 112-50.

[11] Khan MI, Pandith SA, Ramazan S, Shah MA, Malik AH, Reshi ZA. *Rheum moorcroftianum* (Polygonaceae) in Kashmir Himalaya. Phytotaxa 2019; 405(5): 269-75.
[http://dx.doi.org/10.11646/phytotaxa.405.5.6]

[12] Lozina-Lozinskaja A. Sistematiceskij obzor dikorastuscich vidov roda *Rheum* L. Trudy Bot Inst Akad Nauk SSSR. Ser 1936; 1: 67-141.

[13] Kao T, Cheng C-Y. Synopsis of the Chinese Rheum. Acta Phytotaxonom Sinica 1975; pp. 69-82.

[14] Li A. Flora reipublicae popularis Sinicae: tomus 25 (1). Angiospermae, Dicotyledoneae, Polygonaceae. Beijing: Science Press 1998; p. 237.

[15] Yang M, Zhang D, Zheng J, Liu J. Pollen morphology and its systematic and ecological significance in *Rheum* (Polygonaceae) from China. Nord J Bot 2001; 21(4): 411-8.
[http://dx.doi.org/10.1111/j.1756-1051.2001.tb00789.x]

[16] Wang A, Yang M, Liu J. Molecular phylogeny, recent radiation and evolution of gross morphology of the rhubarb genus Rheum (Polygonaceae) inferred from chloroplast DNA trnL-F sequences. Ann Bot 2005; 96(3): 489-98.
[http://dx.doi.org/10.1093/aob/mci201] [PMID: 15994840]

[17] Sun Y, Wang A, Wan D, Wang Q, Liu J. Rapid radiation of Rheum (Polygonaceae) and parallel evolution of morphological traits. Mol Phylogenet Evol 2012; 63(1): 150-8.

[http://dx.doi.org/10.1016/j.ympev.2012.01.002] [PMID: 22266181]

[18] Sanchez A, Schuster TM, Kron KA. A large-scale phylogeny of Polygonaceae based on molecular data. Int J Plant Sci 2009; 170(8): 1044-55.
[http://dx.doi.org/10.1086/605121]

[19] Wan D, Sun Y, Zhang X, *et al.* Multiple ITS copies reveal extensive hybridization within *Rheum* (Polygonaceae), a genus that has undergone rapid radiation. PLoS One 2014; 9(2): e89769.
[http://dx.doi.org/10.1371/journal.pone.0089769] [PMID: 24587023]

[20] Xie Z. Eco-geographical distribution of the species from Rheum L. Polygonaceae in China. China's biodiversity conservation toward the 21st Century.. China Forest Press 2000; pp. 2-78.

[21] Ali SI. Flora of Pakistan: Pakistan agricultural research council. Karachi, Pakistan: Agricultural Research Council 1980; pp. 28-258.

[22] Pandith SA, Dar RA, Lattoo SK, Shah MA, Reshi ZA. *Rheum australe*, an endangered high-value medicinal herb of North Western Himalayas: a review of its botany, ethnomedical uses, phytochemistry and pharmacology. Phytochem Rev 2018; 17(3): 573-609.
[http://dx.doi.org/10.1007/s11101-018-9551-7] [PMID: 32214920]

[23] Rumpunen K, Henriksen K. Phytochemical and morphological characterization of seventy-one cultivars and selections of culinary rhubarb (Rheum spp.). J Hortic Sci Biotechnol 1999; 74(1): 13-8.
[http://dx.doi.org/10.1080/14620316.1999.11511064]

[24] Heaney RP, Weaver CM. Oxalate: Effect on calcium absorbability. Am J Clin Nutr 1989; 50(4): 830-2.
[http://dx.doi.org/10.1093/ajcn/50.4.830] [PMID: 2801588]

[25] Albersmeyer M, Hilge R, Schröttle A, Weiss M, Sitter T, Vielhauer V. Acute kidney injury after ingestion of rhubarb: secondary oxalate nephropathy in a patient with type 1 diabetes. BMC Nephrol 2012; 13(1): 141.
[http://dx.doi.org/10.1186/1471-2369-13-141] [PMID: 23110375]

[26] Chai W, Liebman M. Effect of different cooking methods on vegetable oxalate content. J Agric Food Chem 2005; 53(8): 3027-30.
[http://dx.doi.org/10.1021/jf048128d] [PMID: 15826055]

[27] Sanz P, Reig R. Clinical and pathological findings in fatal plant oxalosis. A review. Am J Forensic Med Pathol 1992; 13(4): 342-5.
[http://dx.doi.org/10.1097/00000433-199212000-00016] [PMID: 1288268]

[28] Tallqvist H, Vaananen I. Death of a child from oxalic acid poisoning due to eating rhubarb leaves. Ann Paediatr Fenn 1960; 6: 144-7.
[PMID: 13836748]

[29] Moore M. Herbal formulas for the clinic and home. Southwest School of Botanical Medicine 1997; p. 97.

[30] Maclean W, Taylor K. Clinical manual of chinese herbal patent medicines. Pangolin Press 2000; pp. 1-235.

[31] Gardener C, Ghronicle G. Gardener's Magazine. Rhubarb: The Wondrous Drug. New Jersey: Princeton Legacy Library, Princeton University Press 2014; pp. 191-321.

[32] Clementi EM, Misiti F. Potential health benefits of rhubarb. Bioactive foods in promoting health. Amsterdam, The Netherlands: Elsevier 2010; pp. 407-23.
[http://dx.doi.org/10.1016/B978-0-12-374628-3.00027-X]

[33] Lai F, Zhang Y, Xie D, *et al.* A systematic review of rhubarb (a Traditional Chinese Medicine) used for the treatment of experimental sepsis. Evid Based Complement Alternat Med 2015; 2015: 1-12.
[http://dx.doi.org/10.1155/2015/131283] [PMID: 26339264]

[34] Ashnagar A, Naseri NG, Nasab HH. Isolation and identification of anthralin from the roots of rhubarb

plant (*Rheum palmatum*). J Chem 2007; 4: 546-9.

[35] Grieve M, Leyel C. The medicinal, culinary, cosmetic and economic properties, cultivation and folklore of herbs, grasses, fungi, shrubs and trees with all their modern scientific uses. A modern herbal Tiger books International, Mackays of Chatham. Chatham, Kent.: PLC 1994; pp. 961-2.

[36] Hoffmann D. The complete illustrated herbal: A safe and practical guide to making and using herbal remedies: Barnes & Noble Books. New York, USA: East Street 1996; pp. 28-77.

[37] Mantani N, Sekiya N, Sakai S, Kogure T, Shimada Y, Terasawa K. Rhubarb use in patients treated with Kampo medicines--a risk for gastric cancer? Yakugaku Zasshi 2002; 122(6): 403-5.
[http://dx.doi.org/10.1248/yakushi.122.403] [PMID: 12087778]

[38] Oi H, Matsuura D, Miyake M, *et al*. Identification in traditional herbal medications and confirmation by synthesis of factors that inhibit cholera toxin-induced fluid accumulation. Proc Natl Acad Sci 2002; 99(5): 3042-6.
[http://dx.doi.org/10.1073/pnas.052709499] [PMID: 11854470]

[39] Castleman M. The healing herbs: The ultimate guide to the curative power of nature's medicine. G and C Merriam, Pensylvania 1991; pp. 128-35.

[40] Rokaya MB, Münzbergová Z, Timsina B, Bhattarai KR. Rheum australe D. Don: A review of its botany, ethnobotany, phytochemistry and pharmacology. J Ethnopharmacol 2012; 141(3): 761-74.
[http://dx.doi.org/10.1016/j.jep.2012.03.048] [PMID: 22504148]

[41] Bilal S, Mir M, Parrah J, *et al*. Rhubarb: the wondrous drug. A review. Int J Pharm Biol Sci 2013; 3: 228-33.

[42] Alaadin AM, Al-Khateeb EH, Jäger AK. Antibacterial activity of the Iraqi *Rheum ribes*. Root. Pharm Biol 2007; 45(9): 688-90.
[http://dx.doi.org/10.1080/13880200701575049]

[43] Hatano T, Uebayashi H, Ito H, Shiota S, Tsuchiya T, Yoshida T. Phenolic constituents of Cassia seeds and antibacterial effect of some naphthalenes and anthraquinones on methicillin-resistant *Staphylococcus aureus*. Chem Pharm Bull 1999; 47(8): 1121-7.
[http://dx.doi.org/10.1248/cpb.47.1121] [PMID: 10478467]

[44] Chen CL, Chen QH. [Biochemical study of Chinese rhubarb. XIX. Localization of inhibition of anthraquinone derivatives on the mitochondrial respiratory chain]. Yao Xue Xue Bao 1987; 22(1): 12-8.
[PMID: 3037851]

[45] Agarwal SK, Singh SS, Verma S, Kumar S. Antifungal activity of anthraquinone derivatives from *Rheum emodi*. J Ethnopharmacol 2000; 72(1-2): 43-6.
[http://dx.doi.org/10.1016/S0378-8741(00)00195-1] [PMID: 10967452]

[46] Suresh Babu K, Srinivas PV, Praveen B, Kishore KS, Murty US, Rao JM. Antimicrobial constituents from the rhizomes of *Rheum emodi*. Phytochemistry 2003; 62(2): 203-7.
[http://dx.doi.org/10.1016/S0031-9422(02)00571-X] [PMID: 12482457]

[47] Lu C, Wang H, Lv W, *et al*. Antibacterial properties of anthraquinones extracted from rhubarb against *Aeromonas hydrophila*. Fish Sci 2011; 77(3): 375-84.
[http://dx.doi.org/10.1007/s12562-011-0341-z]

[48] Hussain H, Al-Harrasi A, Al-Rawahi A, *et al*. A fruitful decade from 2005 to 2014 for anthraquinone patents. pert Opin Ther Pat 2015; 25(9): 1053-64.
[http://dx.doi.org/10.1517/13543776.2015.1050793]

[49] Sydiskis RJ, Owen DG, Lohr JL, Rosler KH, Blomster RN. Inactivation of enveloped viruses by anthraquinones extracted from plants. Antimicrob Agents Chemother 1991; 35(12): 2463-6.
[http://dx.doi.org/10.1128/AAC.35.12.2463] [PMID: 1810179]

[50] May G, Willuhn G. Antiviral effect of aqueous plant extracts in tissue culture. Arzneimittelforschung

1978; 28(1): 1-7.
[PMID: 204315]

[51] Kurokawa M, Ochiai H, Nagasaka K, *et al.* Antiviral traditional medicines against herpes simplex virus (HSV-1), poliovirus, and measles virus *in vitro* and their therapeutic efficacies for HSV-1 infection in mice. Antiviral Res 1993; 22(2-3): 175-88.
[http://dx.doi.org/10.1016/0166-3542(93)90094-Y] [PMID: 8279811]

[52] Wang Z, Wang G, Xu H, Wang P. Anti-herpes virus action of ethanol-extract from the root and rhizome of Rheum officinale Baill. Zhongguo Zhongyao Zazhi 1996; 21(6): 364-366, 384.
[PMID: 9388926]

[53] Miraj S. Therapeutic effects of *Rheum palmatum* L.(Dahuang): A systematic review. Pharma Chem 2016; 8: 50-4.

[54] Zhang T, Wu Z, Du J, *et al.* Anti-Japanese-encephalitis-viral effects of kaempferol and daidzin and their RNA-binding characteristics. PLoS One 2012; 7(1): e30259.
[http://dx.doi.org/10.1371/journal.pone.0030259] [PMID: 22276167]

[55] Kalisz S, Oszmiański J, Kolniak-Ostek J, Grobelna A, Kieliszek M, Cendrowski A. Effect of a variety of polyphenols compounds and antioxidant properties of rhubarb (Rheum rhabarbarum). Lebensm Wiss Technol 2020; 118: 108775.
[http://dx.doi.org/10.1016/j.lwt.2019.108775]

[56] Arvindekar AU, Laddha KS. An efficient microwave-assisted extraction of anthraquinones from *Rheum emodi*: Optimisation using RSM, UV and HPLC analysis and antioxidant studies. Ind Crops Prod 2016; 83: 587-95.
[http://dx.doi.org/10.1016/j.indcrop.2015.12.066]

[57] Mishra SK, Tiwari S, Shrivastava A, *et al.* Antidyslipidemic effect and antioxidant activity of anthraquinone derivatives from *Rheum emodi* rhizomes in dyslipidemic rats. J Nat Med 2014; 68(2): 363-71.
[http://dx.doi.org/10.1007/s11418-013-0810-z] [PMID: 24343839]

[58] Yen G, Duh P-D, Chuang D-Y. Antioxidant activity of anthraquinones and anthrone. Food Chem 2000; 70(4): 437-41.
[http://dx.doi.org/10.1016/S0308-8146(00)00108-4]

[59] Rajkumar V, Guha G, Ashok Kumar R. Antioxidant and anti-cancer potentials of *Rheum emodi* rhizome extracts. Evid Based Complement Alternat Med 2011; 2011: 1-9.
[http://dx.doi.org/10.1093/ecam/neq048] [PMID: 21792364]

[60] Rajkumar V, Guha G, Kumar RA. Apoptosis induction in MDA-MB-435S, Hep3B and PC-3 cell lines by *Rheum emodi* rhizome extracts. Asian Pac J Cancer Prev 2011; 12(5): 1197-200.
[PMID: 21875266]

[61] Lu CC, Yang JS, Huang AC, *et al.* Chrysophanol induces necrosis through the production of ROS and alteration of ATP levels in J5 human liver cancer cells. Mol Nutr Food Res 2010; 54(7): 967-76.
[http://dx.doi.org/10.1002/mnfr.200900265] [PMID: 20169580]

[62] Cha TL, Qiu L, Chen CT, Wen Y, Hung MC. Emodin down-regulates androgen receptor and inhibits prostate cancer cell growth. Cancer Res 2005; 65(6): 2287-95.
[http://dx.doi.org/10.1158/0008-5472.CAN-04-3250] [PMID: 15781642]

[63] Lee HZ, Hsu SL, Liu MC, Wu CH. Effects and mechanisms of aloe-emodin on cell death in human lung squamous cell carcinoma. Eur J Pharmacol 2001; 431(3): 287-95.
[http://dx.doi.org/10.1016/S0014-2999(01)01467-4] [PMID: 11730720]

[64] Arvindekar A, More T, Payghan PV, Laddha K, Ghoshal N, Arvindekar A. Evaluation of anti-diabetic and alpha glucosidase inhibitory action of anthraquinones from *Rheum emodi*. Food Funct 2015; 6(8): 2693-700.
[http://dx.doi.org/10.1039/C5FO00519A] [PMID: 26145710]

[65] Hosseini A, Mollazadeh H, Amiri MS, Sadeghnia HR, Ghorbani A. Effects of a standardized extract of *Rheum turkestanicum* Janischew root on diabetic changes in the kidney, liver and heart of streptozotocin-induced diabetic rats. Biomed Pharmacother 2017; 86: 605-11.
[http://dx.doi.org/10.1016/j.biopha.2016.12.059] [PMID: 28027536]

[66] Yang J, Li L. Effects of Rheum on renal hypertrophy and hyperfiltration of experimental diabetes in rat. Zhongguo Zhong Xi Yi Jie He Za Zhi 1993; 13(5): 286-8.
[PMID: 8219681]

[67] Cheon M-S, Yoon T-S, Choi G-Y, Kim S-J, Lee A. Comparative study of extracts from rhubarb on inflammatory activity in RAW 264.7 cells. Hanguk Yakyong Changmul Hakhoe Chi 2009; 17: 109-14.

[68] Moon MK, Kang DG, Lee JK, Kim JS, Lee HS. Vasodilatory and anti-inflammatory effects of the aqueous extract of rhubarb *via* a NO-cGMP pathway. Life Sci 2006; 78(14): 1550-7.
[http://dx.doi.org/10.1016/j.lfs.2005.07.028] [PMID: 16269157]

[69] Martínez-Salgado C, López-Hernández FJ, López-Novoa JM. Glomerular nephrotoxicity of aminoglycosides. Toxicol Appl Pharmacol 2007; 223(1): 86-98.
[http://dx.doi.org/10.1016/j.taap.2007.05.004] [PMID: 17602717]

[70] Alam MMA, Javed K, Jafri MA. Effect of *Rheum emodi* (Revand Hindi) on renal functions in rats. J Ethnopharmacol 2005; 96(1-2): 121-5.
[http://dx.doi.org/10.1016/j.jep.2004.08.028] [PMID: 15588659]

[71] Ngai HHY, Sit WH, Wan JMF. The nephroprotective effects of the herbal medicine preparation, WH30+, on the chemical-induced acute and chronic renal failure in rats. Am J Chin Med 2005; 33(3): 491-500.
[http://dx.doi.org/10.1142/S0192415X05003089] [PMID: 16047565]

[72] Kim Y-W, Hyun C. Evaluation of therapeutic effect of the extract from Rhubarb (*Rheum officinalis*) in dogs with chronic renal failure. J Vet Clin 2012; 29: 435-40.

[73] Ibrahim M, Khaja MN, Aara A, *et al*. Hepatoprotective activity of *Sapindus mukorossi* and *Rheum emodi* extracts: *in vitro* and *in vivo* studies. World J Gastroenterol 2008; 14(16): 2566-71.
[http://dx.doi.org/10.3748/wjg.14.2566] [PMID: 18442207]

[74] Ali A, Akhtar MS, Habib A, Bashir S. Isolation, identification, and *in vivo* evaluation of flavonoid fractions of chloroform/methanol extracts of *Rheum emodi* roots for their hepatoprotective activity in Wistar rats. Int J Nutr Pharmacol Neurol Dis 2016; 6(1): 28.
[http://dx.doi.org/10.4103/2231-0738.173784]

[75] Neyrinck AM, Etxeberria U, Jouret A, Delzenne NM. Supplementation with crude rhubarb extract lessens liver inflammation and hepatic lipid accumulation in a model of acute alcohol-induced steatohepatitis. Arch Public Health 2014; 72(S1): P6.
[http://dx.doi.org/10.1186/2049-3258-72-S1-P6]

[76] Neyrinck AM, Etxeberria U, Taminiau B, *et al*. Rhubarb extract prevents hepatic inflammation induced by acute alcohol intake, an effect related to the modulation of the gut microbiota. Mol Nutr Food Res 2017; 61(1): 1500899.
[http://dx.doi.org/10.1002/mnfr.201500899] [PMID: 26990039]

[77] Ding Y, Zhao L, Mei H, *et al*. Exploration of Emodin to treat alpha-naphthylisothiocyanate-induced cholestatic hepatitis *via* anti-inflammatory pathway. Eur J Pharmacol 2008; 590(1-3): 377-86.
[http://dx.doi.org/10.1016/j.ejphar.2008.06.044] [PMID: 18590720]

[78] Kosikowska U, Smolarz H, Malm A. Antimicrobial activity and total content of polyphenols of *Rheum* L. species growing in Poland. Open Life Sci 2010; 5(6): 814-20.
[http://dx.doi.org/10.2478/s11535-010-0067-4]

[79] Krafczyk N, Kötke M, Lehnert N, Glomb MAJEFR. Technology, Phenolic composition of rhubarb. Resour Technol 2008; 228: 187.

[80] Pandith SA, Dhar N, Rana S, *et al.* Functional promiscuity of two divergent paralogs of type III plant polyketide synthases. Plant Physiol 2016; 171(4): 2599-619.
[http://dx.doi.org/10.1104/pp.16.00003] [PMID: 27268960]

[81] Rokaya MB, Maršík P, Münzbergová Z. Active constituents in *Rheum acuminatum* and *Rheum australe* (Polygonaceae) roots: A variation between cultivated and naturally growing plants. Biochem Syst Ecol 2012; 41: 83-90.
[http://dx.doi.org/10.1016/j.bse.2011.11.004]

[82] Abrol E, Vyas D, Koul S. Metabolic shift from secondary metabolite production to induction of anti-oxidative enzymes during NaCl stress in *Swertia chirata* Buch.-Ham. Acta Physiol Plant 2012; 34(2): 541-6.
[http://dx.doi.org/10.1007/s11738-011-0851-4]

[83] Agarwal SK, Singh SS, Lakshmi V, Verma S, Kumar S. Chemistry and pharmacology of rhubarb (*Rheum* species)—a review. J Sci Ind Res (India) 2001; 60: 1-9.

[84] Pandith SA, Dhar N, Rana S, *et al.* Characterization and functional promiscuity of two divergent paralogs of Type III plant polyketide synthases from Rheum emodi Wall ex. Meissn. Plant Physiol 2016; 171(4): 2599-619.
[http://dx.doi.org/10.1104/pp.16.00003] [PMID: 27268960]

[85] Stafford HA. Flavonoid metabolism. Boca Raton, Florida: CRC press 1990; pp. 1-264.

[86] Pandey A, Misra P, Choudhary D, *et al.* AtMYB12 expression in tomato leads to large scale differential modulation in transcriptome and flavonoid content in leaf and fruit tissues. Sci Rep 2015; 5(1): 12412.
[http://dx.doi.org/10.1038/srep12412] [PMID: 26206248]

[87] Maaliki D, Shaito AA, Pintus G, El-Yazbi A, Eid AH. Flavonoids in hypertension: a brief review of the underlying mechanisms. Curr Opin Pharmacol 2019; 45: 57-65.
[http://dx.doi.org/10.1016/j.coph.2019.04.014] [PMID: 31102958]

[88] Jucá MM, Cysne Filho FMS, De Almeida JC, *et al.* Flavonoids: Biological activities and therapeutic potential. Nat Prod Res 2020; 34(5): 692-705.
[http://dx.doi.org/10.1080/14786419.2018.1493588] [PMID: 30445839]

[89] Rahaman ST, Mondal S. Flavonoids: A vital resource in healthcare and medicine. Pharm Pharmacol Int J 2020; 8(2): 91-104.
[http://dx.doi.org/10.15406/ppij.2020.08.00285]

[90] Ye M, Han J, Chen H, Zheng J, Guo D. Analysis of phenolic compounds in rhubarbs using liquid chromatography coupled with electrospray ionization mass spectrometry. J Am Soc Mass Spectrom 2007; 18(1): 82-91.
[http://dx.doi.org/10.1016/j.jasms.2006.08.009] [PMID: 17029978]

[91] Greco G, Turrini E, Catanzaro E, Fimognari C. Marine Anthraquinones: Pharmacological and Toxicological Issues. Mar Drugs 2021; 19(5): 272.
[http://dx.doi.org/10.3390/md19050272] [PMID: 34068184]

[92] Singh R, Chauhan SJC. 10-Anthraquinones and other biologically active compounds from the genus Rubia. Chemistry & Biodiversity 1: 1241-64.2004;

[93] Pandith SA, Dar RA, Lattoo SK, Shah MA, Reshi ZA. *Rheum australe*, an endangered high-value medicinal herb of North Western Himalayas: a review of its botany, ethnomedical uses, phytochemistry and pharmacology. Phytochem Rev 2018; 17(3): 573-609.
[http://dx.doi.org/10.1007/s11101-018-9551-7] [PMID: 32214920]

[94] Chien SC, Wu YC, Chen ZW, Yang WC. Naturally occurring anthraquinones: chemistry and therapeutic potential in autoimmune diabetes. Evid Based Complement Alternat Med 2015; 2015: 1-13.
[http://dx.doi.org/10.1155/2015/357357] [PMID: 25866536]

[95]   Fouillaud M, Venkatachalam M, Girard-Valenciennes E, Caro Y, Dufossé L. Anthraquinones and derivatives from marine-derived fungi: Structural diversity and selected biological activities. Mar Drugs 2016; 14(4): 64.
[http://dx.doi.org/10.3390/md14040064] [PMID: 27023571]

[96]   Diaz-Munoz G, Miranda IL, Sartori SK, De Rezende DC, Diaz MA. Anthraquinones: an overview. Studies in natural products chemistry. Amsterdam, The Netherlands: Elsevier 2018; pp. 313-38.

[97]   Ye M, Han J, Chen H, Zheng J, Guo D. Analysis of phenolic compounds in rhubarbs using liquid chromatography coupled with electrospray ionization mass spectrometry. J Am Soc Mass Spectrom 2007; 18(1): 82-91.
[http://dx.doi.org/10.1016/j.jasms.2006.08.009] [PMID: 17029978]

[98]   Malik EM, Müller CE. Anthraquinones as pharmacological tools and drugs. Med Res Rev 2016; 36(4): 705-48.
[http://dx.doi.org/10.1002/med.21391] [PMID: 27111664]

[99]   Wei WT, Lin SZ, Liu DL, Wang ZH. The distinct mechanisms of the antitumor activity of emodin in different types of cancer (Review). Oncol Rep 2013; 30(6): 2555-62.
[http://dx.doi.org/10.3892/or.2013.2741] [PMID: 24065213]

[100]   Dusemund B, Gilbert J, Gott D, *et al.* Food additives and nutrient sources added to food: developments since the creation of EFSA. EFSA J 2012; 10(10): s1006.
[http://dx.doi.org/10.2903/j.efsa.2012.s1006]

[101]   Chen C, Gao J, Wang TS, *et al.* NMR-based metabolomic techniques identify the toxicity of emodin in HepG2 cells. Sci Rep 2018; 8(1): 9379.
[http://dx.doi.org/10.1038/s41598-018-27359-4] [PMID: 29925852]

[102]   Morales M, Ros Barcelo A, Pedreno M. Plant stilbenes: recent advances in their chemistry and biology. Advances in Plant Physiology 2000; 3: 39-70.

[103]   Roupe K, Remsberg C, Yáñez J, Davies N. Pharmacometrics of stilbenes: seguing towards the clinic. Curr Clin Pharmacol 2006; 1(1): 81-101.
[http://dx.doi.org/10.2174/157488406775268246] [PMID: 18666380]

[104]   Baur JA, Sinclair DA. Therapeutic potential of resveratrol: the *in vivo* evidence. Nat Rev Drug Discov 2006; 5(6): 493-506.
[http://dx.doi.org/10.1038/nrd2060] [PMID: 16732220]

[105]   Dávid CZ, Hohmann J, Vasas A. Chemistry and pharmacology of cyperaceae stilbenoids: A review. Molecules 2021; 26(9): 2794.
[http://dx.doi.org/10.3390/molecules26092794] [PMID: 34068509]

[106]   Piotrowska H, Kucinska M, Murias M. Biological activity of piceatannol: Leaving the shadow of resveratrol. Mutat Res Rev Mutat Res 2012; 750(1): 60-82.
[http://dx.doi.org/10.1016/j.mrrev.2011.11.001] [PMID: 22108298]

[107]   Kala CP. Indigenous uses, population density, and conservation of threatened medicinal plants in protected areas of the Indian Himalayas. Conserv Biol 2005; 19(2): 368-78.
[http://dx.doi.org/10.1111/j.1523-1739.2005.00602.x]

[108]   Kabir Dar A, Siddiqui M, Wahid-ul H, Lone A, Manzoor N. Threat Status of *Rheum emodi*-A Study in Selected Cis-Himalayan Regions of Kashmir Valley Jammu & Kashmir India. Med Aromat Plants 2015; 4: 183.

[109]   Tali BA, Ganie AH, Nawchoo IA, Wani AA, Reshi ZA. Assessment of threat status of selected endemic medicinal plants using IUCN regional guidelines: A case study from Kashmir Himalaya. J Nat Conserv 2015; 23: 80-9.
[http://dx.doi.org/10.1016/j.jnc.2014.06.004]

[110]   Panis B, Nagel M, Van den houwe I. Challenges and prospects for the conservation of crop genetic

resources in field genebanks, in *in vitro* collections and/or in liquid nitrogen. Plants 2020; 9(12): 1634.
[http://dx.doi.org/10.3390/plants9121634] [PMID: 33255385]

[111] Roggemans J, Claes MC. Rapid clonal propagation of rhubarb by *in vitro* culture of shoot-tips. Sci Hortic (Amsterdam) 1979; 11(3): 241-6.
[http://dx.doi.org/10.1016/0304-4238(79)90005-0]

[112] Bairu MW, Aremu AO, Van Staden J. Somaclonal variation in plants: causes and detection methods. Plant Growth Regul 2011; 63(2): 147-73.
[http://dx.doi.org/10.1007/s10725-010-9554-x]

[113] Miguel C, Marum L. An epigenetic view of plant cells cultured *in vitro*: somaclonal variation and beyond. J Exp Bot 2011; 62(11): 3713-25.
[http://dx.doi.org/10.1093/jxb/err155] [PMID: 21617249]

[114] Schmidt EDL, Guzzo F, Toonen MAJ, Vries SC. A leucine-rich repeat containing receptor-like kinase marks somatic plant cells competent to form embryos. Development 1997; 124(10): 2049-62.
[http://dx.doi.org/10.1242/dev.124.10.2049] [PMID: 9169851]

[115] Komamine A, Murata N, Nomura K. 2004 SIVB Congress Symposium Proceeding: Mechanisms of somatic embryogenesis in carrot suspension cultures—Morphology, physiology, biochemistry, and molecular biology. in vitro Cell Dev Biol Plant 2005; 41(1): 6-10.
[http://dx.doi.org/10.1079/IVP2004593]

[116] Dodeman VL, Ducreux G, Kreis M. Zygotic embryogenesis versus somatic embryogenesis. J Exp Bot 1997; 48: 1493-509.

[117] Steward FC, Mapes MO, Mears K. Growth and organized development of cultured cells. II. Organization in cultures grown from freely suspended cell. Am J Bot 1958; 45(10): 705-8.
[http://dx.doi.org/10.1002/j.1537-2197.1958.tb10599.x]

[118] Reinert J. Über die Kontrolle der Morphogenese und die Induktion von Adventivembryonen an Gewebekulturen aus Karotten. Planta 1959; 53(4): 318-33.
[http://dx.doi.org/10.1007/BF01881795]

[119] Quiroz-Figueroa FR, Rojas-Herrera R, Galaz-Avalos RM, Loyola-Vargas VM. Embryo production through somatic embryogenesis can be used to study cell differentiation in plants. Plant Cell Tissue Organ Cult 2006; 86(3): 285-301.
[http://dx.doi.org/10.1007/s11240-006-9139-6]

[120] Mathieu M, Lelu-Walter MA, Blervacq AS, David H, Hawkins S, Neutelings G. Germin-like genes are expressed during somatic embryogenesis and early development of conifers. Plant Mol Biol 2006; 61(4-5): 615-27.
[http://dx.doi.org/10.1007/s11103-006-0036-5] [PMID: 16897479]

[121] Yeung EC. Structural and developmental patterns in somatic embryogenesis.in vitro embryogenesis in plants current plant science and biotechnology in agriculture. Dordrecht: Springer 1995; Vol. 20: pp. 205-47.
[http://dx.doi.org/10.1007/978-94-011-0485-2_6]

[122] Lakshmanan P, Taji A. Somatic embryogenesis in leguminous plants. Plant Biol 2000; 2(2): 136-48.
[http://dx.doi.org/10.1055/s-2000-9159]

[123] Jin S, Zhang X, liang S, Nie Y, Guo X, Huang C. Factors affecting transformation efficiency of embryogenic callus of Upland cotton (*Gossypium hirsutum*) with *Agrobacterium tumefaciens*. Plant Cell Tissue Organ Cult 2005; 81(2): 229-37.
[http://dx.doi.org/10.1007/s11240-004-5209-9]

[124] Li ZT, Dhekney S, Dutt M, *et al.* Optimizing Agrobacterium-mediated transformation of grapevine. in vitro Cell Dev Biol Plant 2006; 42(3): 220-7.
[http://dx.doi.org/10.1079/IVP2006770]

[125] Nomura K, Komamine A. Identification and isolation of single cells that produce somatic embryos at a

high frequency in a carrot suspension culture. Plant Physiol 1985; 79(4): 988-91.
[http://dx.doi.org/10.1104/pp.79.4.988] [PMID: 16664558]

[126] Quiroz-Figueroa F, Fuentes-Cerda C, Rojas-Herrera R, Loyola-Vargas V. Histological studies on the developmental stages and differentiation of two different somatic embryogenesis systems of *Coffea arabica*. Plant Cell Rep 2002; 20(12): 1141-9.
[http://dx.doi.org/10.1007/s00299-002-0464-x]

[127] Dudits D, Bögre L, János G. Molecular and cellular approaches to the analysis of plant embryo development from somatic cells *in vitro*. J Cell Sci 1975; 99(3): 473-82.
[http://dx.doi.org/10.1242/jcs.99.3.473]

[128] De Jong AJ, Schmidt EDL, De Vries SC. Early events in higher-plant embryogenesis. Plant Mol Biol 1993; 22(2): 367-77.
[http://dx.doi.org/10.1007/BF00014943] [PMID: 8507837]

[129] Fehér A, Pasternak TP, Dudits D. Transition of somatic plant cells to an embryogenic state. Plant Cell Tissue Organ Cult 2003; 74(3): 201-28.
[http://dx.doi.org/10.1023/A:1024033216561]

[130] Sagare AP, Lee YL, Lin TC, Chen CC, Tsay HS. Cytokinin-induced somatic embryogenesis and plant regeneration in *Corydalis yanhusuo* (Fumariaceae) — a medicinal plant. Plant Sci 2000; 160(1): 139-47.
[http://dx.doi.org/10.1016/S0168-9452(00)00377-0] [PMID: 11164586]

[131] Senger S, Mock HP, Conrad U, Manteuffel R. Immunomodulation of ABA function affects early events in somatic embryo development. Plant Cell Rep 2001; 20(2): 112-20.
[http://dx.doi.org/10.1007/s002990000290] [PMID: 30759896]

[132] Trigiano R, May R, Conger B. Reduced nitrogen influences somatic embryo quality and plant regeneration from suspension cultures of orchardgrass. *in vitro*–. Plant 1992; 28: 187-91.

[133] Sharp WR, Sondahl MR, Caldas LS, Maraffa SB. The physiology of *in vitro* asexual embryogenesis. Hortic Rev 2011; 2: 268-310. [Vegetative propagation].
[http://dx.doi.org/10.1002/9781118060759.ch6]

[134] Williams EG, Maheswaran G. Somatic embryogenesis: factors influencing coordinated behaviour of cells as an embryogenic group. Ann Bot 1986; 57(4): 443-62.
[http://dx.doi.org/10.1093/oxfordjournals.aob.a087127]

[135] Germanà M. Somatic embryogenesis and plant regeneration from anther culture of *Citrus aurantium* and *C. reticulata*. Biologia 2003; 58: 843-50.

[136] Malik MR, Wang F, Dirpaul JM, *et al*. Transcript profiling and identification of molecular markers for early microspore embryogenesis in *Brassica napus*. Plant Physiol 2007; 144(1): 134-54.
[http://dx.doi.org/10.1104/pp.106.092932] [PMID: 17384168]

[137] Aydin Y, Ipekci Z, Talas-Oğraş T, Zehir A, Bajrovic K, Gozukirmizi N. High frequency somatic embryogenesis in cotton. Biol Plant 2004; 48(4): 491-5.
[http://dx.doi.org/10.1023/B:BIOP.0000047142.07987.e1]

[138] Rout GR, Mohapatra A, Jain SM. Tissue culture of ornamental pot plant: A critical review on present scenario and future prospects. Biotechnol Adv 2006; 24(6): 531-60.
[http://dx.doi.org/10.1016/j.biotechadv.2006.05.001] [PMID: 16814509]

[139] Thorpe TA. History of plant tissue culture. Mol Biotechnol 2007; 37(2): 169-80.
[http://dx.doi.org/10.1007/s12033-007-0031-3] [PMID: 17914178]

[140] Seguí-Simarro JM, Nuez F. Embryogenesis induction, callogenesis, and plant regeneration by *in vitro* culture of tomato isolated microspores and whole anthers. J Exp Bot 2007; 58(5): 1119-32.
[http://dx.doi.org/10.1093/jxb/erl271] [PMID: 17237159]

[141] Yang X, Zhang X, Jin S, Fu L, Wang L. Production and characterization of asymmetric hybrids

between upland cotton Coker 201 (Gossypium hirsutum) and wild cotton (G. klozschianum Anderss). Plant Cell Tissue Organ Cult 2007; 89(2-3): 225-35.
[http://dx.doi.org/10.1007/s11240-007-9245-0]

[142] Jiménez VM, Bangerth F. Endogenous hormone concentrations and embryogenic callus development in wheat. Plant Cell Tissue Organ Cult 2001; 67(1): 37-46.
[http://dx.doi.org/10.1023/A:1011671310451]

[143] Haccius B. Question of unicellular origin of non-zygotic embryos in callus cultures. Phytomorphology 1978; 28: 74-81.

[144] Raghavan V. Experimental embryogenesis in vascular plants. Cambridge: Cambridge University Press 1976; p. 611.

[145] Maithani U. *In-vitro* Propagation Studies of *Rheum moorcroftianum* Royle: A Threatened Medicinal plant from Garhwal Himalaya. Int J Curr Microbiol Appl Sci 2015; 4: 596-9.

[146] Ji-yong J. Tissue Culture of Rhubarb. Yinshan Academic Journal 2010; 2: 55-68.

[147] Lepse L. Comparison of *in vitro* and traditional propagation methods of Rhubarb (*Rheum rhabarbarum*) according to morphological features and yield. In: International symposium on acclimatization and establishment of micropropagated plants. Acta Hortic 2007; 265-70.

[148] Sepehr MF, Ghorbanli Z. Formation of catechin in callus cultures and micropropagation of *Rheum ribes* L. Pak J Biol Sci 2005; 8: 1346-50.
[http://dx.doi.org/10.3923/pjbs.2005.1346.1350]

[149] Mun SC, Mun GS. Development of an efficient callus proliferation system for *Rheum coreanum* Nakai, a rare medicinal plant growing in Democratic People's Republic of Korea. Saudi J Biol Sci 2016; 23(4): 488-94.
[http://dx.doi.org/10.1016/j.sjbs.2015.05.017] [PMID: 27298581]

[150] Xu W, Chen G, Li Y, Wang L. Studies on tissue culture technique of *Rheum tanguticum*. Xibei Zhiwu Xuebao 2004; 24: 1734-8.

[151] Lal N, Ahuja PS. Propagation of Indian Rhubarh (*Rheum emodi* Wall.) using shoot-tip and leaf explant culture. Plant Cell Rep 1989; 8(8): 493-6.
[http://dx.doi.org/10.1007/BF00269057] [PMID: 24233537]

[152] Rashid S, Kaloo ZA, Singh S, Bashir I. Callus induction and shoot regeneration from rhizome explants of *Rheum webbianum* Royle- a threatened medicinal plant growing in Kashmir Himalaya. Journal of Scientific and Innovative Research 2014; 3(5): 515-8.
[http://dx.doi.org/10.31254/jsir.2014.3508]

[153] Arya S, Kalia RK, Arya ID. Induction of somatic embryogenesis in *Pinus roxburghii* Sarg. Plant Cell Rep 2000; 19(8): 775-80.
[http://dx.doi.org/10.1007/s002990000197] [PMID: 30754868]

[154] Yantcheva A, Vlahova M, Atanassov A. Direct somatic embryogenesis and plant regeneration of carnation (Dianthus caryophyllus L.). Plant Cell Rep 1998; 18(1-2): 148-53.
[http://dx.doi.org/10.1007/s002990050548]

[155] Hamidah M, Karim AGA, Debergh P. Somatic embryogenesis and plant regeneration in *Anthurium scherzerianum*. Plant Cell Tissue Organ Cult 1997; 48(3): 189-93.
[http://dx.doi.org/10.1023/A:1005834131478]

[156] Hussein S, Ibrahim R, Pick Kiong AL. Adventitious shoots regeneration from root and stem explants of *Eurycoma longifolia* Jack-an important tropical medicinal plants. Int J Agric Res 2010; 5(7): 543-53.
[http://dx.doi.org/10.3923/ijar.2010.543.553]

[157] Horstman A, Bemer M, Boutilier K. A transcriptional view on somatic embryogenesis. Regeneration 2017; 4(4): 201-16.

[http://dx.doi.org/10.1002/reg2.91] [PMID: 29299323]

[158] Moscatiello R, Baldan B, Navazio L. Plant cell suspension cultures. plant mineral nutrients. methods in molecular biology. Totowa, NJ: Springer, Humana Press 2013; Vol. 953: pp. 77-93.
[http://dx.doi.org/10.1007/978-1-62703-152-3_5]

[159] Murashige T, Skoog F. A revised medium for rapid growth and bio assays with tobacco tissue cultures. Physiol Plant 1962; 15(3): 473-97.
[http://dx.doi.org/10.1111/j.1399-3054.1962.tb08052.x]

[160] Gamborg OL, Miller RA, Ojima K. Nutrient requirements of suspension cultures of soybean root cells. Exp Cell Res 1968; 50(1): 151-8.
[http://dx.doi.org/10.1016/0014-4827(68)90403-5] [PMID: 5650857]

[161] Zenk MH. 6. Chasing the enzymes of secondary metabolism: Plant cell cultures as a pot of gold. Phytochemistry 1991; 30(12): 3861-3.
[http://dx.doi.org/10.1016/0031-9422(91)83424-J]

[162] Buitelaar RM, Tramper J. Strategies to improve the production of secondary metabolites with plant cell cultures: a literature review. J Biotechnol 1992; 23(2): 111-41.
[http://dx.doi.org/10.1016/0168-1656(92)90087-P]

[163] Kreis W. 1993; Arzneistoffe aus pflanzlichen zell-und gewebekulturen. Deutsche apotheker zeitung-stuttgart 133: 17-7.

[164] Shuler M. Strategies for improving productivity in plant cell, tissue, and organ culture in bioreactors Bioproducts and Bioprocesses 2. Springer 1993; pp. 235-45.

[165] Wen Su W. Bioprocessing technology for plant cell suspension cultures. Applied Biochemistry and Biotechnology 1995; 50189: 230.

[166] Scragg AH. The problems associated with high biomass levels in plant cell suspensions. Plant Cell Tissue Organ Cult 1995; 43(2): 163-70.
[http://dx.doi.org/10.1007/BF00052172]

[167] Kargi F, Rosenberg MZ. Plant cell bioreactors: present status and future trends. Biotechnol Prog 1987; 3(1): 1-8.
[http://dx.doi.org/10.1002/btpr.5420030102]

[168] Payne G, Shuler M, Brodelius P. 1987; Large-scale culture of plant cells Methods Mol Biol 6: 477-94.
[http://dx.doi.org/10.1385/0-89603-161-6:477]

[169] Panda AK, Mishra S, Bisaria VS, Bhojwani SS. Plant cell reactors—A perspective. Enzyme Microb Technol 1989; 11(7): 386-97.
[http://dx.doi.org/10.1016/0141-0229(89)90132-4]

[170] Doran PM. Design of reactors for plant cells and organs Bioprocess design and control. Berlin, Heidelberg: Springer 1993; pp. 115-68.
[http://dx.doi.org/10.1007/BFb0007198]

[171] Zimmerman JL. Somatic embryogenesis: a model for early development in higher plants. Plant Cell 1993; 5(10): 1411-23.
[http://dx.doi.org/10.2307/3869792] [PMID: 12271037]

[172] Michael A, De Klerk G-J. Plant propagation by tissue culture: The background. Berlin, Heidelberg: Springer 2008; pp. 62-153.

[173] Mustafa NR, De Winter W, Van Iren F, Verpoorte R. Initiation, growth and cryopreservation of plant cell suspension cultures. Nat Protoc 2011; 6(6): 715-42.
[http://dx.doi.org/10.1038/nprot.2010.144] [PMID: 21637194]

[174] Bhatia S, Sharma K, Dahiya R, Bera T. Modern applications of plant biotechnology in pharmaceutical sciences. New York, USA: Academic Press 2015; pp. 1-125.

[175] Strosse H, Domergue R, Panis B, Escalant JV, Côte F. Banana and plantain embryogenic cell suspensions: Bioversity International. The International Network for the improvement of Bnana and Plantain. France: Montpellier 2003; pp. 5-25.

[176] Soomro R, Memon RA. Establishment of callus and suspension culture in *Jatropha curcas*. Pak J Bot 2007; 39: 2431-41.

[177] Ascough GD, Fennell CW. The regulation of plant growth and development in liquid culture. S Afr J Bot 2004; 70(2): 181-90.
[http://dx.doi.org/10.1016/S0254-6299(15)30234-9]

[178] Liu C, Moon K, Honda H, Kobayashi T. *In situ* regeneration of rice (*Oryza sativa* L.) callus immobilized in polyurethane foam. J Biosci Bioeng 2001; 91(1): 76-80.
[http://dx.doi.org/10.1016/S1389-1723(01)80115-8] [PMID: 16232950]

[179] Biondi S. Requirements for a tissue culture facility.In proceedings/ plant tissue culture Methods and applications in agriculture. 1-20.New York USA: Academic Press. Campinas 1981; pp.

[180] Wang Y, Jeknić Z, Ernst RC, Chen THH. Efficient plant regeneration from suspension-cultured cells of tall bearded iris. HortScience 1999; 34(4): 730-5.
[http://dx.doi.org/10.21273/HORTSCI.34.4.730]

[181] Harris R. *In vitro* propagation of *Vitis*. Vitis 1982; 21: 22-32.

[182] Harris R, Mason EBB. Two machines for *in vitro* propagation of plants in liquid media. Can J Plant Sci 1983; 63(1): 311-6.
[http://dx.doi.org/10.4141/cjps83-032]

[183] Levin R, Gaba V, Tal B, Hirsch S, DeNola D. Automated plant tissue culture for mass propagation. Bio/Technology 1988; 6: 1035-40.

[184] Malik S, Sharma N, Sharma UK, *et al.* Qualitative and quantitative analysis of anthraquinone derivatives in rhizomes of tissue culture-raised *Rheum emodi* Wall. plants. J Plant Physiol 2010; 167(9): 749-56.
[http://dx.doi.org/10.1016/j.jplph.2009.12.007] [PMID: 20144491]

[185] Hussain MS, Fareed S, Ansari S, Rahman MA, Ahmad IZ, Saeed M. Current approaches toward production of secondary plant metabolites. J Pharm Bioallied Sci 2012; 4(1): 10-20.
[http://dx.doi.org/10.4103/0975-7406.92725] [PMID: 22368394]

[186] Karuppusamy S. A review on trends in production of secondary metabolites from higher plants by *in vitro* tissue, organ and cell cultures. J Med Plants Res 2009; 3: 1222-39.

[187] Varma V. Advancements in the production of secondary metabolites. Journal of Natural Products 2010; 3.

[188] Khan T, Khan MA, Karam K, Ullah N, Mashwani ZR, Nadhman A. Plant *in vitro* culture technologies; a promise into factories of secondary metabolites against COVID-19. Front Plant Sci 2021; 12: 610194.
[http://dx.doi.org/10.3389/fpls.2021.610194] [PMID: 33777062]

[189] Wink M. Why do lupin cell cultures fail to produce alkaloids in large quantities? Plant Cell Tissue Organ Cult 1987; 8(2): 103-11.
[http://dx.doi.org/10.1007/BF00043147]

[190] Vasil IK, Constabel F, Schell J. Cell culture and somatic cell genetics of plants. New York, USA: Academic Press 1984; pp. 14-98.

[191] Bajaj Y, Furmanowa M, Olszowska O. Biotechnology of the micropropagation of medicinal and aromatic plants. Medicinal and aromatic plants I. Berlin, Heidelberg: Springer 1988; pp. 60-103.
[http://dx.doi.org/10.1007/978-3-642-73026-9_3]

[192] Furuya T, Ayabe S, Noda K. Chrysophanol and emodin from callus tissue of rhubarb (Rheum

palmatum). Phytochemistry 1975; 14(5-6): 1457.
[http://dx.doi.org/10.1016/S0031-9422(00)98666-7]

[193] Rai P. The production of anthraquinones in callus cultures of *Rheum palmatum*. Lloydia 1978; 41: 114-6.

[194] Ohshima Y, Takahashi K, Shibata S. Tissue culture of rhubarb and isolation of sennosides from the callus. Planta Med 1988; 54(1): 20-4.
[http://dx.doi.org/10.1055/s-2006-962322] [PMID: 3375332]

[195] Kurosaki F, Nagase H, Nishi A. Stimulation of anthraquinone production in rhubarb tissue culture by an ethylene-generating reagent. Phytochemistry 1992; 31(8): 2735-8.
[http://dx.doi.org/10.1016/0031-9422(92)83621-5]

[196] Farzami Sepehr M, Ghorbanli M. Effects of nutritional factors on the formation of anthraquinones in callus cultures of *Rheum ribes*. Plant Cell Tissue Organ Cult 2002; 68(2): 171-5.
[http://dx.doi.org/10.1023/A:1013837232047]

[197] Charlwood BV, Rhodes MJ. Secondary products from plant tissue culture. Clarendon Broad Street: Clarendon Press 1990; pp. 1-41.

[198] Ramachandra Rao S, Ravishankar GA. Plant cell cultures: Chemical factories of secondary metabolites. Biotechnol Adv 2002; 20(2): 101-53.
[http://dx.doi.org/10.1016/S0734-9750(02)00007-1] [PMID: 14538059]

[199] Wink M, Alfermann AW, Franke R, *et al.* Sustainable bioproduction of phytochemicals by plant *in vitro* cultures: anticancer agents. Plant Genet Resour 2005; 3(2): 90-100.
[http://dx.doi.org/10.1079/PGR200575]

[200] DiCosmo F, Misawa M. Plant cell and tissue culture: Alternatives for metabolite production. Biotechnol Adv 1995; 13(3): 425-53.
[http://dx.doi.org/10.1016/0734-9750(95)02005-N] [PMID: 14536096]

[201] Hussain MS, Fareed S, Ansari S, Rahman MA, Ahmad IZ, Saeed M. Current approaches toward production of secondary plant metabolites. J Pharm Bioallied Sci 2012; 4(1): 10-20.
[http://dx.doi.org/10.4103/0975-7406.92725] [PMID: 22368394]

[202] Davioud E, Kan C, Hamon J, Tempé J, Husson HP. Production of indole alkaloids by *in vitro* root cultures from *Catharanthus trichophyllus*. Phytochemistry 1989; 28(10): 2675-80.
[http://dx.doi.org/10.1016/S0031-9422(00)98066-X]

[203] Smetanska I. Production of secondary metabolites using plant cell cultures. Adv Biochem Eng Biotechnol 2008; 111: 187-228.
[http://dx.doi.org/10.1007/10_2008_103] [PMID: 18594786]

[204] Zhao J, Davis LC, Verpoorte R. Elicitor signal transduction leading to production of plant secondary metabolites. Biotechnol Adv 2005; 23(4): 283-333.
[http://dx.doi.org/10.1016/j.biotechadv.2005.01.003] [PMID: 15848039]

[205] Namdeo A. Plant cell elicitation for production of secondary metabolites: a review. Pharmacogn Rev 2007; 1: 69-79.

[206] Romagnoli LG, Knorr D. Effects of ferulic acid treatment on growth and flavor development of cultured *vanilla planifolia* cells. Food Biotechnol 1988; 2(1): 93-104.
[http://dx.doi.org/10.1080/08905438809549678]

[207] Ling AP, Ong S, Sobri H. Strategies in enhancing secondary metabolites production in plant cell cultures. Med Aromat Plant Sci Biotechnol 2011; 5: 94-101.

[208] Maziah Mahmood, Mahmood M, Fadzillah N, Daud S. Effects of precursor supplementation on the production of triterpenes by *Centella asiatica* callus culture. Pak J Biol Sci 2005; 8(8): 1160-9.
[http://dx.doi.org/10.3923/pjbs.2005.1160.1169]

[209] Kim OT, Kim MY, Hong MH, Ahn JC, Hwang B. Stimulation of asiaticoside accumulation in the

whole plant cultures of *Centella asiatica* (L.) Urban by elicitors. Plant Cell Rep 2004; 23(5): 339-44.
[http://dx.doi.org/10.1007/s00299-004-0826-7] [PMID: 15316748]

[210] Karppinen K, Hokkanen J, Tolonen A, Mattila S, Hohtola A. Biosynthesis of hyperforin and adhyperforin from amino acid precursors in shoot cultures of *Hypericum perforatum*. Phytochemistry 2007; 68(7): 1038-45.
[http://dx.doi.org/10.1016/j.phytochem.2007.01.001] [PMID: 17307206]

[211] Jin JH, Shin JH, Kim JH, Chung IS, Lee HJ. Effect of chitosan elicitation and media components on the production of anthraquinone colorants in madder (*Rubia akane* Nakai) cell culture. Biotechnol Bioprocess Eng 1999; 4(4): 300-4.
[http://dx.doi.org/10.1007/BF02933757]

[212] Staniszewska I, Królicka A, Maliński E, Łojkowska E, Szafranek J. Elicitation of secondary metabolites in *in vitro* cultures of *Ammi majus* L. Enzyme Microb Technol 2003; 33(5): 565-8.
[http://dx.doi.org/10.1016/S0141-0229(03)00180-7]

[213] Kasparova M, Siatka T. Production of anthracene derivatives in elicited tissue cultures of Rheum palmatum L. Ceska a Slovenska farmacie: casopis Ceske farmaceuticke spolecnosti a Slovenske farmaceuticke spolecnosti 1999; 256-61.

[214] Kasparova M, Siatka T. Effect of salicylic acid on production of anthracene derivatives in a culture of Rheum palmatum L. *in vitro*. Ceska a Slovenska farmacie: casopis Ceske farmaceuticke spolecnosti a Slovenske farmaceuticke spolecnosti 2002; 51: 177-81.

[215] Farzami M, Ghorbant M. Formation of catechin in callus cultures and micropropagation of *Rheum ribes* L. Pak J Biol Sci 2005; 8: 1346-50.
[http://dx.doi.org/10.3923/pjbs.2005.1346.1350]

[216] Doran PM. Biotechnology of hairy root systems. Berlin, Heidelberg: Springer 2013; pp. 1-157.
[http://dx.doi.org/10.1007/978-3-642-39019-7]

[217] Chilton MD, Tepfer DA, Petit A, David C, Casse-Delbart F, Tempé J. *Agrobacterium rhizogenes* inserts T-DNA into the genomes of the host plant root cells. Nature 1982; 295(5848): 432-4.
[http://dx.doi.org/10.1038/295432a0]

[218] Gelvin SB. Agrobacterium in the genomics age. Plant Physiol 2009; 150(4): 1665-76.
[http://dx.doi.org/10.1104/pp.109.139873] [PMID: 19439569]

[219] Guillon S, Trémouillaux-Guiller J, Pati PK, Rideau M, Gantet P. Harnessing the potential of hairy roots: Dawn of a new era. Trends Biotechnol 2006; 24(9): 403-9.
[http://dx.doi.org/10.1016/j.tibtech.2006.07.002] [PMID: 16870285]

[220] De Meijer EPM, Bagatta M, Carboni A, *et al*. The inheritance of chemical phenotype in *Cannabis sativa* L. Genetics 2003; 163(1): 335-46.
[http://dx.doi.org/10.1093/genetics/163.1.335] [PMID: 12586720]

[221] Palazon J, Pinol M, Cusido R, Morales C, Bonfill M. Application of transformed root technology to the production of bioactive metabolites. Recent Research Developments -. Plant Physiol 1997; 1: 125-43.

[222] Sevón N, Oksman-Caldentey KM. *Agrobacterium rhizogenes*-mediated transformation: root cultures as a source of alkaloids. Planta Med 2002; 68(10): 859-68.
[http://dx.doi.org/10.1055/s-2002-34924] [PMID: 12391546]

[223] Hashimoto T, Yun DJ, Yamada Y. Production of tropane alkaloids in genetically engineered root cultures. Phytochemistry 1993; 32(3): 713-8.
[http://dx.doi.org/10.1016/S0031-9422(00)95159-8]

[224] Shanks J, Morgan J. Plant 'hairy root' culture. Curr Opin Biotechnol 1999; 10(2): 151-5.
[http://dx.doi.org/10.1016/S0958-1669(99)80026-3] [PMID: 10209145]

[225] Bakkali AT, Jaziri M, Foriers A, Vander Heyden Y, Vanhaelen M, Homès J. Lawsone accumulation in

normal and transformed cultures of henna, *Lawsonia inermis*. Plant Cell Tissue Organ Cult 1997; 51(2): 83-7.
[http://dx.doi.org/10.1023/A:1005926228301]

[226] Kittipongpatana N, Hock RS, Porter JR. Production of solasodine by hairy root, callus, and cell suspension cultures of *Solanum aviculare* Forst. Plant Cell Tissue Organ Cult 1998; 52(3): 133-43.
[http://dx.doi.org/10.1023/A:1005974611043]

[227] Thimmaraju R, Bhagyalakshmi N, Narayan MS, Ravishankar GA. Food-grade chemical and biological agents permeabilize red beet hairy roots, assisting the release of betalaines. Biotechnol Prog 2003; 19(4): 1274-82.
[http://dx.doi.org/10.1021/bp0201399] [PMID: 12892491]

[228] Thimmaraju R, Bhagyalakshmi N, Narayan MS, Ravishankar GA. Kinetics of pigment release from hairy root cultures of *Beta vulgaris* under the influence of pH, sonication, temperature and oxygen stress. Process Biochem 2003; 38(7): 1069-76.
[http://dx.doi.org/10.1016/S0032-9592(02)00234-0]

[229] Wongsamuth R, Doran PM. Production of monoclonal antibodies by tobacco hairy roots. Biotechnol Bioeng 1997; 54(5): 401-15.
[http://dx.doi.org/10.1002/(SICI)1097-0290(19970605)54:5<401::AID-BIT1>3.0.CO;2-I] [PMID: 18634133]

[230] Medina-Bolívar F, Cramer C. Production of recombinant proteins by hairy roots cultured in plastic sleeve bioreactors. Recombinant Gene Expression. Berlin, Heidelberg: Springer 2004; pp. 351-63.

[231] Liu C, Towler MJ, Medrano G, Cramer CL, Weathers PJ. Production of mouse interleukin-12 is greater in tobacco hairy roots grown in a mist reactor than in an airlift reactor. Biotechnol Bioeng 2009; 102(4): 1074-86.
[http://dx.doi.org/10.1002/bit.22154] [PMID: 18988263]

[232] Woods RR, Geyer BC, Mor TS. Hairy-root organ cultures for the production of human acetylcholinesterase. BMC Biotechnol 2008; 8(1): 95.
[http://dx.doi.org/10.1186/1472-6750-8-95] [PMID: 19105816]

[233] Gurusamy PD, Schäfer H, Ramamoorthy S, Wink M. Biologically active recombinant human erythropoietin expressed in hairy root cultures and regenerated plantlets of *Nicotiana tabacum* L. PLoS One 2017; 12(8): e0182367.
[http://dx.doi.org/10.1371/journal.pone.0182367] [PMID: 28800637]

[234] Pham NB, Schäfer H, Wink M. Production and secretion of recombinant thaumatin in tobacco hairy root cultures. Biotechnol J 2012; 7(4): 537-45.
[http://dx.doi.org/10.1002/biot.201100430] [PMID: 22125283]

[235] Luchakivskaia IuS, Olevinskaia ZM, Kishchenko EM, Spivak NIa, Kuchuk NV. Obtaining of hairy-root, callus and suspension carrot culture (*Daucus carota* L.) able to accumulate human interferon alpha-2b. Tsitol Genet 2012; 46(1): 18-26.
[PMID: 22420216]

[236] Cardon F, Pallisse R, Bardor M, *et al*. *Brassica rapa* hairy root based expression system leads to the production of highly homogenous and reproducible profiles of recombinant human alpha-L-iduronidase. Plant Biotechnol J 2019; 17(2): 505-16.
[http://dx.doi.org/10.1111/pbi.12994] [PMID: 30058762]

[237] Chang Z, Guo D, Shen X, Wang S, Zheng J. Anthraquinone production and analysis in the hairy root cultures of Rheum palmatum L. Yao Xue Xue Bao 1998; 33(11): 869-72.
[PMID: 12016951]

[238] Chang Z, Shen X, Da G, Lu K, Wang S. Anthroquinone production in hairy root cultures of *Rheum wittrochii* lundstr. J Beijing Med Univ 1998; 30(5): 413-5.

[239] Ron M, Kajala K, Pauluzzi G, *et al*. Hairy root transformation using *Agrobacterium rhizogenes* as a

tool for exploring cell type-specific gene expression and function using tomato as a model. Plant Physiol 2014; 166(2): 455-69.
[http://dx.doi.org/10.1104/pp.114.239392] [PMID: 24868032]

[240] Michno JM, Wang X, Liu J, Curtin SJ, Kono TJY, Stupar RM. CRISPR/Cas mutagenesis of soybean and *Medicago truncatula* using a new web-tool and a modified Cas9 enzyme. GM Crops Food 2015; 6(4): 243-52.
[http://dx.doi.org/10.1080/21645698.2015.1106063] [PMID: 26479970]

[241] Du H, Zeng X, Zhao M, *et al.* Efficient targeted mutagenesis in soybean by TALENs and CRISPR/Cas9. J Biotechnol 2016; 217: 90-7.
[http://dx.doi.org/10.1016/j.jbiotec.2015.11.005] [PMID: 26603121]

[242] Kirchner TW, Niehaus M, Debener T, Schenk MK, Herde M. Efficient generation of mutations mediated by CRISPR/Cas9 in the hairy root transformation system of *Brassica carinata*. PLoS One 2017; 12(9): e0185429.
[http://dx.doi.org/10.1371/journal.pone.0185429] [PMID: 28937992]

CHAPTER 9

In Vitro Plant Regeneration from Nodal Segments and Biochemical Fidelity Analysis of *Operculina Turpethum*, a Threatened Medicinal Plant of Odisha

Kumari Monalisa[1,2,#], Shashikanta Behera[1,#], Shasmita[1], Debasish Mohapatra[1], Anil K. Biswal[3] and Soumendra K. Naik[1,2,*]

[1] *Department of Botany, Ravenshaw University, Cuttack-753003, Odisha, India*

[2] *Center of Excellence in Environment and Public Health, Ravenshaw University, Cuttack-753 003, Odisha, India*

[3] *Department of Botany, Maharaja Sriram Chandra Bhanja Deo University, Takatpur, Baripada - 757001, Odisha, India*

Abstract: Trivrit [*Operculina turpethum* (L.) Silva Manso], belonging to the family Convolvulaceae, is a perennial, herbaceous and creeping vine. It is a medicinal plant which is widely used in traditional systems of Indian medicine. The roots, undamaged bark, stem and leaves possess immense medicinal properties and are used in the treatment of various ailments, including bronchitis, skin diseases, tuberculosis, cough, asthma, rheumatism, jaundice, ulcer, gastrointestinal disturbances, *etc*. The plant is enlisted as threatened species in different states of India, particularly in Odisha, due to indiscriminate destruction of forests, shrinkage of natural habitats, and unsustainable harvesting and collection for medicinal uses. Thus, there is an urgency for its protection and conservation. To scale up the production of *O. turpethum*, aiming at its conservation, micropropagation can be an alternative in order to circumvent the limitations of conventional propagation of the plant. Keeping this in view, an efficient protocol for plant regeneration of *O. turpethum* by axillary shoot proliferation from nodal segments was optimized. Multiple shoots were induced from mature nodal explants by axillary shoot proliferation on Murashige and Skoog's (1962) (MS) medium augmented with different types and concentrations of plant growth regulators. The highest number of shoots (13.3) proliferated on MS + 3.0 mg/L meta-Topolin. *In vitro* regenerated shoots were rooted on ½ MS medium containing 0.5 mg/L indole-3-butyric acid. *In vitro* regenerated plants with well-developed roots were successfully acclimatized in the small pots containing sterile garden soil and sand (1:1), followed by transfer to the large pot containing garden soil. Finally, plants were successfully

---

[*] **Corresponding author Soumendra K. Naik:** Department of Botany, Ravenshaw University, Cuttack-753003, Odisha, India; & Center of Excellence in Environment and Public Health, Ravenshaw University, Cuttack-753003, Odisha, India; E-mail sknuu@yahoo.com

[#] Kumari Monalisa and Shashikaanta Behera have equal contribution

Mohammad Anis & Mehrun Nisha Khanam (Eds.)
All rights reserved-© 2024 Bentham Science Publishers

established in the field. The biochemical fidelity, in terms of secondary metabolites, was checked for tissue culture raised-field established plant vis-à-vis mother plant.

Keywords: Axillary shoot proliferation, Biochemical fidelity, Medicinal plant, Nodal explants.

INTRODUCTION

Operculina turpethum (L.) Silva Manso (Trivrit), belonging to the family Convolvulaceae, is a large, perennial, and herbaceous climber that exudates milky juice [1, 2]. It is commonly known as transparent wood rose as well as Indian jalap in English. This plant is regarded as an important medicinal herb in the Ayurvedic system of medicine. Trivrit has two varieties known as *Aruna or Shweta* [*O. turpethum* (L.) Silva Manso (syn. *Ipomoea turpethum)*] with whitish coloured root and Shyama (*Ipomoea petaloideschois)* has blackish coloured root [3]. *O. turpethum* root, stem, and leaf are rich in medicinal properties and are used for the treatment of various ailments [4]. Due to the presence of resins containing glycosides like turpethinic acid A, B, C, D, E, and scopoletin [5, 6], the roots of *O. turpethum* with its undamaged bark show thermogenic, purgative, antidiabetic, antipyretic, anti-inflammatory, antioxidant, anti-cancer, antimicrobial, antiproliferative, and stimulant properties [7, 8, 9]. The oil extracts from the root bark are used for the treatment of skin disorders and diseases (particularly of scaly nature) and also play a vital role in the treatment of cough and asthma [10, 11]. In addition to this, the root is also used as the chief ingredient for the treatment of ulcer and related gastrointestinal disturbances [12, 13]. Stem extracts of *O. turpethum* show antioxidant, antibacterial, hepato-protective, and anti-clastogenic effects due to the presence of a chemical EH4 (N-p coumarytyraminse) [14, 15]. Besides these, the alcoholic extracts derived from the fresh fruits of *O. turpethum* are reported to have antibacterial activities [16].

O. turpethum has been enlisted as a threatened (endangered or vulnerable) medicinal plant in different states of India, including Odisha, due to overexploitation, habitat destruction, and unsustainable trading of its stem and roots [17, 18]. The conventional methods for multiplication of *O. turpethum* are done through vegetative propagation and seeds which are time-consuming and require some pre-treatment like mechanical scarification that possesses restrictions and limitations [19]. Therefore, conventional methods of propagation are inadequate to meet the demand for raw materials. Because of above-mentioned problems, there is a need for special techniques (*e.g.*, development of reliable *in vitro* plant regeneration protocol) for the conservation and protection of this threatened medicinal plant. Two earlier reports on plant regeneration of this plant using nodal segments and cotyledonary nodes as explants are not sufficient.

Therefore, the objective of this study was to develop an efficient, reliable *in vitro* plant regeneration protocol by axillary shoot proliferation of nodal explants and assessment of their biochemical fidelity in terms of phytochemical analysis.

MATERIALS AND METHODS

Collection and Surface Sterilization of the Explants

A healthy *O. turpethum* plant maintained at the Department of Botany, Ravenshaw University, Cuttack, Odisha, India, was used as the source of explant. Healthy, young, and tender shoots of the plant were collected from the vine. The nodal segments of 1.0-1.5 cm (excluding 3 nodes from the tip portion) were used for the plant regeneration experiment.

The nodal segments were washed under running tap water for about 25 mins, followed by treatment with 5% (v/v) aqueous solution of a liquid detergent, 'Teepol' for 20 mins (Reckitt Benckiser Ltd., HP, India). Then explants were rinsed with distilled water 4-5 times. Before inoculation, the nodal segments were surface sterilized under aseptic conditions inside a laminar airflow cabinet with 0.1% (w/v) aqueous solution of mercuric chloride ($HgCl_2$, Himedia, India) for 6 mins. After surface sterilization, nodal segments were rinsed thoroughly 5-6 times with sterile double distilled water.

Culture Medium and Culture Conditions

The surface sterilized mature nodal explants were inoculated in different culture media including Murashige and Skoog's [20] (MS) medium or MS medium supplemented with N^6-benzylaminopurine (BA; 0.5-5.0 mg/L), meta-Topolin (mT; 0.5-5.0 mg/L), Zeatin (Z; 0.5-5.0 mg/L), and Kinetin (KIN; 0.5-5.0 mg/L) for the regeneration of multiple shoots by axillary bud proliferation.

For root induction, the 3.0-4.0 cm long shoots were excised from the primary cultures and cultured individually in the culture tubes (Borosil, India) containing half- (½) or full-strength MS medium alone or ½ MS or MS medium supplemented with Indole-3-butyric acid (IBA; 0.5-2.0 mg/L).

All media were supplemented with 3% (w/v) sucrose and gelled with 0.7% (w/v) agar. The pH of the medium was adjusted to 5.8 ± 0.1 prior to autoclave at 121 °C for 17 mins. All the cultures were maintained at 25 ± 1 °C with a photoperiod of 16h/ 8h under illumination of 35-50 µmol m^{-2} s^{-1} photon flux density provided by cool white fluorescent tubes (Phillips, India).

Acclimatization of Plantlets of *O. turpethum*

In vitro rooted plantlets were removed from the rooting medium, and agar was cleaned from the roots carefully under running tap water. Plantlets were planted in plastic pots containing sterile garden soil:sand (1:1). Potted plantlets were covered with polyethylene bags for maintaining humidity and then kept in the culture room. The plantlets were watered as per the requirement. After 6-8 days from the date of transfer, holes were punched in the polyethylene bags to reduce humidity. After 2-4 days of punching holes, the polyethylene bags were removed, and then after 2 days, plantlets were taken out from the culture room and kept under shade for 7 days. Finally, after 21 days, plantlets were transferred to larger pots containing garden soil and kept out door under full sun.

Phytochemical Analysis

Preparation of the Sample

For phytochemical analysis, roots, stem, and leaves of tissue culture regenerated-field established plants and of the mother plant were collected and washed under running tap water, followed by rinsing in double distilled water. The plant parts were then cut into small pieces and dried in the shade at room temperature to get a constant weight.

Quantification of alkaloid, flavonoid, tannin, saponin, and phenolics

Phytochemicals like tannin, saponin, flavonoid, and phenolics were estimated as per the standard protocol [21], and alkaloid was estimated by the protocol followed [22]. The total alkaloids, saponins, and flavonoids quantity was represented as mg/g in dry weight of roots, stem, leaves. Quantification for phenolics and tannins was done based on gallic acid equivalents (GAE), tannic acid equivalents (TAE) standard curve and finally, the data were expressed in mg standard equivalent weight/g of dry weight of the sample.

Data Recording and Statistical Analysis

For *in vitro* shoot proliferation experiments, each treatment had five flasks, and each flask contained two nodal segments as explant. For the rooting experiment, five tubes and each tube with one explant were taken. Both experiments were repeated thrice. The mean number of shoots per explant, mean number of roots per explant, and mean length of shoots and roots (in cm) were recorded after 6 weeks of culture. Data were analysed using Analysis of Variance (ANOVA) for a completely randomized design (CRD). Duncan's new multiple range test (DMRT) [23] was used to designate the means with a significant effect at $P \leq 0.05$.

RESULTS

Evaluation of Growth Regulators for Axillary Shoot Proliferation

The influence of the different types of plant growth regulators on axillary shoot proliferation from nodal explants has been tested, and the result is documented in Table 1. The explants responded for shoot initiation in all the media tested within one week of culture. After six weeks of culture, about 56.6% of explants exhibited shoot regeneration on MS medium alone, with an average of only 1.0 shoot per explant and an average shoot length of 1.8 cm (Fig. 1A). The maximum percentage (95.6%) of explants showed shoot regeneration on MS supplemented with 3.0 mg/L mT. About 91.1% and 88.9% of explants responded for shoot regeneration on MS medium supplemented with 2.0 mg/L BA and MS medium augmented with 4.0 mg/L Z, respectively Table 1. Multiple shoot proliferation was observed on MS medium supplemented with 2.0 mg/L BA (Fig. 1B) with an average of 4.5 shoots per explant at six weeks of culture, having a shoot length of 4.1 cm (Fig. 1C). After six weeks of culture, the highest number of shoot initiation was observed on MS medium supplemented with 3.0 mg/L mT with an average of 13.3 number of shoots per explant having shoot length of 4.8 cm (Fig. 1D, E). KIN has not been found to have a more stimulating effect on shoot proliferation from nodal segments compared with other plant growth regulators tested.

Table 1. Effect of cytokinins on shoot proliferation from a mature nodal segment of *O. turpethum*.

MS Basal Medium Supplemented with Plant Growth Regulators (mg/L)				% Shoot Regeneration	Shoots/Explants	Mean Length of Shoot (cm)
BA	mT	Z	KIN	-	-	-
0.0	-	-	-	55.6 ± 1.0 [r]	1.0 ± 0.0 [t]	1.8 ± 0.2 [o-q]
0.5	-	-	-	66.7 ± 1.0 [m]	2.0 ± 0.4 [qr]	1.9 ± 0.1 [op]
1.0	-	-	-	77.8 ± 1.7 [gh]	2.5 ± 0.2 [o-q]	2.7 ± 0.2 [j-l]
1.5	-	-	-	86.7 ± 1.5 [d]	3.2 ± 0.3 [k-m]	3.3 ± 0.2 [gh]
2.0	-	-	-	91.1 ± 1.1 [b]	4.5 ± 0.3 [h]	4.1 ± 0.4 [cd]
2.5	-	-	-	82.2 ± 1.4 [f]	2.9 ± 0.0 [l-o]	3.0 ± 0.3 [h-j]
3.0	-	-	-	60.0 ± 0.0 [p]	2.8 ± 0.3 [m-p]	2.3 ± 0.2 [mn]
5.0	-	-	-	57.8 ± 1.4 [q]	1.6 ± 0.2 [rs]	1.7 ± 0.2 [p-r]
-	0.5	-	-	71.1 ± 1.9 [k]	3.8 ± 0.3 [ij]	3.2 ± 0.3 [g-i]
-	1.0	-	-	77.8 ± 1.4 [gh]	7.6 ± 0.7 [d]	3.9 ± 0.1 [de]
-	2.0	-	-	84.4 ± 2.0 [e]	9.5 ± 0.5 [c]	4.5 ± 0.3 [ab]

(Table 1) cont.....

MS Basal Medium Supplemented with Plant Growth Regulators (mg/L)				% Shoot Regeneration	Shoots/Explants	Mean Length of Shoot (cm)
BA	mT	Z	KIN	-	-	-
-	3.0	-	-	95.6 ± 0.6 [a]	13.3 ± 0.5 [a]	4.8 ± 0.1 [a]
-	4.0	-	-	88.9 ± 1.6 [c]	10.2 ± 0.3 [b]	4.3 ± 0.3 [bc]
-	5.0	-	-	75.6 ± 1.0 [i]	6.7 ± 0.3 [e]	3.9 ± 0.1 [de]
-	-	0.5	-	62.2 ± 2.0 [o]	2.0 ± 0.0 [qr]	2.5 ± 0.0 [k-m]
-	-	1.0	-	66.7 ± 1.2 [m]	2.8 ± 0.5 [m-p]	2.5 ± 0.2 [k-m]
-	-	2.0	-	75.6 ± 1.0 [i]	3.3 ± 0.1 [j-l]	3.0 ± 0.0 [h-j]
-	-	3.0	-	82.2 ± 2.0 [f]	5.3 ± 0.3 [g]	3.5 ± 0.4 [fg]
-	-	4.0	-	88.9 ± 1.7 [c]	6.0 ± 0.5 [f]	3.8 ± 0.2 [d-f]
-	-	5.0	-	80.0 ± 1.0 [g]	4.2 ± 0.2 [hi]	3.5 ± 0.3 [fg]
-	-	-	0.5	60.0 ± 0.5 [p]	1.0 ± 0.0 [t]	2.0 ± 0.0 [no]
-	-	-	1.0	64.4 ± 0.7 [n]	2.0 ± 0.2 [qr]	2.3 ± 0.3 [mn]
-	-	-	2.0	73.3 ± 0.5 [j]	3.2 ± 0.3 [k-m]	3.0 ± 0.0 [h-j]
-	-	-	3.0	77.8 ± 1.4 [gh]	3.5 ± 0.5 [jk]	3.0 ± 0.1 [h-j]
-	-	-	4.0	73.3 ± 1.2 [j]	3.0 ± 0.0 [k-n]	2.8 ± 0.1 [jk]
-	-	-	5.0	68.9 ± 1.6 [l]	2.0 ± 0.0 [qr]	2.5 ± 0.0 [k-m]

In a column, different letters in superscripts indicate statistically significant differences between the means (P≤0.05; Duncan's new multiple range test).

Rooting of *in vitro* Regenerated Shoots

In vitro regenerated shoots were tested for rooting on MS and ½ MS medium alone and on MS and ½ MS medium supplemented with different concentrations of auxin (IBA; 0.5-2.0 mg/L). The percentage of rooting of shoots varied among the different rooting media tested Table **2**. The addition of auxin was essential for the induction of roots as growth regulator free (MS and ½ MS) media failed to induce root(s) from the *in vitro* regenerated shoots. Cent percent of *in vitro* regenerated shoots responded for rooting on ½ MS supplemented with 0.5 mg/L IBA with an average of 8.3 roots produced per shoot having a root length of 5.2 cm (Fig. **1F**).

Table 2. Rooting of *in vitro* regenerated shoots of *O. turpethum*.

Rooting Medium		Rooting (%)	Mean no. of Roots/Shoots	Mean Root Length (cm)
Basal Medium	IBA (mg/L)			
MS	0.0	0.0 ± 0.0 [f]	0.0 ± 0.0 [i]	0.0 ± 0.0 [i]
	0.5	100 ± 0.0 [a]	2.2 ± 0.1 [h]	4.5 ± 0.2 [b]
	1.0	100 ± 0.0 [a]	5.0 ± 0.3 [e]	3.7 ± 0.1 [cd]
	1.5	95.6 ± 2.0 [b]	6.5 ± 0.3 [c]	2.3 ± 0.3 [f]
	2.0	77.8 ± 1.4 [d]	7.0 ± 0.3 [b]	2.0 ± 0.0 [gh]
½ MS	0.0	0.0 ± 0.0 [f]	0.0 ± 0.0 [i]	0.0 ± 0.0 [i]
	0.5	100 ± 0.0 [a]	8.3 ± 0.5 [a]	5.2 ± 0.3 [a]
	1.0	84.4 ± 3.4 [c]	5.8 ± 0.5 [d]	3.8 ± 0.2 [c]
	1.5	77.8 ± 0.9 [d]	4.3 ± 0.5 [f]	3.3 ± 0.2 [e]
	2.0	73.3 ± 1.6 [e]	3.6 ± 0.2 [g]	2.2 ± 0.3 [fg]

In a column, different letters in superscripts indicate statistically significant differences between the means ($P \leq 0.05$; Duncan's new multiple range test).

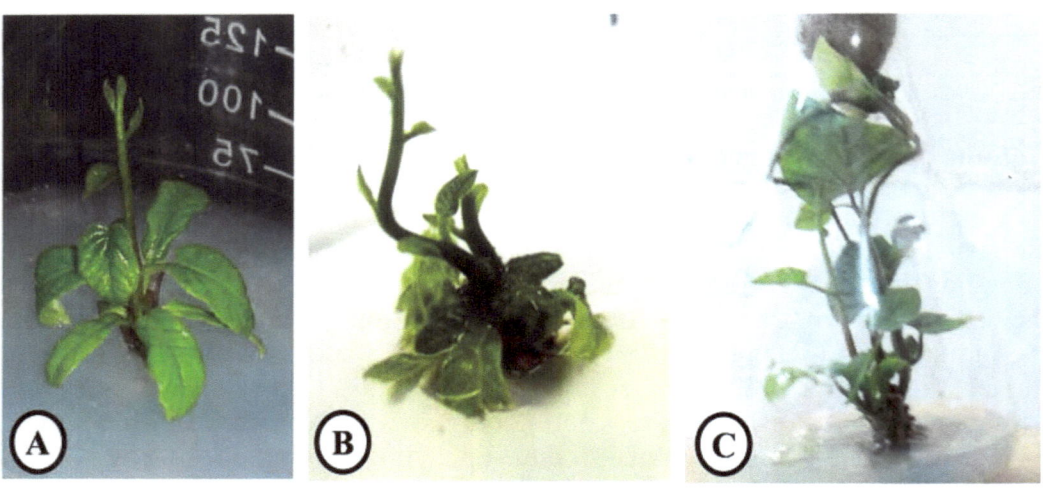

(Fig. 1) contd.....

Fig. (1). Micropropagation by axillary shoot proliferation from mature nodal explants of *Operculina turpethum*. (**A**) Shoot regeneration from the nodal segment on MS medium. (**B**) Multiple shoot initiation on MS medium supplemented with BA (2.0 mg/L). (**C**) Shoot elongation on MS medium + BA (2.0 mg/L) at 6 weeks of culture. (**D**) Multiple shoot initiation on MS + mT (3.0 mg/L). (**E**) Shoot elongation on MS + mT (3.0 mg/L) at 6 weeks of culture. (**F**) Rooting of *in vitro* regenerated shoot on ½ MS medium supplemented with IBA (0.5 mg/L). (**G**) Acclimatization of *in vitro* regenerated plantlet in a small pot containing sterile sand:garden soil (1:1). (**H**) Acclimatized plant growing in a large pot containing garden soil.

Acclimatization of *in vitro* Regenerated Plantlets

About 65% of plantlets were successfully acclimatized and survived when transferred to the mixture of sterile garden soil and sand (Fig. **1G**). Upon transfer to the clay pots (16 × 17 cm) containing garden soil, 90% of the acclimatized plantlets survived (Fig. **1H**). Regenerated plants showed normal growth and were morphologically similar to the mother plant.

Phytochemical Analysis

Qualitative analysis of phytochemicals suggested that there has been no variation in the quantity of phytochemicals between the mother plant and tissue culture raised plants Table **3**. Out of different phytochemicals (tannins, saponins, flavonoids, alkaloids, and phenolics) studied, tannin was found to be in a maximum amount in all the plant parts, *i.e.*, roots, stem, leaves tested. Among all the plant parts (*e.g.*, roots, stem, and leaves) tested, the highest quantity of all phytochemicals has been observed in the root Table **3**.

Table 3. Phytochemicals analysis of tissue cultured plant derived from mature node.

Phytochemicals (mg/g)	Mother Plant	Tissue Culture Plant
	Root	
Alkaloids (mg/ g DW)	3.20 ± 0.10	3.03 ± 0.06
Flavonoids (mg/ g DW)	23.31 ± 0.37	21.35 ± 0.60
Tannins (mg TAE/ g DW)	122.70 ± 1.86	120.35 ± 1.12
Saponins (mg/ g DW)	89.50 ± 2.17	85.81 ± 1.06
Phenolics (mg GAE/ g DW)	10.37 ± 0.47	10.13 ± 0.50
	Stem	
Alkaloids (mg/ g DW)	2.79 ± 0.24	2.63 ± 0.22
Flavonoids (mg/ g DW)	19.07 ± 0.78	17.78 ± 0.3
Tannins (mg TAE/ g DW)	97.24 ± 1.37	98.96 ± 1.56
Saponins (mg/ g DW)	78.90 ± 1.01	90.03 ± 1.00
Phenolics (mg GAE/ g DW)	7.88 ± 0.42	6.53 ± 0.34
	Leaf	
Alkaloids (mg/ g DW)	2.87 ± 0.30	2.76 ± 0.41
Flavonoids (mg/ g DW)	12.09 ± 0.43	10.77 ± 0.48
Tannins (mg TAE/ g DW)	83.53 ± 1.85	81.65 ± 0.99
Saponins (mg/ g DW)	78.10 ± 0.95	77.63 ± 0.77
Phenolics (mg GAE/ g DW)	6.17 ± 0.29	6.26 ± 0.30

DISCUSSION

Different factors such as explant type, the path of regeneration, choice of basal medium, type of plant growth regulators used, *etc.* , affect the *in vitro* clonal propagation (micropropagation). Usually, the parts of the plant containing vegetative meristematic buds are preferred explants for *in vitro* plant regeneration as they are less prone to clonal instability than unorganized callus [24] or seedling explants. Thus, the nodal segments were taken as explant in the present experiment for *in vitro* plant regeneration. The nodal explants have already been used in a number of plant species, such as *Bacopa monnieri* [25], *Ipomoea batatas* [26, 27], *Ipomoea mauritiana* [28], *Pongamia pinnata* [29, 30], *Punica granatum* [31] and *Symplocos racemosa* [32].

In this experiment, MS was used as the basal medium as this is the most commonly used basal medium for plant tissue culture. MS medium without any plant growth regulators was not sufficient for multiple shoot proliferation of *O. turpethum*. Therefore, different types and concentrations of cytokinins were used in this experiment to produce multiple shoots. In this present study, mT (3.0 mg/L) was found to be the most effective cytokinin among other cytokinins (BA, KIN, and Z) tested for *in vitro* multiple shoot regeneration from the nodal explant. On this medium, the maximum number of *in vitro* shoot as well as shoot length, was regenerated from the nodal segment within six weeks of culture. meta-Topolin plays a crucial role in the development of a higher frequency of shoots in comparison to other cytokinins at equal concentrations. This might be due to its ability to metabolize much faster than other cytokinins as well as due to its typical chemical structure (*i.e.*, -OH group on its side chain that helps in the construction of O-glucoside conjugate) [33]. Corroborating our study, it was observed in several plants, such as *Curcuma amada* [34], *Daphne mezereum* [35], *Hedychium coronarium* [36], *Stylosanthes amata* cv. Verano [37], *Pterocarpus marsupium* [38], *Gluta usitata* [39], *Rose* cv. Frisco [40], *Ansellia africana* [41, 42], and *Dendrocalamus asper* [43] MS medium supplemented with a single cytokinin (mT) was optimum for *in vitro* shoot regeneration.

Complete plant regeneration requires rooting of the *in vitro* regenerated shoots. MS or ½ MS devoid of any auxin was enough for rooting in a number of plant species, including *Bacopa monnieri* [44, 45], *Curcuma amada* [46], *Curcuma soloensis* [47], *Zingiber pettiolatum* [48], and *Hedychium bousigonianum* [49]. The addition of auxin(s) is usually essential for root induction in a number of plant species. In this study, it was observed that the addition of IBA to the basal media was essential for the rooting of *in vitro* regenerated shoots of *O. turpethum*. Without successful acclimatization of plants in the soil, the *in vitro* multiplication protocol is a failure. In this study, the regenerated plants were successfully acclimatized.

There is a need to check the phytochemical quantity of tissue cultured plants before their use for commercial or industrial purposes. Thus, in this study, the phytochemical analysis of different bioactive compounds (*e.g.*, alkaloids, flavonoids, tannins, saponins, and phenolics) was studied in different parts like roots, stem, and leaves both in mother and tissue culture raised plants. It was observed that the tissue culture raised plants were phytochemically comparable with the mother plant. Similar work was previously reported in *Hedychium coronarium* and *Curcuma amada*.

CONCLUSION

Operculina turpethum possesses various medicinal and pharmaceutical properties due to the presence of several important bioactive compounds. This plant is enlisted as threatened plant species in several states of India by reason of its overexploitation. In this study, an efficient micropropagation protocol was developed for *O. turpethum* using nodal segments. The highest number of shoots (13.3) proliferated on MS supplemented with 3.0 mg/L mT. *In vitro* shoots were rooted on ½ MS medium + 0.5 mg/L IBA. *In vitro* regenerated plants were successfully acclimatized and subsequently successfully established in the field. The biochemical fidelity was checked for tissue culture raised-field established plant vis-à-vis mother plant. The current protocol has the potential to circumvent the limitations of conventional methods of propagation of *O. turpethum*. The large-scale production of the medicinally valued plant species, *O. turpethum*, through this method, could provide (i) planting materials for the reintroduction of plants in their natural habitat and (ii) raw materials to pharmaceutical companies. As a consequence, this protocol will help in the conservation of *O. turpethum*.

ACKNOWLEDGEMENT

Authors acknowledge to Department of Botany, Ravenshaw University, Cuttack, Odisha, India for providing necessary facilities.

REFERENCES

[1] Austin DF. *Operculina* turpethum (Convolvulaceae) as a medicinal plant in Asia. Econ Bot 1982; 36(3): 265-9.
[http://dx.doi.org/10.1007/BF02858545]

[2] Kirtikar KA, Basu BA. Indian medicinal plants. Periodical Expert's Book Agency 2000.New Delhi

[3] Murty KPS. Bhavprakasha of bhavmishra. Varanasi: Chaukhamba Shri Krishna Das 2008; 1.

[4] Kohli KR, Nipanikar SU, Kadbhane KP. A comprehensive review on *trivrit* (*Operculina turpethum* syn. *Ipomoea turpethum*). Int J Pharm Bio-sci 2010; 1(4): 443-52.

[5] Rastogi RP, Mehrotra BN. Compendium of Indian medicinal plants. Lucknow: CDRI 1979; 2.

[6] Jain S, Saxena VK. Nonsaponifiable matter from the stem *Operculina turpethum*. Acta Cienc Indica Chem 1987; 13(3): 1171-2.

[7] Anbuselvam C, Vijayavel K, Balasubramanian MP. Protective effect of *Operculina turpethum* against 7,12-dimethyl benz(a)anthracene induced oxidative stress with reference to breast cancer in experimental rats. Chem Biol Interact 2007; 168(3): 229-36.
[http://dx.doi.org/10.1016/j.cbi.2007.04.007] [PMID: 17531963]

[8] Alam MJ, Alam I, Sharmins SA, Rahman MM, Anisuzzaman M, Alam MF. Micropropagation and antimicrobial activity of *Operculina turpethum* (syn. *Ipomoea turpethum*), an endangered medicinal plant. Plant Omics 2010; 3(2): 40-6.

[9] Shankaraiah P, Srinivasa RC. Comparative antidiabetic activity of induced diabetic rats. Int Curr Pharm J 2012; 1(9): 2272-8.

[10] Khare AK, Srivastava MC, Tewari JP, Puri JN, Singh S. A preliminary study of anti-inflammatory activity of *Ipomoea turpethumNisoth*. Indian J Drug 1982; 19: 224-6.

[11] Ahmed A, Howlader MS, Dey SK, Hira A, Hossain MH, Uddin MMN. Phytochemical screening and antibacterial activity of different fractions of *Operculina turpethum* root and leaf. Am j sci ind 2013; 4(2): 167-72.
[http://dx.doi.org/10.5251/ajsir.2013.4.2.167.172]

[12] Rajashekar MB. Pharmacological screening of root of *Operculina turpethum* and its formulations. Acta Pharm Sci 2006; 48: 11-7.

[13] Tripathi B. Charakasamhita of agnivesha elaborated by charaka & drudhabala. Varanasi: Chaukhamba Surbharti Publishers 2008; II.

[14] Rashid MH, Gafur MA, Sadik MG, Rahman AA. Antibacterial and cytotoxic activities of extracts and isolated compounds of *Ipomoea turpethum*. Pak J Biol Sci 2002; 5(5): 597-9.

[15] Ahmad R, Ahmed S, Khan NU, Hasnain A. Operculina turpethum attenuates n-nitrosodimethylamine induced toxic liver injury and clastogenicity in rats. Chem Biol Interact 2009; 181(2): 145-53.
[http://dx.doi.org/10.1016/j.cbi.2009.06.021] [PMID: 19589336]

[16] Anonymous . The wealth of India: A dictionary of Indian raw materials and industrial products. New Delhi: CSIR 1966.

[17] Biswal AK, Nair MV. Threatened plants of orissa and priority species for conservation. In: Rawat GS, Ed. Special habitats and threatened plants of india, envis bulletin: Wildlife and protected areas. Dehradun: Wildlife Institute 2008; 11: p. 1.

[18] Sharma V, Singh M. *Operculina turpethum* as a panoramic herbal medicine: A review. Int J Pharm Sci Res 2012; 3(1): 21-5.

[19] Sebastinraj J, Britto SJ, Kumar SRS. Micropropagation of *Operculiina turpethum* Linn. Silva Manso. using cotyledonary node explants. Acd J Plant Sci 2013; 6(2): 77-81.

[20] Murashige T, Skoog F. A revised medium for rapid growth and bioassays with tobacco tissue cultures. Physiol Plant 1962; 15(3): 473-97.
[http://dx.doi.org/10.1111/j.1399-3054.1962.tb08052.x]

[21] Behera S, Kamila PK, Rout KK, Barik DP, Panda PC, Naik SK. An efficient plant regeneration protocol of an industrially important plant, *Hedychium coronarium* J. Koenig and establishment of genetic & biochemical fidelity of the regenerants. Ind Crops Prod 2018; 126: 58-68.
[http://dx.doi.org/10.1016/j.indcrop.2018.09.058]

[22] Jain P, Sharma HP, Basri F, Priya K, Singh P. Phytochemical analysis of *Bacopa monnieri* L. Wetts and their antifungal activities. Indian J Tradit Knowl 2016; 16(2): 310-8.

[23] Gomez KA, Gomez AA. Statistical procedure for agricultural research. New York: Wiley 1984.

[24] Shenoy VB, Vasil IK. Biochemical and molecular analysis of plants derived from embryogenic tissue

cultures of napier grass *Pennisetum purpureum* K. Schum. Theor Appl Genet 1992; 83(8): 947-55.
[http://dx.doi.org/10.1007/BF00232955] [PMID: 24202918]

[25] Behera S, Nayak N. An efficient micropropagation protocol of *Bacopa monnieri* L. Pennell through two-stage culture of nodal segments and *ex vitro* acclimatization. J Appl Biol Biotechnol 2015; 3(3): 16-21.

[26] Yang X. Rapid production of virus-free plantlets by shoot tip culture *in vitro* of purple-coloured sweet potato *Ipomoea batatas* L. Lam. Pak J Bot 2010; 42(3): 2069-75.

[27] Mvuria JM, Ombori O. Low-cost macronutrients in the micropropagation of selected sweet potato [*Ipomoea batatas* L. Lam] varieties. J Agric Environ Sci 2014; 3(1): 89-101.

[28] Islam MS, Bari MA. Rapid *in vitro* multiplication, callogenesis and indirect shoot regeneration in *Ipomoea mauritiana*- a rare medicinal plant in Bangladesh. Med Aromat Plants 2013; 2(6): 1-3.
[http://dx.doi.org/10.4172/2167-0412.1000138]

[29] Sujatha K, Hazra S. Micropropagation of mature *Pongamia pinnata* Pierre. In Vitro. Cell Dev Bio - Plant . 2007; 43: pp. 608-13.

[30] Kesari V, Ramesh AM, Rangan L. High frequency direct organogenesis and evaluation of genetic stability for *in vitro* regenerated *Pongamia pinnata*, a valuable biodiesel plant. Biomass Bioenergy 2012; 44: 23-32.
[http://dx.doi.org/10.1016/j.biombioe.2012.03.029]

[31] Naik SK, Pattnaik S, Chand PK. *In vitro* propagation of pomegranate (*Punica granatum* L. cv. Ganesh) through axillary shoot proliferation from nodal segments of mature tree. Sci Hortic 1999; 79(3-4): 175-83.
[http://dx.doi.org/10.1016/S0304-4238(98)00218-0]

[32] Behera S, Barik DP, Naik SK. Micropropagation of *Symplocos racemosa* Roxb., a threatened medicinal tree of India. Curr Sci 2017; 113(4): 555-8.

[33] Pramanik B, Sarkar S, Bhattacharyya S, Gantait S. *Meta*-topolin-induced enhanced biomass production *via* direct and indirect regeneration, synthetic seed production, and genetic fidelity assessment of *Bacopa monnieri* L. Pennell, a memory-booster plant. Acta Physiol Plant 2021; 43(7): 107.
[http://dx.doi.org/10.1007/s11738-021-03279-1]

[34] Behera S, Monalisa K, Meher RK, *et al.* Phytochemical fidelity and therapeutic activity of micropropagated *Curcuma amada* Roxb. A valuable medicinal herb. Ind Crops Prod 2022; 176: 114401.
[http://dx.doi.org/10.1016/j.indcrop.2021.114401]

[35] Nowakowska K, Pacholczak A, Tepper W. The effect of selected growth regulators and culture media on regeneration of *Daphne mezereum* L. 'Alba'. Rend Lincei Sci Fis Nat 2019; 30(1): 197-205.
[http://dx.doi.org/10.1007/s12210-019-00777-w]

[36] Behera S, Kar SK, Rout KK, Barik DP, Panda PC, Naik SK. Assessment of genetic and biochemical fidelity of field-established *Hedchium coronarium* J. Koenig regenerated from axenic cotyledonary node on *meta*-topolin supplemented medium. Ind Crops Prod 2019; 134: 206-15.
[http://dx.doi.org/10.1016/j.indcrop.2019.03.051]

[37] Ngoenngam L, Pongtongkam P, Aranannant J, Poeaim S, Poeam A. *In vitro* effect of gamma irradiation and plant growth regulators *pgrs* for induction and development of *Stylosanthes hamate* cv. Verano. Agric Technol Thail 2019; 15: 63-74.

[38] Ahmad A, Anis M. Meta-topolin improves *in vitro* morphogenesis, rhizogenesis and biochemical analysis in *Pterocarpus marsupium* Roxb.: A potential drug-yielding tree. J Plant Growth Regul 2019; 38(3): 1007-16.
[http://dx.doi.org/10.1007/s00344-018-09910-9]

[39] Rakrawee R, Kittibanpacha K, Chareonsap PP, Poeaim A. Efficiency of antioxidant and absorbent on

browning and the optimal factors of plant regeneration from young seed of *Gluta usitata* (217 Mae Ka) by tissue culture. Agric Technol Thail 2018; 14: 911-22.

[40] Mahmood S, Reza MR, Hossain MG, Hauser B. Response of cytokinins on *in vitro* shoot multiplication of Rose cv *Frisco*. Res Rev J Agric Sci Technol 2018; 5: 8-12.

[41] Bhattacharyya P, Kumar V, Van Staden J. *In vitro* encapsulation based short term storage and assessment of genetic homogeneity in regenerated *Ansellia africana* (Leopard orchid) using gene targeted molecular markers. Plant Cell Tissue Organ Cult 2018; 133(2): 299-310.
[http://dx.doi.org/10.1007/s11240-018-1382-0]

[42] Bhattacharyya P, Kumar V, Van Staden J. Assessment of genetic stability amongst micropropagated *Ansellia africana*, a vulnerable medicinal orchid species of africa using scot markers. S Afr J Bot 2017; 108: 294-302.
[http://dx.doi.org/10.1016/j.sajb.2016.11.007]

[43] Ornellas TS, Werner D, Holderbaum DF, Scherer RF, Guerra MP. Effects of vitrofural, bap and meta-topolin in the *in vitro* culture of *Dendrocalamus asper*. Acta Hortic 2017; (1155): 285-92.
[http://dx.doi.org/10.17660/ActaHortic.2017.1155.41]

[44] Asha KI, Devi AI, Dwivedi NK, Nair RA. *In vitro* regeneration of brahmi (*Bacopa monnieri* (Linn.) Pennell - an important medicinal herb through nodal segment culture. Res Plant Biol 2013; 3(1): 1-7.

[45] Koul A, Sharma A, Gupta S, Mallubhotla S. Cost effective protocol for micropropagation of *Bacopa monnieri* using leaf explants. Int J Sci Res 2014; 3(4): 210-2.

[46] Nayak S. High frequency *in vitro* production of microrhizomes of *Curcuma amada*. Indian J Exp Biol 2002; 40(2): 230-2.
[PMID: 12622191]

[47] Zhang S, Liu N, Sheng A, Ma G, Wu G. Direct and callus-mediated regeneration of *Curcuma soloensis* Valeton (Zingiberaceae) and *ex vitro* performance of regenerated plants. Sci Hortic 2011; 130(4): 899-905.
[http://dx.doi.org/10.1016/j.scienta.2011.08.038]

[48] Prathanturarug S, Angsumalee D, Pongsiri N, Suwacharangoon S, Jenjittikul T. *In vitro* propagation of *Zingiber petiolatum* (Holttum) Theilade, a rare zingiberaceous plant from Thailand. *In Vitro*. Cell Dev Biol Plant 2004; 40(3): 317-20.
[http://dx.doi.org/10.1079/IVP2003505]

[49] Shakhanokho HF. Somatic embryogenesis in *Hedychium bousigonianum*. Hortic Sci 2009; 44(5): 1487-90.

CHAPTER 10

Tissue and Cell Culture of Tea (*Camellia sp.*)

Abhishek Mazumder[1], Urvashi Lama[2], Meghali Borkotoky[3], Sangeeta Borchetia[3], Shabana Begam[1] and Tapan Kumar Mondal[1,*]

[1] *ICAR-National Institute for Plant Biotechnology (ICAR-NIPB), New Delhi, Pusa, India*
[2] *Department of Botany, Sovarani Memorial College, Jagatballavpur, Howrah, West Bengal, India*
[3] *Tocklai Tea Research Institute, Jorhat, Assam, India*

Abstract: Tea(*Camellia sp.*) is a non-alcoholic drink consumed across the globe. Upon consumption, it provides refreshment and enormous health benefits. Tea possesses antioxidant compounds which prevent human health from several diseases and disorders as well. Micropropagation and somatic embryogenesis are two distinct cell and tissue culture methods which have been utilized for a long time for the production of secondary metabolites having economical and industrial values. Micropropagation is a clonal propagation method accomplished by selection of explants and establishment of culture in basal media followed by shoot multiplication, development of callus, rhizogenesis, hardening and acclimatization by transferring plantlets from the laboratory to an open environment in the greenhouse or in the field. Somatic embryogenesis is the development of embryos from somatic cells, not from the zygotic cells. It consists of induction, multiplication, development and maturation of the embryo. Globular, heart and torpedo, these three distinguishable developmental stages are visible in somatic embryogenesis. Numerous genes associated with cell division, organ formation and specific cellular processes related to somatic embryogenesis have been identified. Tea possesses several secondary metabolites which have versatile functions. Caffeine, theobromine and theophylline are typical secondary metabolites which impart characteristic taste and flavour to tea. In addition, polyphenols, catechins, proanthocyanin and flavonoids act as antioxidant compounds and possess several health benefits. Various cell and tissue culture methods have been adopted for the biosynthesis of secondary metabolites on laboratory and industrial scales. These methods can be adopted on a larger scale, from experimental laboratory investigation to the industrial setup for the discovery of novel metabolic compounds for their potential applications as medicines and in commercial sectors.

Keywords: Antioxidants, Callus, Micropropagation, Somatic embryogenesis, Secondary metabolites, Tea.

** **Corresponding author Tapan Kumar Mondal:** ICAR-National Institute for Plant Biotechnology (ICAR-NIPB), New Delhi, Pusa, India; Email: mondltk@yahoo.co.in*

Mohammad Anis & Mehrun Nisha Khanam (Eds.)
All rights reserved-© 2024 Bentham Science Publishers

INTRODUCTION

Tea is a popular beverage consumed worldwide. It is prepared with leaves of *Camellia sinensis* (L.) O. Kuntze. Besides holding medicinal properties and being economically important, it is a living representation of ancient cultural practices that originated in China and gradually spread throughout the globe. Owing to its large demand, micropropagation serves as a better alternative to traditional propagation methods for commercial use. Micropropagation technique was developed in 1946 by Ball, who is regarded as the father of micropropagation [1 - 4]. It is also worthwhile to mention the contribution of a pioneer scientist in tissue culture [5] and the initiation of a systematic study on micropropagation of tea [6].

MICROPROPAGATION OF TEA

In regard to woody and perennial plants like tea, which exhibit recalcitrancy and have long gestation periods, micropropagation offers a rapid, cost-effective production [7, 8]. There have been several works on the micropropagation of tea in the past [9 - 11]. The main objective of micropropagation is to achieve plants with desirable characteristics, increased productivity and with developed resistance to drought, pest, salinity, acidity, alkalinity, frost and other factors that limit plant growth. It can be achieved by – (i) enhancing axillary bud breaking, (ii) production of adventitious buds directly or indirectly *via* the callus, and (iii) somatic embryogenesis directly or indirectly from explants [12 - 14].

In this book, we will classify the micropropagation of tea into four broad stages which are discussed below:

Stage I: Selection of explants and establishment of culture through explants.

Stage II: Shoot multiplication: Initiation and multiplication of callus.

Stage III: Rhizogenesis.

Stage IV: Hardening & Acclimatization: Transfer from *in vitro* to *ex vitro* condition.

Stage I: Selection of Explants and Establishment of Culture Through Explants

Explant material forms the basis of all micropropagation works. Factors like the availability of the explant material throughout the year, its origin, and its type are to be considered. Apart from this, sterilization of explants, as well as aseptic laboratory conditions, are necessary to establish *in vitro* propagation successfully. For *in vitro* culture of tea, different kinds of explants such as meristems, shoot tips

[15, 16], parts of stem *viz.* nodal segments [17], stem segments, epidermal layers of stem segments, stem segments without epidermal layer for shoot regeneration, zygotic embryos, mature and immature cotyledons for the induction of adventitious buds [18, 19] were used. Generally, shoot tips were used for tea micropropagation through either apical meristem culture or shoot apex culture. Shoot tip and axillary bud culture were reported as effective, easier, simpler and quicker in securing the growth of shoots [20]. While comparing the effect of both methods in C. *sinensis* var. *assamica* and C. *sinensis* var. *sinensis*, it was found that the shoot tip culture was more effective than axillary bud culture in the former variety, while axillary bud culture was more effective than the shoot tip culture in the latter variety.

Stage II: Initiation, Multiplication, and Elongation of Shoots

By the late 1980s of the twentieth century, enhancement of the rate of multiplication was well established in the micropropagation of tea. Several types of basal media were used, *viz.* Heller's medium, MS (half or full strength) [21], B5 medium [22], and WPM [23], in which MS was considered the most common and efficient basal media for shoot proliferation and multiplication, while half-strength MS was reported to be as effective as the full strength MS [24 - 27]. Basal media supplemented with PGRs, commonly 1 mg/l BAP and IBA, IAA, Kinetin, NAA, GA3, or 2,4-D, were used in varying composition and concentrations by several researchers. In a study, it was reported that shoot tips and nodal segments of tea, when cultured in a media composition of MS with BAP (3.0 mg/l) and IBA (0.05 mg/l), showed the highest shoot elongation [28]. They also showed that seeds without seed cover gave an early response to shooting formation compared to seeds containing seed coat. TDZ was reported to be a significant cytokinin like growth factor for micropropagation of tea with high proliferation rates [29]. Several researchers [30 - 33] have found that Picloram {2,4,5-trichlorophenoxyacetic acid (2,4,5-T)} can significantly aid in the shoot elongation of tea plants. A protocol for somatic embryogenesis, a micropropagation technique, was developed [34] in which efficient embryo induction from mature cotyledons was found in MS media with BAP (3 mg/l) and NAA (0.1 mg/l) while media with BAP (2 mg/l) and NAA (3 mg/l) induced callus of leaf explants effectively.

Growth adjuvants for tissue culture of tea include coconut milk [35, 36], yeast extract [37], casein acid hydrolysate, serine and glutamine as nitrogen sources [38] and sucrose as carbon sources [39]. Also, 3% (w/v) sucrose as a carbon source for micropropagation of tea shoots were used.

Stage III: Rhizogenesis

Production of a healthy root system is a prerequisite for successful hardening and acclimatization of *in vitro* grown plantlets. While both *in vitro* and *ex vitro* rhizogenesis were reported in tea micropropagation, *in vitro* rooting of tea was influenced by the type of cultivar [40], auxin and salt concentration in the basal medium, the duration of auxin treatment and the physical condition of the culture and its environment whereas *ex vitro* rooting of tea was dependent on pH of the hardening media and relative humidity of the hardening chamber.

A low concentration of auxin was generally maintained in the medium for the induction of roots in tea. Treatment with IBA (0.5-1 mg/l) for 30 min, although consuming much time to induce rooting, was often preferred over NAA [40 - 42].

Ex vitro rooting (97% rooting) was found to be better than *in vitro* rooting methods [43], such as agar solidified medium, liquid medium with filter paper bridges [44] or liquid shake culture when the shoot ends were dipped in IBA (50 mg/l) for 2 h before transplantation. Physical conditions like low light or darkness with a low pH favoured rhizogenesis in tea.

Stage IV: Hardening & Acclimatization: Transfer from *in vitro* to *ex vitro* condition

Acclimatizing *in vitro* grown plantlets in an external environment to laboratory conditions stands as a critical and tricky phase of micropropagation. The first step in this phase is the hardening of *in vitro* grown plantlets – conventional hardening, biological hardening, *in vitro* hardening, and micrografting.

Conventionally, they are hardened on soil containing additives like peat moss, vermiculite, perlite, cow dung, vermicompost, soilrite, *etc.* , with varying ratios for about 6 months, preferably under shade, providing the plantlets protection from direct sunlight exposure, rainfall or frost. Low temperature, low sunlight intensity, and high humidity are factors that serve as intermediary conditions for plantlets from *in vitro* conditions to *ex-vitro*. They are then transferred to poly tubes or pots varying in size (small to large). This was supported by similar other works [45]. The standard method is transferring the rooted plantlets to potting mixtures containing an equal ratio (1:1) of either – (i) soil and vermiculite [46] or (ii) soil and peat under high humidity using humid chambers, misting or fogging units [25]. As it has been observed, the later mixture was more commonly used in recent works. Generally, the microshoots were kept in MS media supplemented with 500 mg/l IBA for 30 minutes before transplantation [47, 48].

Besides the abiotic stress, micropropagated plantlets also undergo biotic stress from microorganisms present in the rooting media. It was found that the fungus *Fusarium oxysporum* was mainly responsible for the mortality of micropropagated tea plants. To combat the stress, microbial inoculants or cultures of fungi such as *Trichoderma* VAM, *Piriformospora indica* could be used [49] while bacteria such as *Bacillus subtilis* and *Pseudomonas corrugate* were useful for hardening before transplantation.

Micrografting comes as a cost-effective and time-saving substitute to other types of hardening methods of micropropagated plants, although the age of the rootstock is the most important factor. However, the graft compatibility did not depend on the addition of a plant growth regulator (PGR) [50], which therefore was cost-effective for commercial use. The hardening time of a tea cultivar is significantly reduced from six months to almost one year when it is micrografted.

Studies of Micropropagated Raised Plants on Field

Information on field transplanted plants is quite limited. Analysis of morphological, physiological, genetic, and biochemical parameters constitutes a detailed study of transplanted plants. It was demonstrated in a study that the addition of cobalt metal ion elicitors to the nutrient medium had resulted in the synthesis of the secondary metabolite cinnamic acid in *Camellia sinensis* by 11.9% [51, 52].

To assess the genetic diversity in micropropagation of tea, two markers, *viz.* inter simple sequence repeats (ISSR) [53 - 58] and start codon targeted (SCoT) marker [59], have been used. They have also been used to assess genetic stability during micropropagation [60 - 65]. Apart from ISSR, markers like RAPD and RFLP fingerprints were also considered [66] to evaluate the genetic integrity of micropropagated plants of three diploid and triploid elite tea clones representing *C. sinensis* (China type) and *C. assamica* (Assam, India type) [67].

Due to the heterozygosity of the tea genome, micropropagation of true-to-type individuals and the genetic instability are more genotype-dependent than dependent on culture conditions [68, 69].

Problems of Micropropagation in Tea: Explant Oxidative Browning and Microbial Contamination

Explant oxidative browning and microbial contamination of *C. sinensis* var. *sinensis* are common and severe problems in the micropropagation of tea. Owing to the presence of high content of phenolic compounds in tea, tea explants are easily prone to polyphenol leaching, thereby getting oxidised to form some toxic compounds, ultimately resulting in the lowering of the pH of the culture medium. Remarkable works from researchers [70 - 72] were done to prevent polyphenol oxidation. The compound 2-aminoindane [1] 2-phosphonic acid (AIP) has been used as an inhibitor of phenylalanine ammonia lyase (PAL), an enzyme required in the production of polyphenols, and different antibiotics to control explant browning and necrosis. These compounds and 6-benzylaminopurine (BAP) and thidiazuron at different concentrations were supplemented in the regular plant growth medium. Moreover, the use of antibiotics, namely timentin, (150 mg/l) and gentamycin (30 mg/l), eliminated the surface and endophytic microbes associated with the explants of C. *sinensis* var. *sinensis* effectively [73].

SOMATIC EMBRYOGENESIS

Somatic embryogenesis is a process developed by exploiting the plant's cellular totipotency, *i.e.*, the ability of plant cells to give rise to a completely new plant. It can be defined as the development of embryos under appropriate artificial conditions or *in vitro* from somatic cells instead of germ-line cells. It is a complex developmental reprogramming, where the somatic cells are converted to embryogenic competent cells [74]. Thus, somatic embryos can also be termed non-zygotic embryos. Apart from conventional zygotic embryos, some plants give rise to embryos *via* different means under natural conditions, like parthenogenetic embryos, which are formed from an unfertilized egg and androgenic embryos or from male gametes or micro pollen [75]. But neither of these is the case with tea. Besides, the cost of *in vitro* propagation techniques for tea production surpasses the commercial utility of the tea industry. In such circumstances, somatic embryogenesis is a very good alternative which can play a significant role in both the production and genetic improvement of tea [76]. In tea, the regeneration of tea plantlets was reported to be successful for the first time [77]. Direct somatic embryogenesis without an intervening callus phase will maintain genetic fidelity. In tea, somatic embryogenesis has also been exploited for clonal propagation [78], and genetic transformation of tea [79, 80]. The various stages, factors, and molecular mechanisms which govern the somatic embryogenesis of tea have been summarized.

Different Stages of Somatic Embryogenesis

Induction

Cells of zygotic embryos of plants possess embryogenic competence and are termed as PEDCs (pre-embryogenic determined cells), but other than the zygotic embryo, the somatic cells from explants like leaves, hypocotyls *etc.* , need external factors to become embryogenic and are termed as IEDCs (induced embryogenic determined cells). This process of inducing embryonic competence in somatic cells is called induction. Plant growth hormones such as auxin (2,4-D, NAA) and cytokinins are commonly used as inducers. Many other factors, like pH, electrical gradient, heavy metals, *etc.*, can be factors affecting the induction of SE, as suggested by previous studies. Depending on the explant, the requirement of growth hormones may differ [81]. The protocol of somatic embryogenesis from three different explants of *C sinensis* had been demonstrated [82], each of which requires different composition of media and conditions for induction. For example, the nodal tea buds require a bud sprouting medium (BSM) first, then the sprouting axillary buds require a bud culture maintenance medium (BMM), and nodal cuttings from the shoots of this culture are then introduced into the preinduction media (PIM) containing MS macro with 0.5mg/L BA and 0.1mg/L IBA for 6-8 weeks and then subcultured into fresh PIM. Later in PIM, it was incubated for 16 weeks for induction, whereas cotyledons from mature tea seeds were cut into thin slices and introduced into PIM for induction. These cultures were incubated for 90 days or 12 weeks in the dark without sub culturing for induction [76]. Different growth regulators were reported [83] in cotyledon derived calli of *C sinensis*. It was reported that 2-8 mg/L of 2,4-D was necessary for callus induction [84].

Multiplication or Maturation of Callus

The unorganized mass of cells is a mixture of embryogenic and non-embryogenic cells [85]. These are transferred into multiple flasks with fresh media, or into suspensions for single cell culture to further divide and mature. Here, auxin plays an important role in some plants. Besides inducing embryogenesis, auxins can also cause cell elongation. Due to its continuous exposure, non-embryogenic cells elongate and lead to their breakdown. Once it occurs, the first stage of the somatic embryo begins, which is pre-globular stage. Cotyledons of tea clones after forming callus required basal medium supplemented with IBA and BA for 10-12 weeks for maturation and later development of the embryo [86]. But secondary embryogenesis and callus from cotyledon explants on MS nutrient medium supplemented with 3mg/L 2,4-D and 0.2mg/L kinetin were reported [30]. Formation of adventitious embryos directly on the callus free surface of 17% cotyledonary explants of *C. sinensis* inoculated on MS medium supplemented with 0-10 mg/L BA, 0-2 mg/L IBA and 0-2 mg/L naphthaleneacetic acid (NAA) was also reported [87]. They also mentioned similar observations in *C. japonica*.

Development of Embryo

Similar to zygotic embryogenesis, the development of somatic embryos also occurs in different stages as follows:

Globular stage

After the maturation of the embryogenic callus, the pre-globular stage of the embryo is formed. These pre-globular embryos are also known as globules. At this first stage of the embryo, auxin free media is needed for some plants and some plants like maize require auxin such as 2,4-D [88]. In this process, the disruption of cells from each other stops, which leads to the development of globular embryos from globules [89]. Here the first differentiation step of the formation of protoderm outside the globule occurs. This leads to the further development of globular embryos. Towards their later stages, the embryos become capable of synthesizing their own auxin and morphogenesis beyond this stage, which depends on the proper polar transport of auxin. The continuous division leads to a more triangular shape of the embryo. In this stage, the radial patterning is first established and continues to the later stages creating the ring layers of structures found in stems and roots. The outer layer, which is the protoderm forms the epidermis; the next layer from the protoderm forms the ground tissue meristem, and the central layer of elongated cells are called procambium, which becomes vascular tissue and pericycles in the roots [90].

Heart Stage

The heart stage occurs through rapid cell division on each side of the triangular globular embryo. The center of this outgrowth becomes the shoot apical meristem, and the outgrowths on each side will eventually become cotyledons. This is the stage where bilateral symmetry in the embryo is established. Some plants require treatment with abscisic acid to reach the heart stage and for morphogenesis beyond the heart stage. The apical basal axio patterning is even clearer at this stage with three clearly defined regions. The apical region gives rise to the cotyledons and shoot apical meristems, the middle or central region gives rise to the hypocotyl, root and the majority of the root meristem, and lastly, hypophysis also gives rise to the root meristem [91].

Torpedo Stage

Morphogenesis from the heart to torpedo occurs *via* cell elongation throughout the embryo as well as the continuous growth of cells in the periphery of the embryo from which cotyledons start to emerge surrounding the shoot apical meristem. Cotyledons play an important role in the regeneration capacity of the embryos, because of their possible tendency to regulate auxin flow, fixing physiological gradients within the embryo for a prompt conversion at the inception of germination [92].

Maturation of Embryo

Unlike zygotic embryos, somatic embryos do not go through the maturation phase. Normally, this phase includes the accumulation of reserved food materials and proteins that impart desiccation tolerance to the embryos. This stage can be achieved by *in vitro* SE with ABA, which is reported to increase desiccation tolerance in somatic embryos of plants like carrot, celery, soybean, alfalfa *etc.*

Germination

Germination or conversion into plantlets has been generally very poor in the case of somatic embryos. One major reason is the incomplete or improper development of somatic embryos. Even normal looking embryos may not have undergone complete development. Another factor is the intrinsic polarity, as mentioned earlier. Intrinsic polarity can also have some effect during germination. Due to this reason, in plants like maize, the somatic embryo's attachment to the callus is important until the formation of a proper shoot meristem, only such embryos develop into complete embryos with a higher probability of germination. Another factor affecting germination is the highly asynchronous development of somatic embryos, as new centers of embryogeny arise from embryos. Therefore, it is essential to obtain uniformity in the initial population by sieving the inoculum. But in tea, it is generally highly asynchronous and not controllable. Repetitive embryogenesis by budding off new globular somatic embryos was reported [93].

The authors standardize the media and conditions for the growth of somatic embryos without intervening callus formation, giving rise to a long-term direct embryogenesis system in tea, from mature seeds of TS450. The embryonal axes were separated from the cotyledons and placed in half MS media with 2mg/L NAA, 1mg/L GA_3 and 0.5mg/L BAP for embryogenesis. Very few somatic embryos were first observed on the surface of cotyledonary explants four weeks after initiation of culture as small globules (Fig. **1A**). Within 5-6 weeks, secondary globular embryos were formed on the surface of the older embryos. With repeated subculture in MS media with 2mg/L NAA and 0.5mg/L BAP, the embryos multiplied and developed into heart-shaped, torpedo-shaped and cotyledonary-shaped embryos (Fig. **1B**) without any intervening callus phase. Globular somatic embryos have a high potential for repetitive embryogenesis. The conversion percentage was 32.5% when transferred to MS medium with high BAP concentration (5mg/L) (Fig. **1C** and **1D**).

Fig. (1). Somatic embryogenesis in Tea. (**1A**) Initiation of somatic embryos from TS450. (**1B**) Embryo multiplication (**1C & 1D**) Regeneration from somatic embryos.

The explants media compositions for different stages of somatic embryogenesis reported are presented in Table **1**.

Table 1. Factors affecting somatic embryogenesis of tea.

S. No.	Explants of *C. sinensis*	Medium				References
-	-	Induction	Maturation	Germination	Multiplication	-
1.	Immature cotyledon	MS+BA (0.5) + BA (0.5)	-	MS + BA (0.5) + NAA (5)	-	[86]
2.	Whole cotyledon	Modified MS+BA (10) + YE (2)	-	Modified MS+ BA (10) with 2% sucrose	-	[46]

(Table 1) cont.....

S. No.	Explants of *C. sinensis*	Medium				References
3.	Mature seed	½ MS macro+Full micro MS+AHS(100)+Gln(100)	½ MS macro + full micro MS + AHS 100 + Gln (100)	½ MS macro+ Full micro MS+AHS (100) + Gln (100)	-	[93]
4.	Whole cotyledon	MS+BA (4) + IBA (2)	-	MS + BA (10) + IBA (0.5)	-	[87]
5.	Mature deembryonated cotyledon	MS+BA(2)+IBA(0.2)	Modified MS + BA (2) + IBA (0.2) + Gln (1) + K$_2$SO$_4$	Modified MS+BA (2) + IBA (0.2) + Gln (1)	-	[78]
6.	Mature cotyledon	MS+kin (10) + IAA (1)	-	-	-	[83]

Molecular Mechanisms of Somatic Embryogenesis (SE) in Tea

There has been extensive research investigating the role of genes involved in SE processes, which has shown that three categories of genes are expressed: (i) genes involved in cell division, (ii) genes involved in organ formation and (iii) genes specific for the process of SE [94]. The latter group of genes has been the primary target of several studies in model species, *e.g.*, leafy cotyledon (LEC) genes in *Arabidopsis* [95], WUSCHELL homeobox (WOX) in *Picea abies* [96] and WUS in Ginseng [97]. Although it is possible to identify embryogenic from non-embryogenic cultures by visual inspection, reliable expression markers associated with early SE that enable discrimination among cultures containing a high percentage of embryogenic cells when the culture is a mixture of embryogenic and non-embryogenic cells would be extremely valuable. Apart from their use in early embryogenic culture discrimination, molecular markers would also be invaluable to study the underlying molecular mechanisms regulating the change of somatic cells to embryogenic cells. Among several putative marker genes, two transcription factors, WOX2, which is one of the 15 WOX family proteins, and HAP3 (heme-activated protein 3), which is encoded by the *LEC* gene, are well known to play a key role in controlling many aspects of plant SE [98, 99]. *LEC* genes are the most well-known. There are two classes of this gene both encoding regulatory proteins that plays an important role in the induction of somatic embryo development. They have also been found to be interacting directly with hormone-responsive genes. *LEC2* gene can activate *YUC2* and *YUC4* genes that encode auxin biosynthesis enzymes, thus increasing endogenous auxin levels in the absence of exogenous auxin to induce somatic embryogenesis [100]. BBM is expressed in the root meristem. Root meristems also express BBM target genes that play a general role in maintaining cells in an undifferentiated state. Expression of BBM is enriched in the quiescent center, a group of four to seven cells that give rise to root initial/meristem cells. WUS genes in all stages of embryogenesis are expressed in a small group of cells underneath the stem cells maintaining and reprograming cell fate. It also bypasses the auxin requirement. SERK is a special group of protein kinases which was first isolated from carrots and was expressed in embryogenic competent cells of suspension cultures up to the globular-shaped stage of embryogenesis. AGL15 was identified as a component of the SERK1 protein complex, both expressed in response to auxin treatment. Moreover, direct induction of AGL15 by LEC2 is also reported in previous studies [100]. In the case of plants like *Arabidopsis*, it was reported that *YUC* gene family members encode flavin monooxygenase leading to the initiation of somatic embryo differentiation through the YUC flavin monooxygenase mediated synthesis of auxin [101].

Among *Camellia* species, tea (*C. sinensis*) is the most studied for the induction of somatic embryogenesis. However, despite being the most extensively studied among *Camellia* species for the induction of somatic embryogenesis, molecular aspects of woody plants like tea are yet to be explored. Somatic embryogenesis (SE), a complex process of clonal propagation by which plants can form embryos without meiosis and fertilization, shares some developmental and physiological similarities with zygotic embryogenesis as it involves common factors of hormonal, transcriptional, developmental and epigenetic control [102]. SE induction, as well as organogenesis and plant regeneration, depends on the addition of plant growth regulators (PGR), such as auxins and cytokinins. During SE induction, auxin response factor (ARF) genes display specific expressions being up- or down-regulated, suggesting that auxin signaling is central in the process. Moreover, *YUCCA* and *AUX/IAA* genes that are involved in auxin biosynthesis and response, respectively, are transcriptionally regulated during SE by LAFL transcription factors [103]. Transcriptional regulation performs an essential role in somatic embryogenesis. In *Arabidopsis*, the ectopic expression of some transcription factors, such as *LEC* gene, *BBM*, *WUS/WOX* genes or *AGL15*, can increase the efficiency of SE induction and lead to the formation of somatic embryos without adding hormones [104]. As the mechanisms underlying the control of somatic embryogenesis are multiple, involve complex gene regulatory networks, hormonal and epigenetic controls, and remain poorly understood, hence it has enough scope for further exploration.

CELL CULTURE AND SECONDARY METABOLITE BIOSYNTHESIS

A variety of primary metabolites like carbohydrates, amino acids, fatty acids, chlorophylls, cytochromes, and metabolic intermediates of the anabolic and catabolic pathways are present in all plant species, including tea (*Camelia sp.*). In addition, plants contain a large array of compounds called secondary metabolites, which have no apparent, direct metabolic function. A few plant species possess a number of specific metabolites which perform definite ecological functions like insect attraction for transferring pollen from one plant to another plant, attracting animals for fruit consumption which helps in the dispersal of seeds. Secondary metabolites can also act as natural pesticides to protect plants from harmful insects and pathogens. In tea plants, a few characteristic secondary metabolites add distinct flavour and aroma and help in maintaining good cup quality. Moreover, tea plant possesses a number of specific secondary metabolites which have antioxidant activity. Polyphenolic compounds *viz.* Catechins (green tea) and theaflavins (black tea) act as strong antioxidant molecules in tea plants. Flavonoids present in black tea have shown radioprotective effects and assist in radiation therapy for combating cancer [105]. Additionally, the risk of heart disease gets reduced in people consuming 3-4 cups of black tea in a day. Polyphenol, namely L-theanine, provides overall protection to the nervous system. Moreover, polyphenols can hinder oxidative damage, apoptosis, and pulmonary emphysema in human lungs induced due to smoking cigarettes [106]. Enormous polyphenols and caffeine present in tea have been experimentally shown to have antidiabetic activity, and therefore, tea and its extracts have gained attention to prepare antidiabetic medicines [107]. Overall, tea has enormous beneficial effects on improving human health and protecting them from several diseases and disorders.

Secondary Metabolite Production in Tea Through Cell Culture Method (*Camellia sp.*)

Tea is a nonalcoholic beverage which is consumed by two-thirds of the global population. The genus *Camellia* includes several species, namely, *C. sinensis*, *C. assamica*, *C. taliensis*, *C. gymnogyna* and *C. tachangensis*, *etc*. Out of several species, two of them, *C. sinensis* and *C. assamica*, are cultivated as plantation crops for the production of tea throughout the world [108]. Tea, processed from the plant "*C. sinensis* L. cv. Kuntze", consists of four different kinds of basic teas, *viz.* green tea, black tea, oolong tea and white tea. The taste and flavour of tea are imparted by several methylxanthines, *viz.* caffeine, theobromine, and theophylline and several alkaloids. Suspension culture of tea cells/tissues has been utilized for the production of secondary metabolites for commercial exploitation by the tea industry in the market [109, 110]. In 1964, for the first time in the world, caffeine was prepared from tea leaves in the purified form [111].

Polyphenol content in cell culture depends on the incorporation of precursor compounds like shikimic acid and quinic acid. The phenolic acid formation has been found to be restricted by the formation of gallic acid [112]. In a callus culture experiment, it had been found that the production of soluble phenolic compounds, flavans and lignins had been enhanced in the presence of Naphthalene acetic acid (NAA) instead of 2,4-Dichlorophenoxy acetic acid (2,4-D) . Production of phenolic compounds and flavans can be enhanced in auxin-containing cell culture medium when kinetin is applied. Whereas the application of abscisic acid had been found to play an inhibitory role on cell growth in the culture resulting in no phenolic compound production [113]. For the production of catechins and proanthocyanin, glucose (5%) is the best carbon source. In the stem-derived callus culture, a significantly elevated level of theanine formation had been witnessed after the addition of ethylamine HCl and benzylaminopurine (BAP) [109]. In the cell culture of tea, catechins were produced up to 30% [114]. Production of theanine in cell culture has also been assessed based on the major inorganic components of the medium. In cell culture experiments, primary amines were found to be connected with γ-carboxylic acid in glutamic acid [115]. A variety of chemically active compounds have been synthesized in *in-vitro* cell cultures of tea where peroxidase activity is high [116, 117]. Moreover, the production of caffeine and theobromine has been performed through callus and root suspension culture of tea. Additionally, soluble polymeric polyphenol compound lignin had been found to be synthesized on the surface of the callus culture of tea plant after exposing the medium to UV-B radiation. Moreover, other phenolic compounds had been found to be deposited in the cell wall and intercellular spaces in the same experiment [118]. Additionally, the production of flavans and lignin has been found to be happened in the callus culture from different tissues, *viz.* root, stem and leaf, in the presence of cadmium (Cd). Flavans had been found to be produced less than control in leaf calli, whereas flavans had been produced in almost equal amounts to that in control in root calli and in increased amounts in stem calli. On the other hand, root and stem calli formed an enhanced amount of lignin than control, but in leaf calli, lignin was produced in the same amount than that of control. Therefore, the formation of flavans and lignin in the Cd induced callus culture depends on the source tissue of calli and the concentration of Cd present in the culture medium [119]. In addition, the suspension culture of tea had synthesized catechins, caffeine, and polyphenols in elevated amounts when shikimic acid had been supplemented in the medium in an *in vitro* environment [120].

Catechins are one of the most important groups of flavonoids, which have antibacterial, antiallergic, and antioxidative properties [121, 122]. Although a few studies had been able to produce catechins in callus culture, the growth rate of the cells was very low and the content of flavonoids was less than the original tea plant. In another investigation, researchers utilized liquid Gamborg B5 medium and were successful in making a large-scale synthesis of catechins. During the culture period, light irradiation was standardized to obtain increased cell growth and catechin production. Flavonoid (catechin and proanthocyanidin) production was enhanced significantly in this experiment (150 mg/g dry cell weight). This amount was greater than that present in the intact plant. On the basis of the cell life cycle and expression of the major enzyme phenylalanine ammonia lyase (PAL) involved in the biosynthetic pathway of flavonoids, a kinetic model has been postulated for the production of flavonoids in tea in this study [123].

CONCLUSION

While *in vitro* propagation of tea is constrained, considerable success has also been achieved in the area of micropropagation of tea. It has been used to protect and produce endangered plants. Micropropagation of tea or any other plant can be used as a tool to reduce the problems of our global society [124]. Cost-effective protocols of micropropagation of tea plants with high leaf productivity, quality and resistance to biotic and abiotic stresses are still required for both quality and quantity production of tea. The complex gene regulatory networks associated with somatic embryogenesis need to be unraveled for better utilization of genes and regulatory elements for utilization of this method for commercial production of secondary metabolites. Overall, several cell culture experiments have been fruitful in the synthesis of secondary metabolites, *viz.* flavonoids, isoflavonoids, procyanidins, alkaloids, polyphenols, caffeine, theobromine, theophylline, and many more which have been commercially utilized. Some of these metabolites add characteristic flavour and act on our nervous system, whereas many of these compounds have medicinal properties which can prevent diabetes, cardiovascular diseases, lung infection, and cancer. In summary, understanding the mechanisms, gene functions and regulatory networks of cell and tissue culture methods, especially micropropagation and somatic embryogenesis, can provide useful information in the future for effective, economical production of secondary metabolites for commercialization and their utilization in human society.

REFERENCES

[1]   Gautheret RJ. Plant tissue culture: A history. Bot Mag Tokyo 1983; 96(4): 393-410.
[http://dx.doi.org/10.1007/BF02488184]

[2]   Gautheret RJ. History of plant tissue and cell culture: A personal account. Cell culture and somatic cell genetics of plants. New York: Academic Press 1985; Vol. 2: pp. 1-59.

[3]   Dagla HR. Plant tissue culture. Resonance 2012; 17(8): 759-67.
[http://dx.doi.org/10.1007/s12045-012-0086-8]

[4] Kaur A, Malhotra PK, Manchanda P, Gosal SS. Micropropagation and somatic embryogenesis in sugarcane. Biotechnologies of Crop Improvement 2018; 1: 57-91.
[http://dx.doi.org/10.1007/978-3-319-78283-6_2]

[5] Forrest GI. Studies on the polyphenol metabolism of tissue cultures derived from the tea plant (*Camellia sinensis* L.). Biochem J 1969; 113(5): 765-72.
[http://dx.doi.org/10.1042/bj1130765] [PMID: 5821008]

[6] Kato M. Regeneration of plantlets from tea stem callus. Japanese Journal of Breeding 1985; 35(3): 317-22.
[http://dx.doi.org/10.1270/jsbbs1951.35.317]

[7] Borchetia S, Das SC, Handique PJ, Das S. High multiplication frequency and genetic stability for commercialization of the three varieties of micropropagated tea plants (Camellia spp.). Sci Hortic 2009; 120(4): 544-50.
[http://dx.doi.org/10.1016/j.scienta.2008.12.007]

[8] Mukhopadhyay M, Mondal TK, Chand PK. Biotechnological advances in tea (*Camellia sinensis* [L.] O. Kuntze): a review. Plant Cell Rep 2016; 35(2): 255-87.
[http://dx.doi.org/10.1007/s00299-015-1884-8] [PMID: 26563347]

[9] Kato M. Camellia sinensis L. (Tea): *In vitro* regeneration. In: Bajaj YSP, Ed. Biotechnology in agriculture and forestry. Berlin, Heidelberg: Springer-Verlag 1989; 7: pp. 83-98.

[10] Vieitez AM, Vieitez ML, Ballester A, Vieitez E. Micropropagation of *Camellia spp*. In: Bajaj YSP, Ed. Biotechnology in agriculture and forestry. Berlin, Heidelberg: Springer-Verlag 1992; 19: pp. 361-87.

[11] Dood AW. Tissue culture of tea (*Camellia sinensis* (L.) O. Kuntze) :A review. Int J Trop Agric 1994; 12: 212-47.

[12] Murashige T. Plant propagation through tissue culture. Annu Rev Plant Physiol 1974; 25(1): 135-66.
[http://dx.doi.org/10.1146/annurev.pp.25.060174.001031]

[13] Murashige T. The impact of plant tissue culture on agriculture.Frontiers of plant tissue culture 1978.Intl Assoc Plant Tissue Culture. Univ. of Calgary Printing Services 1978; pp. 15-26.

[14] Thorpe TA. History of plant tissue culture. Mol Biotechnol 2007; 37(2): 169-80.
[http://dx.doi.org/10.1007/s12033-007-0031-3] [PMID: 17914178]

[15] Gvasaliya MV. Spontaneous and induced cultivars and forms of tea [Camellia sinensis (L.) Kuntze] in the humid subtropics of Russia and Abkhazia, propagation and conservation *in vitro*. PhD thesis. Krasnodar: Kuban State Agrarian University 2015.

[16] Samarina L, Gvasaliya M, Koninskaya N, *et al*. A comparison of genetic stability in tea [*Camellia sinensis* (L.) Kuntze] plantlets derived from callus with plantlets from long-term *in vitro* propagation. Plant Cell Tissue Organ Cult 2019; 138(3): 467-74.
[http://dx.doi.org/10.1007/s11240-019-01642-2]

[17] Vieitez AM, Barciela J, Ballester A. Propagation of *Camellia japonica* cv. Alba Plena by tissue culture. J Hortic Sci 1989; 64(2): 177-82.
[http://dx.doi.org/10.1080/14620316.1989.11515942]

[18] Iddagoda N, Kataeva NN, Butenko RG. *In vitro* clonal micropropagation of tea (*Camellia sinensis* L.) 1. Defining the optimum condition for culturing by means of a mathematical design technique. Indian J Plant Physiol 1988; 31: 1-10.

[19] Jha T, Sen SK. Micropropagation of an elite Darjeeling tea clone. Plant Cell Rep 1992; 11(2): 101-4.

[20] Widhianata H, Taryono . Organogenesis responses of tea [*Camellia sinensis* (L.) O. Kuntze] var. *assamica* and *sinensis*.. AIP Conf Proc 2019; 2099: 020026.
[http://dx.doi.org/10.1063/1.5098431]

[21] Murashige T, Skoog F. A revised medium for rapid growth and bioassay with tobacco tissue cultures. Physiol Plant 1962; 15(3): 473-97.
[http://dx.doi.org/10.1111/j.1399-3054.1962.tb08052.x]

[22] Gamborg OL, Miller RA, Ojima K. Nutrient requirements of suspension cultures of soybean root cells. Exp Cell Res 1968; 50(1): 151-8.
[http://dx.doi.org/10.1016/0014-4827(68)90403-5] [PMID: 5650857]

[23] Llyod G, McCown B. Commercially feasible micropropagation of mountain laurel, *kalmia lalifolia* by use of shoot tip culture. Comb Proc Int Pl Prop Soc 1980; 30: 421-7.

[24] Agarwal B, Singh U, Banerjee M. *In vitro* clonal propagation of tea (*Camellia sinensis (L.) O. Kuntze*). Plant Cell Tissue Organ Cult 1992; 30(1): 1-5.
[http://dx.doi.org/10.1007/BF00039995]

[25] Banerjee M, Agarwal B. *In vitro* rooting of tea, *Camellia sinensis* (L.) O. Kuntze. Indian J Exp Biol 1990; 28: 936-9.

[26] Kim YD, Yun JG, Seo YR, Karigar CS, Choi MS. Influence of mineral salts on shoot growth and metabolite biosynthesis in tea tree (*Camellia sinensis* L.). J.. Weonye Gwahag Gisulji 2015; 33(1): 106-13.
[http://dx.doi.org/10.7235/hort.2015.14055]

[27] Phukan MK, Mitra GC. Regeneration of tea shoots from nodal explants in tissue culture. Curr Sci 1984; 53: 874-6.

[28] Begum A, Ahmad I, Prodhan SH, Azad AK, Sikder MBH, Ara R.) Study of *in vitro* propagation of tea [*Camellia sinensis (L). O. Kuntze*] through different explants. J Global Biosci 2015; 7: 2878-87.

[29] Das SC, Barman TS. Tea shoot regeneration from embryo callus.New Trends in Biotechnology. New Delhi, Bombay: Oxford and IBH Publishing Co. Pvt. Ltd. 1992; pp. 81-6.

[30] Arulpragasam PV, Latiff R. Studies on the tissue culture on tea (*camellia sinensis* (l.) o. kuntze). 1. development of a culture method for the multiplication of shoots. Sri Lank J Tea Sci 1986; 55: 44-7.

[31] Nakamura Y. Shoot tip culture of tea cultivar yabukita. Tea Res J 1987; 1987(65): 1-7.
[http://dx.doi.org/10.5979/cha.1987.1]

[32] Nakamura Y. *In vitro* rapid plantlet culture from axillary buds of tea plant (C. sinensis (L.) O. Kuntze). Bull Shizuoka Tea Expt Stat 1987; 13: 23-7.

[33] Nakamura Y. Differentiation of adventitious buds and its varietal difference in stem segment culture of *camellia sinensis* (l.) o. kuntze. Tea Res J 1989; 1989(70): 41-9.
[http://dx.doi.org/10.5979/cha.1989.70_41]

[34] Abeywardana DASR, Ranaweera KK, Ranatunga MAB, Warnasooriya WMRSK. Somatic embryogenesis protocol for tea *[Camellia sinensis* (L.) O. Kuntze] *IRSyRUSI*. 3rd International Research Symposium Part 1 , Faculty of Management Studies. Rajarata University of Sri Lanka. 2015; pp. 80-9.

[35] Rajkumar R, Ayyappan P. Micropropagation of *Camellia sinensis* (L.) O kuntze. J Plant Crops 1992; 20: 252-6.

[36] Sarathchandra TM, Upali PD, Wijeweardena RGA. Studies on the tissue culture of tea {*Camellia sinensis* (L.) O. Kuntze}. Somatic embryogenesis in stem and leaf callus cultures. Sri Lanka J Tea Sci 1988; 52: 50-4.

[37] Phukan MK, Mitra GC. Nutrient requirements for growth and multiplication of tea plants *in vitro*.. Bangladesh J Bot 1990; 19: 65-71.

[38] Chen Z, Liao H. Obtaining plantlet through anther culture *of* tea plants. Zhongguo Chaye 1982; 4: 6-7.

[39] Nakamura Y. Effect of sugar on formation of adventitious buds and growth of axillary buds in tissue culture of tea. *bull shizuoka tea expt.*. Stat 1990; 15: 1-5.

[40] Murali KS, Pandidurai V, Manivel L, Rajkumar R. Clonal variation in multiplication of tea through tissue culture. J Plant Crops 1996; 24: 517-22.

[41] Bidarigh S, Azarpour E. The study effect of cytokinin hormone types on length shoot *in vitro* culture of tea (*camellia sinensis* L.). World Appl Sci J 2011; 13: 1726-9.

[42] Bidarigh S, Hatamzadeh A, Azarpour E. The study effect of iba hormone levels on rooting in microcuttings of tea (*camellia sinensis* L.). World Appl Sci J 2012; 20: 1051-4.

[43] Gunasekare MTK, Evans PK. *In vitro* rooting of microshoots of tea (*Camellia sinensis* L.). Sri Lanka J Tea Sci 2000; 66: 5-15.

[44] Jain SM, Das SC, Barman TS. Enhancement of root induction from *in vitro* regenerated shoots of tea (*Camellia sinensis* L.). Proc Indian Natl Sci Acad 1993; 59: 623-8.

[45] Tian-Ling L. Regeneration of plantlets in cultures of immature cotyledons and young embryos of *Camellia oleifera* Abel. Acta Biol Exp Sin 1982; 15: 393-403.

[46] Arulpragasam PV, Latiff R, Seneviratne P. Studies on tissue culture of tea (*Camellia sinensis* (L.) O. Kuntze). 3. Regeneration of plants from cotyledon callus. Sri Lank J Tea Sci 1988; 57: 20-3.

[47] Banerjee S. Genetic diversity analysis and development of *in vitro* regeneration system in selected clones of tea (Camellia sinensis L.) plant, Ph.D. Thesis. University of Dhaka 2019.

[48] Rajasekaran P, Mohankumar P. Rapid micropropagation of tea (*Camellia spp*). J Plant Crops 1992; 20: 248-52.

[49] Singh A, Sharma J, Rexer KH, Varma A. Plant productivity determinants beyond minerals, water and light: *Piriformospora indica*- A revolutionary plant growth promoting fungus. Curr Sci 2000; 79: 1548-54.

[50] Mondal TK, Parathiraj S, Mohan Kumar P. Micrografting: A technique to shorten the hardening time of micropropagated shoots of tea {Camellia sinensis (L) O. Kuntze}. Sri Lank J Tea Sci 2005; 70: 5-9.

[51] Sutini W, Augustien N, Purwanto DA, Muslihatin W. The production of cinnamic acid secondary metabolites through *in vitro* culture of callus camellia sinensis l. with the elicitor of cobalt metal ions. AIP Conference Proceedings 2019; 2120: 030028.

[52] Akin B. Tissue culture techniques of medicinal and aromatic plants: history, cultivation and micropropagation. JSR-A, 2020; 2687-6167.

[53] R D, v R, S M, N M, S R, S N. RAPD, ISSR and RFLP fingerprints as useful markers to evaluate genetic integrity of micropropagated plants of three diploid and triploid elite tea clones representing camellia sinensis (china type) and c. assamica ssp. assamica (assam-india type). Plant Cell Rep 2002; 21(2): 166-73.
[http://dx.doi.org/10.1007/s00299-002-0496-2]

[54] Ji PZ, Li H, Gao LZ, Zhang J, Chen GZQ, Huang XQ. ISSR diversity and genetic differentiation of ancient tea (*camellia* sinensis var. assamica) plantations from china: implications for precious tea germplasm conservation. Pak J Bot 2011; 43: 281-91.

[55] Lin ZH, Chen RB, Chen CS, *et al.* Preliminary application of issr markers in the genetic relationship analysis of tea plants. Chaye Kexue 2007; 27: 45-50.

[56] Liu BY, Li YY, Tang YC, Wang LY, Cheng H, Wang P-S. Assessment of genetic diversity and relationship of tea germplasm in yunnan as revealed by ISSR markers. Zuo Wu Xue Bao 2010; 36: 391-400.
[http://dx.doi.org/10.3724/SP.J.1006.2010.00391]

[57] Yao MZ, Chen L, Liang YR. Genetic diversity among tea cultivars from china, japan and kenya revealed by issr markers and its implication for parental selection in tea breeding programmes. Plant Breed 2008; 127(2): 166-72.
[http://dx.doi.org/10.1111/j.1439-0523.2007.01448.x]

[58] Yliu B, Wang LY, Yli Y, *et al.* Genetic diversity in tea (*camellia sinensis*) germplasms as revealed by ISSR markers Indian. J Agric Sci 2009; 79: 715-21.

[59] Waheed A. Molecular characterization of different tea clones "Camellia sinensis L" grown at NTHRI, Shinkiari. Int J Biosci 2017; 142-51.

[60] Al-Qurainy F, Nadeem M, Khan S, *et al.* Rapid plant regeneration, validation of genetic integrity by issr markers and conservation of *reseda pentagyna* an endemic plant growing in Saudi Arabia. Saudi J Biol Sci 2018; 25(1): 111-6.
[http://dx.doi.org/10.1016/j.sjbs.2017.07.003] [PMID: 29379366]

[61] Bhattacharyya P, Kumaria S, Diengdoh R, Tandon P. Genetic stability and phytochemical analysis of the *in vitro* regenerated plants of *dendrobium nobile* lindl., an endangered medicinal orchid. Meta Gene 2014; 2: 489-504.
[http://dx.doi.org/10.1016/j.mgene.2014.06.003] [PMID: 25606433]

[62] Bhattacharyya P, Kumar V, Van Staden J. Assessment of genetic stability amongst micropropagated *ansellia africana*, a vulnerable medicinal orchid species of africa using scot markers. S Afr J Bot 2017; 108: 294-302.
[http://dx.doi.org/10.1016/j.sajb.2016.11.007]

[63] Kumari M, Agnihotri D, Chanotiya CS, Mathur AK, Lal RK, Mathur A. Chemical and genetic stability of methyl chavicol-rich indian basil (*ocimum basilicum* var. cim-saumya) micropropagated *in vitro.*. S Afr J Bot 2017; 113: 186-91.
[http://dx.doi.org/10.1016/j.sajb.2017.08.018]

[64] Rathore MS, Mastan SG, Yadav P, Bhatt VD, Shekhawat NS, Chikara J. Shoot regeneration from leaf explants of *withania coagulans* (stocks) dunal and genetic stability evaluation of regenerates with RAPD and ISSR markers. S Afr J Bot 2016; 102: 12-7.
[http://dx.doi.org/10.1016/j.sajb.2015.08.003]

[65] Saha S, Adhikari S, Dey T, Ghosh P. RAPD and ISSR based evaluation of genetic stability of micropropagated plantlets of *Morus alba* L. variety S-1. Meta Gene 2016; 7: 7-15.
[http://dx.doi.org/10.1016/j.mgene.2015.10.004] [PMID: 26693403]

[66] Devarumath RM, Doule RB, Kawar PG, Naikebawane SB, Nerkar YS. Field performance and RAPD analysis to evaluate genetic fidelity of tissue culture raised plantsvis-à-vis conventional setts derived plants of sugarcane. Sugar Tech 2007; 9(1): 17-22.
[http://dx.doi.org/10.1007/BF02956908]

[67] Alizadeh M, Krishna H, Eftekhari M, Modareskia M, Modareskia M. Assessment of clonal fidelity in micropropagated horticultural plants. J Chem Pharm Res 2015; 7(12): 977-90.

[68] Devanand PS, Chen J, Henny RJ, Chao CCT. Assessment of genetic relationships among philodendron cultivars using AFLP markers. J Am Soc Hortic Sci 2004; 129(5): 690-7.
[http://dx.doi.org/10.21273/JASHS.129.5.0690]

[69] Thomas J, Rajkumar R, Mandal A. Metabolite profiling and characterization of somaclonal variants in tea (*camellia* spp.) for identifying productive and quality accession. Phytochemistry 2006; 67(11): 1136-42.
[http://dx.doi.org/10.1016/j.phytochem.2006.03.020] [PMID: 16714038]

[70] Anon . Research and development department. tata tea ltd. Munnar 1999.

[71] Creze J, Beauchesne MG. Camellia cultivation *in vitro*. Int Camellia J 1980; 12: 31-4.

[72] Pandidurai V, Murali KS, Manivel L, Rajkumar R. Factors affecting *in vitro* shoot multiplication and root regeneration in tea. J Plant Crops 1996; 24: 603-9.

[73] Alagarsamy K, Shamala LF, Wei S. Influence of media supplements on inhibition of oxidative browning and bacterial endophytes of *camellia sinensis* var. *sinensis.*. 3 Biotech 2018; 8(8): 356.
[http://dx.doi.org/10.1007/s13205-018-1378-9] [PMID: 30105181]

[74] Park SY, Klimaszewska K, Park JY, Mansfield SD. Lodgepole pine: The first evidence of seed-based somatic embryogenesis and the expression of embryogenesis marker genes in shoot bud cultures of adult trees. Tree Physiol 2010; 30(11): 1469-78.
[http://dx.doi.org/10.1093/treephys/tpq081] [PMID: 20935320]

[75] Kohlenbach HW. Basic aspects of differentiation and plant regeneration from cell and tissue cultures. Pro Life Sci 1977; 355-66.
[http://dx.doi.org/10.1007/978-3-642-66646-9_30]

[76] Monsanto A, Campus A, Middleton, Akula C. Protocols for somatic embryogenesis and plantlet formation from three explants in tea (Camellia sinensis (L.) O. KUNTZE C C) E E. . 2005.

[77] Wu CT. Studies on the tissue culture of tea plant. J Agric Assoc China 1976; 93: 30-42.

[78] Mondal TK, Bhattacharya A, Ahuja PS. Induction of synchronous secondary embryogenesis of tea (*camellia sinensis*). J Plant Physiol 2001; 158: 945-51. a
[http://dx.doi.org/10.1078/0176-1617-00179]

[79] Mondal T, Bhattacharya A, Ahuja P, Chand P. Transgenic tea [*camellia sinensis* (l.) o. kuntze cv. kangra jat] plants obtained by agrobacterium-mediated transformation of somatic embryos. Plant Cell Rep 2001; 20(8): 712-20.
[http://dx.doi.org/10.1007/s002990100382]

[80] Mondal TK, Bhattacharya A, Sood A, Ahuja PS. An efficient protocol for somatic embryogenesis and its use in developing transgenic tea (camellia sinensis (l.) o. kuntze) for field transfer. Plant Biotechnology and In Vitro Biology in the 21st Century 1999; 181-4.

[81] Dantu PK, Tomar PK. Somatic embryogenesis in *cellular and biochemical science*. In: Tripathi G, Ed. (New Delhi: I.K. International House Pvt Ltd. 2010; pp. 892-908.

[82] Akula A, Akula C. Protocols for somatic embryogenesis and plantlet formation from three explants in tea (Camellia sinensis (l.) o. kuntze.Protocol for somatic embryogenesis in woody plants Forestry Sciences. Dordrecht: Springer 2005; 77.
[http://dx.doi.org/10.1007/1-4020-2985-3_15]

[83] Wu CT, Huang TK, Chen GR, Chen SY. A review on the tissue culture of tea plants and on the utilization of callus derived plantlets. In: Rao AN, Ed. Proc Costed Symp Singapore. 104-6.

[84] Vieitez AM. Somatic embryogenesis in *Camellia spp*. In: Jain SM, Gupta PK, Newton RJ, Eds. Somatic Embryogenesis in Woody Plants Forestry Sciences. Dordrecht: Springer 1995; 44-46: pp. 235-76.
[http://dx.doi.org/10.1007/978-94-011-0491-3_14]

[85] Varhaníková M, Uvackova L, Skultety L, Pretova A, Obert B, Hajduch M. Comparative quantitative proteomic analysis of embryogenic and non-embryogenic calli in maize suggests the role of oxylipins in plant totipotency. J Proteomics 2014; 104: 57-65.
[http://dx.doi.org/10.1016/j.jprot.2014.02.003] [PMID: 24530378]

[86] Abraham GC, Raman K. Somatic embryogenesis in tissue culture of immature cotyledons of tea (*Camellia sinensis*). In: Somers DA, Gengenbach BG, Biesboor DD, Hackett WP, Green CE, Eds. 6th Inter Congr on Plant Tissue and Cell Culture. 294.

[87] Kato M. Micropropagation through cotyledon culture in *camellia japonica l. and c.sinensis l. ikushugaku zasshi*. 1986; 36: 31-8.

[88] Bronsema FBF, Van Oostveen WJF, Van Lammeren AAM. Influence of 2,4-D, TIBA and 3,5-D on the growth response of cultured maize embryos. Plant Cell Tissue Organ Cult 2001; 65(1): 45-56.
[http://dx.doi.org/10.1023/A:1010605519845]

[89] Bhojwani SS, Dantu PK. Somatic embryogenesis. Plant tissue culture: An introductory text. academia 2013; pp. 75-92.
[http://dx.doi.org/10.1007/978-81-322-1026-9_7]

[90]   Dantu PK, Tomar PK. Somatic embryogenesis in cellular and biochemical science. In: Tripathi G, Ed. New Delhi: I.K. International House Pvt Ltd 2010; pp. 92-908.

[91]   Bajaj YPS. Somatic embryogenesis and its applications for crop improvement. Bio in Agri Fores 1995; 105-25.

[92]   Tomiczak K, Mikuła A, Sliwinska E, Rybczyński JJ. Autotetraploid plant regeneration by indirect somatic embryogenesis from leaf mesophyll protoplasts of diploid *gentiana decumbens* L.f. *In Vitro*. Cell Dev Biol Plant 2015; 51(3): 350-9.
[http://dx.doi.org/10.1007/s11627-015-9674-0] [PMID: 26097374]

[93]   Akula A, Becker D, Bateson M. High-yielding repetitive somatic embryogenesis and plant recovery in a selected tea clone, 'tri-2025', by temporary immersion. Plant Cell Rep 2000; 19(12): 1140-5.
[http://dx.doi.org/10.1007/s002990000239] [PMID: 30754847]

[94]   Komamine A, Murata N, Nomura K. 2004 SIVB Congress symposium proceeding: mechanisms of somatic embryogenesis in carrot suspension cultures—morphology, physiology, biochemistry, and molecular biology. In Vitro Cell Dev Biol Plant 2005; 41(1): 6-10.
[http://dx.doi.org/10.1079/IVP2004593]

[95]   Gaj MD, Zhang S, Harada JJ, Lemaux PG. Leafy cotyledon genes are essential for induction of somatic embryogenesis of arabidopsis. Planta 2005; 222(6): 977-88.
[http://dx.doi.org/10.1007/s00425-005-0041-y] [PMID: 16034595]

[96]   Palovaara J, Hakman I. Conifer WOX-related homeodomain transcription factors, developmental consideration and expression dynamic of wox2 during *picea abies* somatic embryogenesis. Plant Mol Biol 2008; 66(5): 533-49.
[http://dx.doi.org/10.1007/s11103-008-9289-5] [PMID: 18209956]

[97]   Kiselev KV, Turlenko AV, Zhuravlev YN. PgWUS expression during somatic embryo development in a panax ginseng 2c3 cell culture expressing the rolc oncogene. Plant Growth Regul 2009; 59(3): 237-43.
[http://dx.doi.org/10.1007/s10725-009-9409-5]

[98]   Palovaara J, Hallberg H, Stasolla C, Hakman I. comparative expression pattern analysis of wuschel-related homeobox 2 *wox2* and wox8/9 in developing seeds and somatic embryos of the gymnosperm *picea abies.*. New Phytol 2010; 188(1): 122-35.
[http://dx.doi.org/10.1111/j.1469-8137.2010.03336.x] [PMID: 20561212]

[99]   Yazawa K, Takahata K, Kamada H. Isolation of the gene encoding carrot leafy cotyledon1 and expression analysis during somatic and zygotic embryogenesis. Plant Physiol Biochem 2004; 42(3): 215-23.
[http://dx.doi.org/10.1016/j.plaphy.2003.12.003] [PMID: 15051045]

[100] Trigiano RN, Gray DJ. Plant tissue culture, development, and biotechnology. CRC Press 2011.

[101] Cheng Y, Dai X, Zhao Y. Auxin synthesized by the YUCCA flavin monooxygenases is essential for embryogenesis and leaf formation in arabidopsis. Plant Cell 2007; 19(8): 2430-9.
[http://dx.doi.org/10.1105/tpc.107.053009] [PMID: 17704214]

[102] Salaün C, Lepiniec L, Dubreucq B. Genetic and molecular control of somatic embryogenesis. Plants 2021; 10(7): 1467.
[http://dx.doi.org/10.3390/plants10071467] [PMID: 34371670]

[103] Wójcik AM, Wójcikowska B, Gaj MD. Current perspectives on the auxin-mediated genetic network that controls the induction of somatic embryogenesis in plants. Int J Mol Sci 2020; 21(4): 1333.
[http://dx.doi.org/10.3390/ijms21041333] [PMID: 32079138]

[104] Tian R, Paul P, Joshi S, Perry SE. Genetic activity during early plant embryogenesis. Biochem J 2020; 477(19): 3743-67.
[http://dx.doi.org/10.1042/BCJ20190161] [PMID: 33045058]

[105] Pal S, Saha C. A review on structure–affinity relationship of dietary flavonoids with serum albumins. J Biomol Struct Dyn 2013; 32(7): 1132-47.
[http://dx.doi.org/10.1080/07391102.2013.811700]

[106] Yan Z, Zhong Y, Duan Y, Chen Q, Li F. Antioxidant mechanism of tea polyphenols and its impact on health benefits. Anim Nutr 2020; 6(2): 115-23.
[http://dx.doi.org/10.1016/j.aninu.2020.01.001] [PMID: 32542190]

[107] Fu QY, Li QS, Lin XM, *et al.* Antidiabetic effects of tea. Molecules 2017; 22(5): 849.
[http://dx.doi.org/10.3390/molecules22050849] [PMID: 28531120]

[108] Zhang L, Ttai Y, Wang Y, Meng Q. The proposed biosynthesis of procyanidins by the comparative chemical analysis of five Camellia species using LC-MS. Sci Rep 2017; 7: 46131.
[http://dx.doi.org/10.1038/srep46131]

[109] Matsuura T, Kakuda T, Kinoshita T, Takeuchi N, Sasaki K. Production of theanine by callus culture of tea. Proceedings of the International Symposium on Tea Science. Shizuka, Japan. 1991; pp. 432-6.

[110] Orihara Y, Furuya T. Production of theanine and other? -glutamyl derivatives by camellia sinensis cultured cells. Plant Cell Rep 1990; 9(2): 65-8.
[http://dx.doi.org/10.1007/BF00231550] [PMID: 24226431]

[111] Spedding DJ, Wilson AT. Caffeine Metabolism. Nature 1964; 204(4953): 73.
[http://dx.doi.org/10.1038/204073a0] [PMID: 14240119]

[112] Koretskaya TF, Zaprometov MN. Phenolic compounds in the tissue culture of *Camellia sinensis* and effect of light on their formation. Физиол раст 1975; 22: 941-6.

[113] Zaprometov MN, Zagoskina MV. One more evidence for chloroplast involvement in the biosynthesis of phenolic compounds. Plant Physiol 1979; 34: 165-72.

[114] Hao C, Wang Y, Yang S. Effects of macroelements on the growth of tea callus and the accumulation of catechins. J Tea Sci 1994; 14: 31-6.

[115] Furuya T, Orihara Y, Tsuda Y. Caffeine and theanine from cultured cells of *Camellia sinensis*.. Phytochemistry 1990; 29(8): 2539-43.
[http://dx.doi.org/10.1016/0031-9422(90)85184-H]

[116] Takemoto M, Aoshima Y, Stoynov N, Kutney JP. Establishment of *camellia sinensis* cell culture with high peroxidase activity and oxidative coupling reaction of dibenzylbutanolides. Tetrahedron Lett 2002; 43(39): 6915-7.
[http://dx.doi.org/10.1016/S0040-4039(02)01631-3]

[117] Takemoto M, Suzuki Y, Tanaka K. Enantioselective oxidative coupling of 2-naphthol derivatives catalyzed by Camellia sinensis cell culture. Tetrahedron Lett 2002; 43(47): 8499-501.
[http://dx.doi.org/10.1016/S0040-4039(02)02089-0]

[118] Zagoskina NV, Dubravina GA, Alyavina AK, Goncharuk EA. Effect of ultraviolet (UV-B) radiation on the formation and localization of phenolic compounds in tea plant callus cultures. Russ J Plant Physiol 2003; 50(2): 270-5.
[http://dx.doi.org/10.1023/A:1022945819389]

[119] Zagoskina NV, Goncharuk EA, Alyavina AK. Effect of cadmium on the phenolic compounds formation in the callus cultures derived from various organs of the tea plant. Russ J Plant Physiol 2007; 54(2): 237-43.
[http://dx.doi.org/10.1134/S1021443707020124]

[120] Muthaiya MJ, Nagella P, Thiruvengadam M, Mandal AA. Enhancement of the productivity of tea (*Camellia sinensis*) secondary metabolites in cell suspension cultures using pathway inducers. J Crop Sci Biotechnol 2013; 16(2): 143-9.
[http://dx.doi.org/10.1007/s12892-012-0124-9]

[121] Middleton E Jr. Biological properties of plant flavonoid: An overview. Int J Pharmacogn 1996; 34:

344-8.
[http://dx.doi.org/10.1076/phbi.34.5.344.13245]

[122] Valcic S, Timmermann BN, Alberts DS, *et al.* Inhibitory effect of six green tea catechins and caffeine on the growth of four selected human tumor cell lines. Anticancer Drugs 1996; 7(4): 461-8.
[http://dx.doi.org/10.1097/00001813-199606000-00011] [PMID: 8826614]

[123] Shibasaki-Kitakawa N, Takeishi J, Yonemoto T. Improvement of catechin productivity in suspension cultures of tea callus cells. Biotechnol Prog 2003; 19(2): 655-8.
[http://dx.doi.org/10.1021/bp025539a] [PMID: 12675612]

[124] Sutini W, Guniarti G, Purwanto A. The role *in vitro* culture methods of camellia sinensis l. plant in the global society transformation. 4th international seminar of research month nst proceedings pages. 277-83.

CHAPTER 11

In Vitro Strategies for Isolation and Elicitation of Psoralen, Daidzein and Genistein in Cotyledon Callus of *Cullen Corylifolium* (L.) Medik

Tikkam Singh[1], Renuka Yadav[1] and Veena Agrawal[1,*]

[1] *Department of Botany, University of Delhi, Delhi-110007, India*

Abstract: In recent times, natural herbal products/biomolecules are gaining immense impetus, over modern synthetic allopathic medicines, for curing serious human ailments as the former are proving their better efficacy, causing no or minimum side effects. Consequently, many pharmaceutical industries are coming forward for exploring novel drugs based on medicinal plants. *Cullen corylifolium* (L.) Medik., a well-known traditional medicinal herb of China and India, is extensively used in Ayurvedic medicine to cure several skin diseases such as psoriasis, leprosy and leucoderma. Besides, it also has properties like antioxidant, anti-cancer, anti-inflammatory, hepatoprotective, anti-diabetic, anti-mycobacterial, and anti-helminthic due to the occurrence of a number of important furanocoumarins and isoflavonoids. Furanocoumarins and isoflavonoids are biosynthesized *via* the phenylpropanoid pathway in the plant parts of *C. corylifolium* and are extensively used as anticancerous agents. The prominent marker compounds occurring in *C. corylifolium* are psoralen, genistein and daidzein produced mainly in the green seeds. These are highly expensive and occur in very low amounts. *In vitro* cell, tissue and organ culture can be used as an alternative, controllable, sustainable and eco-friendly tool for rapid multiplication of cells for the synthesis and elicitation of bioactive compounds. In addition, various strategies such as precursors feeding, hairy root culture, biotic and abiotic elicitors, cell suspension cultures, cloning and overexpression of genes involved in biosynthetic pathways of secondary metabolites. are also available for the enhancement of bioactive secondary metabolites. The present review aims at the screening of high-yielding elite plant parts, biosynthetic pathways of psoralen, daidzein and genistein, and various strategies employed for their elicitation and isolation in *C. corylifolium*.

Keywords: Biosynthetic Pathway, *Cullen corylifolium*, Callus, Daidzein, Elicitation, Genistein, Green seed cotyledons, Isolation, Key enzyme genes, Psoralen.

---

* **Corresponding author Veena Agrawal:** Department of Botany, University of Delhi, Delhi-110007, India; E-mail: drveena_du@yahoo.co.in

Mohammad Anis & Mehrun Nisha Khanam (Eds.)
All rights reserved-© 2024 Bentham Science Publishers

INTRODUCTION

Medicinal plants have been reckoned as potential sources of important biomolecules [1] that play a very crucial role in treating several serious ailments and saving numerous precious lives. In recent times, the natural compounds, due to their synergistic activity, less or no toxicity and more compatibility, are gaining much attention in the pharmaceutical industries for novel drug designing and several other therapeutic uses. Incidentally, such compounds occur in very low amounts in plant cells, and their enhancement involves a series of biosynthetic pathways, key enzyme genes and different abiotic and biotic stressors. With the onset of modern phytochemical and analytical tools, few past decades witnessed an immense progression in the discovery of natural products-based drugs, target diseases and their mechanism of action. Considering variable climatic conditions, the Indian subcontinent is a habitat for diverse flora that contributes immensely to herbal medicines at national and international markets. The ever-increasing demand for herbal products is encountered by collections from natural habitats, making them susceptible being threatened and even extinct in some cases. However, the plant cell, tissue and organ culture technique offers a sustainable approach for germplasm conservation as well as enhanced synthesis of bioactive compounds. It is thus imperative to develop a cost-effective and efficient in vitro protocol to cope up with the needs of the pharmaceutical and nutraceutical industries. Besides, in *vitro* elicitation of secondary metabolites facilitates the yield of higher amounts of pharmaceutically desirable compounds with limited resources without sacrificing the entire plants. Callus derived from various plant parts is considered the best source material for the incessant and rapid proliferation of cells for the elicitation of biomolecules [2]. Enhancement of biologically active secondary metabolites can be achieved following different methods, *e.g.*, through precursor feeding, hairy root culture, using biotic and abiotic elicitors, cell suspension cultures, cloning and overexpression of genes involved in biosynthetic pathways of secondary metabolites. Biotic, abiotic and phytohormonal elicitors may induce the synthesis of bioactive compounds [3 - 5]. Though the mechanism of action of elicitors is not much explored, it is believed that elicitors interact with the receptors present on the plasma membrane, which activate the intracellular signal transduction system, including G proteins, calcium ions (Ca^{2+}) and other secondary messenger molecules with the mitogen-activated protein kinases (MAPKs) pathway for the biosynthesis of bioactive compounds [6, 7].

Cullen corylifolium (L.) Medik. (syn. *Psoralea corylifolia* L.) (2n = 22) is a traditional medicinal herb of China and India and extensively used in Ayurvedic medicine [8]. It belongs to the family Fabaceae which comprises almost 751 genera and 19,500 species [9]. The genus *Cullen* consists of 32 species. It is known by various vernacular names such as purple fleabane or West Indian satinwood (English), Babchi or Bakuchi (Hindi), Kusthahantri or Sitavari (Sanskrit), Babechi or Bawachi (Urdu), *etc*. It is widely distributed in the North East tropical Africa, South West Arabian Peninsula, and tropical and subtropical Asia. In India, it is dispersed in Punjab, Rajasthan, Uttar Pradesh, Madhya Pradesh, Chhattisgarh and Tamil Nadu. *Cullen corylifolium* (2n = 22), is an annual herb growing up to the height of 30–180 cm. Fruit is one seeded, elongated and smooth. Mature seeds are glabrous, pitted, compressed, dark brown and non-endospermic. It blooms from July to August and seeds ripen from September to October. Interestingly, it is extensively utilized in Ayurvedic medicine to cure several skin ailments, such as leukoderma, psoriasis and leprosy [10 - 12]. Due to the presence of coumarin, furanocoumarin, flavononoid, isoflavonoid, flavone, meroterpene, chalcone and coumestan, this medicinally important plant possesses potent antioxidant, anti-cancer, anti-inflammatory, hepatoprotective, anti-diabetic, anti-mycobacterial, and anti-helminthic properties [13 - 17]. Furanocoumarins and isoflavonoids are synthesized in the different locations of *C. corylifolium* and are widely employed as anticancerous agents. Biosynthesis of anticancerous agents such as furanocoumarins and isoflavonoids occurs in different locations of *C. corylifolium*. The major bioactive marker compounds of *C. corylifolium,* such as psoralen, genistein and daidzein, are biosynthesized mainly in green seeds *via the* phenylpropanoid pathway [18].

BIOSYNTHETIC PATHWAYS OF PSORALEN, DAIDZEIN AND GENISTEIN

Psoralen, a furanocoumarin, is the major marker compound of *C. corylifolium* synthesized from umbelliferon *via* the phenylpropanoid pathway [19]. Phenylalanine acts as a precursor molecule to form cinnamate by the action of *phenylalanine ammonia lyase* (*PAL*), followed by the formation of p-coumaric acid, which is further converted into umbelliferone (7-hydroxycoumarin) through ortho-hydroxylation. The crucial step of biosynthesis is prenylation, which is catalyzed by *umbelliferone dimethylallyl transferase* and converts umbelliferone to demethylsuberin [20]. The *marmesin synthase* and *psoralen synthase* are two different cytochrome P450 (CYP450) enzymes that finally catalyse the conversion of demethylsuberosin to psoralen [21, 22]. A schematic representation of psoralen biosynthesis is shown in Fig. (**1**).

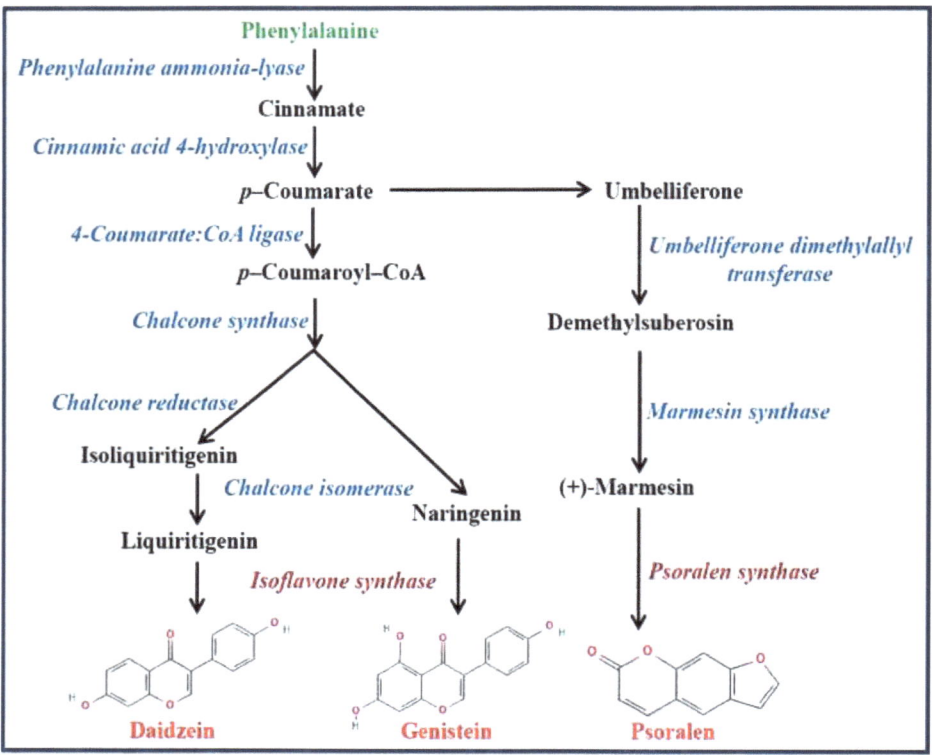

Fig. (1). Schematic representation of biosynthesis of psoralen, daidzein and genistein *via* the phenylpropanoid pathway (Adapted from Parast *et al.* [16], Deng and Lu [26] and Jian *et al.* [22]).

Isoflavones are a subgroup of flavonoids mainly synthesized in the family Fabaceae [23]. Predominant isoflavones occurring in *C. corylifolium* are daidzein and genistein. *Isoflavone synthase* (*IFS*) belongs to the CYP93C subfamily of the P450 superfamily, which transfers flavanone intermediates from the biosynthetic pathways of other flavonoids to isoflavones. *IFS* is the key enzyme which converts flavanones to isoflavones *via* transfer of B-ring from C2 to C3 and is associated with C2=C3 double bond formation. Similarly, *IFS* converts naringenin and legume-specific liquiritigenin into their corresponding isoflavones, genistein and daidzein [24 - 26]. The biosynthetic pathway for daidzein and genistein is represented in Fig. (**1**).

SCREENING FOR HIGH PSORALEN, DAIDZEIN AND GENISTEIN YIELDING PLANT PART OF *C. CORYLIFOLIUM*

Biosynthesis and accumulation of secondary metabolites in plants are mainly influenced by biotic (herbivores, pathogens, *etc.*) and abiotic (light, humidity, temperature, soil, water and pH) factors, including geographical location, temporal and seasonal variations [27], chemotypes [28, 29] genotypes [30], and various stages of plant development [31]. The growth and development of plants can be affected by these factors, as well as their ability to synthesize secondary metabolites, resulting in changes in their overall metabolic profile [32]. In this perspective, the selection of an elite plant part yielding maximum content of desirable compounds is a necessary and foremost step prior to venturing their isolation and elicitation. Incidentally, only a few studies pertaining to the screening of plant parts yielded higher content of psoralen. Various plant parts of *C. corylifolium*, *e.g.*, seeds, inflorescence, buds, bracts, leaf, node and roots, have been analysed for psoralen content employing HPLC. Among all mentioned above, seeds yield a higher quantity of psoralen [33]. It is believed that seeds are the primary source of essential compounds, including bioactive molecules necessary for plant growth and development, defence, and other functions. Detailed investigation was done where they had procured seeds from different locations in India, such as Ghaziabad (Uttar Pradesh), Haridwar (Uttarakhand), Khari-Baoli (Delhi), Kolhapur (Maharashtra), Mansur (Madhya Pradesh), Manesar (Haryana), Sirsa (Haryana) and Sirohi (Rajasthan), which proved that elite chemotype yielding higher psoralen content was from Sirohi region of Rajasthan which produced a maximum quantity of psoralen (4763.00 µg/ g fresh wt.). Before proceeding to the elicitation experiment, screened callus was derived from different sources such as root, node, leaf and cotyledons of green seeds of *P. corylifolia* and analysed through HPLC for optimum yield of psoralen. The results revealed that calluses derived from cotyledons of green seeds had maximum psoralen content, and such calluses have selected for further elicitation using biotic and abiotic elicitors. Besides, there are no such reports pertaining to daidzein and genistein, which need to be further explored.

STRATEGIES FOR EXTRACTION AND ISOLATION OF BIOACTIVE COMPOUNDS IN *C. CORYLIFOLIUM*

For the phytochemical profiling of medicinal plants, extraction is the first crucial step to extract the bioactive compounds from crude plant material for further isolation, identification and characterization. However, before extraction, some basic steps include pre-washing, shade or freeze drying of plant material, and grinding to obtain a homogenous sample. Coarsely powdered plant sample increases the kinetics and surface area in various solvent systems. Various analytical techniques such as maceration, sonification, heating under reflux and Soxhlet extraction are currently being employed for the extraction of biomolecules using different inorganic solvents of variable polarities [34, 35]. Polar solvents (methanol, ethyl acetate and chloroform) and non–polar solvents (hexane, toluene and dichloromethane) are normally being utilized for the extraction of hydrophilic and hydrophobic biomolecules, respectively. Plant crude extract contains various bioactive compounds of varying polarities; however, their isolation still remains a considerable challenge for further characterization and identification. To obtain pure bioactive compounds, a number of analytical tools, *i.e.*, Thin Layer Chromatography (TLC), Column Chromatography, Flash Chromatography, Sephadex Chromatography and High-Performance Liquid Chromatography (HPLC), are currently being employed for the separation and isolation of bioactive components present in crude extract of plants. Further, these isolated bioactive compounds may be identified, validated and characterized by using different techniques such as HPLC, Gas Chromatography–Mass Spectrometry (GC–MS), Liquid Chromatography–Mass Spectrometry (LC–MS), Fourier Transform–Infrared spectroscopy (FTIR) and Ultraviolet–Visible (UV–Vis) spectra High Resolution Mass Spectrometry (HRMS) and Nuclear Magnetic Resonance Spectroscopy (^{1}H-NMR) [36]. A report has used a combination of inorganic solvents, *i.e.*, Toluene: Ethyl acetate (75:25, *v/v*) for the *in vitro* isolation of biomolecules psoralen, genistein and daidzein from the crude extract of seeds of *C. corylifolium* through TLC, column chromatography (Fig. **2**). ^{1}H-NMR and HR-MS techniques have been employed for the characterization of abovesaid compounds (Fig. **2**). There are several phytochemical studies on green seeds or fruit of *C. corylifolium* which have documented the occurrence of several other biologically active compounds, *viz.* bavachin, bakuchiol, bakuchicin, angelicin (isopsoralen), neobavaisoflavone, bavachinin, corylifolinin, corylifols, corylifol D, corylifol E, aryl coumarin, daidzin, coryfolin, corylin, corylifol A, corylifol B, corylifol C, hydroxy bukuchiol, bavachalcone, bavachinone A, bavachinone B, bavacoumestan C, coryaurone A, hydroxypsoralenol A, hydroxypsoralenol B, isobavachalcone, isobavachin, isopsoralen, psoralidin, psoracorylifol D, psoracoumestan, xanthoangelol, bakuisoflavone, bakuflavanone, astragalin, bavacoumestan D, daidzein and genistein [37 - 53].

SCALING UP OF BIOACTIVE COMPOUNDS IN *C. CORYLIFOLIUM* THROUGH VARIOUS STRATEGIES

Plant cell cultures are also known as "Plant Cell Factories", which represent a promising, sustainable, cost-effective and alternative strategy to classical approaches for the production of biomolecules [54, 55]. This biotechnological approach offers a continuous supply of bioactive molecules of pharmaceutical interest by means of large-scale cultures from selected high yielding plant genotype, chemotype or elite plant part [56]. In recent times, natural compounds, due to their synergistic activity, less or no toxicity and more compatibility, are gaining much attention in the industries like pharmaceutical and nutraceutical for novel drug formulation and several other therapeutic benefits [57, 58]. Interestingly, these biologically active compounds are synthesised at specific sites for specific plants, depending on their growth and development stages, environmental factors, and micro- and macronutrient availability [59]. In addition, such conditions may alter the medicinal properties of the active constituent, resulting in poor quality and quantity of the bioactive compounds. Thus, in order to produce desired secondary metabolites that are not affected by environmental conditions, an alternative and sustainable technique must be exploited. As an alternative to traditional methods, plant cell and organ cultures are currently being utilized for the production of bioactive molecules. Plant cell factory offers a continuous and recurrent supply of biomolecules through large-scale culture and possesses the following advantages over the conventional method [60 - 65]:

i. Bioactive compounds of interest can be harvested anywhere.

ii. Uniform quality and continuous production of bioactive compounds.

iii. Production of contamination (bacteria, virus and fungi) free plant material.

iv. Independent of various seasonal, temporal and other environmental conditions.

v. Short growth cycles of a few weeks rather than months/years.

vi. Conservation of endangered plant species without disrupting their natural habitat.

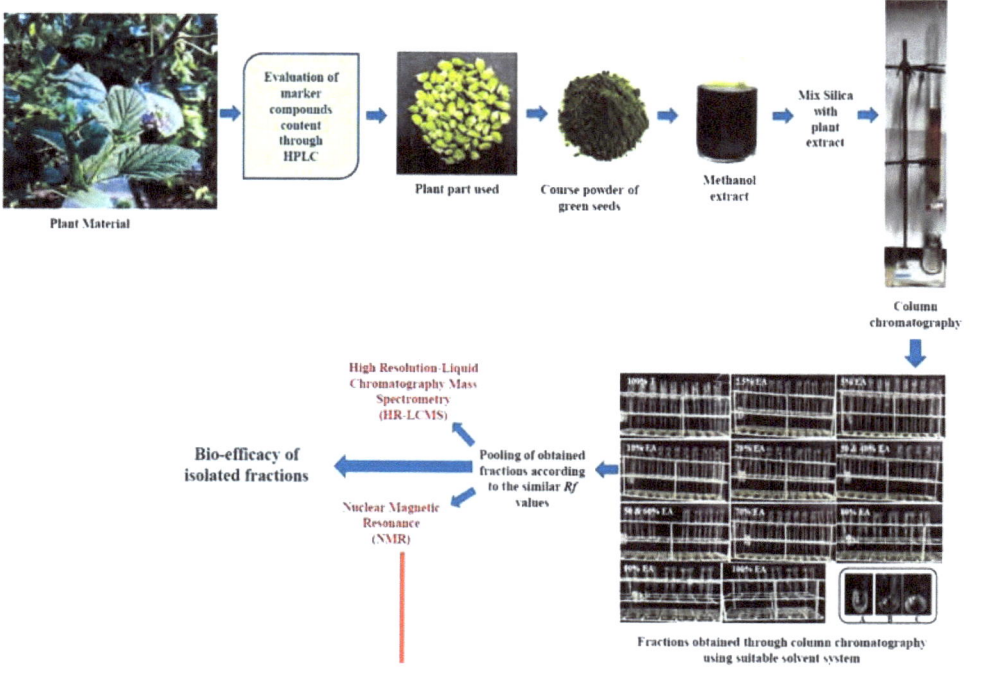

(Fig. 2) contd.....

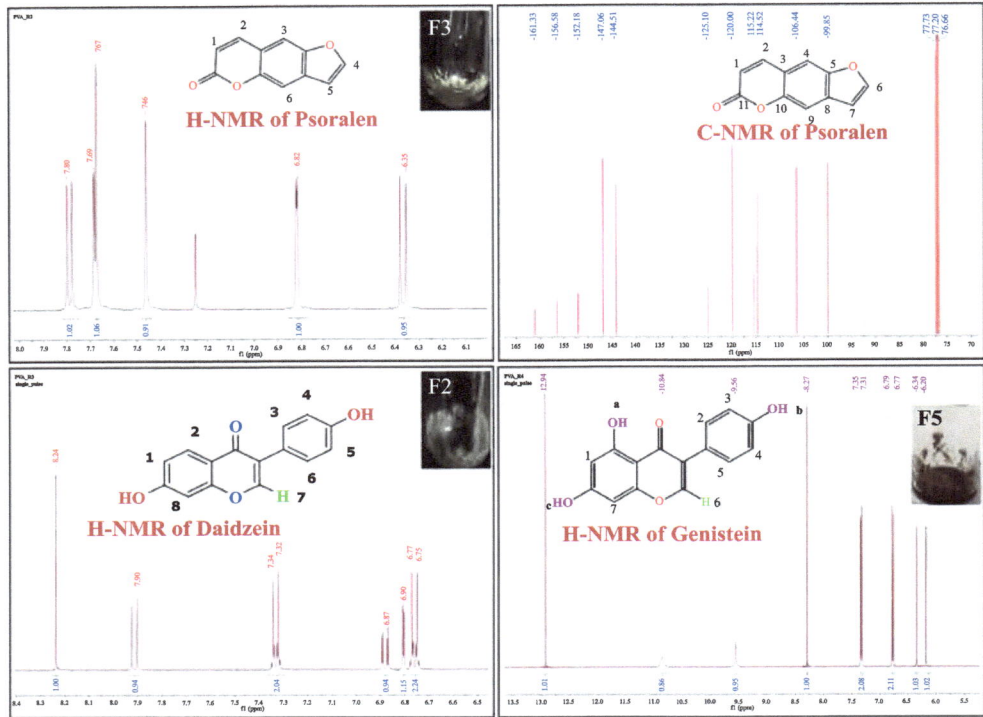

Fig. (2). Schematic representation of extraction, isolation and identification of markers compounds such as psoralen, daidzein and genistein from the green seeds of *C. corylifolium* using various techniques such as column chromatography, HR-LCMS, and NMR. F2, F3 and F5 represent crystals of daidzein, psoralen and genistein, respectively.

A literature survey revealed that psoralen, daidzein and genistein possess several pharmaceutical and therapeutic potential. Such potential of these biomolecules makes them highly valuable to the pharmaceutical industry, thus, the yield of these biomolecules needs to be increased *via* numerous elicitation techniques such as precursor feeding, hairy root cultures, *in vitro* elicitation using biotic and abiotic elicitors, cell suspension cultures, bioreactors and gene cloning of key enzymes (Fig. 3). There are only a few reports pertaining to elicitation of psoralen, daidzein and genistein in *in vitro* cultures or *in vivo* plants of *C. corylifolium* using various biotic and abiotic elicitors or precursor molecules enlisted in Table **1**.

Precursor Feeding

Precursor feeding has been widely utilized to increase the yield of bioactive compounds of interest in plant cell and organ cultures. It enhances the yield by converting the precursor molecules into secondary metabolites or by causing stress to plant cells and organ cultures [66]. It emphasises the addition of precursor molecules of the desired secondary metabolites to the culture medium, later absorbed by the plant cell and organ cultures. Various precursors of the phenylpropanoid pathway have been employed by several researchers to improve the yield of psoralen, daidzein and genistein in *in vitro* cultures of *C. corylifolium*. Phenylalanine has been seen to promote the activity of phenylalanine ammonia lyase (PAL), which ultimately catalyses to cinnamic acid. Parast and co-workers utilized cinnamic acid, umbelliferone and NADPH for enhancement of psoralen content in cotyledon callus cultures of *P. corylifolia* where they recorded the optimum increment in psoralen yield with cinnamic acid. Cinnamic acid may involve in the induction of the genes participating in the psoralen biosynthesis, modulating their expression, thereby accumulating psoralen in *P. corylifolia* [67]. Another group of researchers has also investigated *P. corylifolia* for its daidzein and genistein contents using hairy roots and cell suspension cultures. They fed the hairy roots and cell suspension cultures with precursor phenylalanine which consequently enhanced the yield of daidzein and genistein by 1.3-fold over control [68, 69].

Hairy Root Cultures

Hairy roots of a plant are derived from the infection of *Rhizobium rhizogenes* (previously known as *Agrobacterium rhizogenes*), a gram-negative soil bacterium that produces a variety of biomolecules. Hairy roots emerge from wounded plant parts and show neoplastic growth even on a growth regulator free medium. It is believed that such hairy roots are capable of synthesizing and secreting complex active glycoproteins from a wide range of organisms. Besides chromosomal DNA, *R. rhizogenes* cells also contain extrachromosomal DNA, *i.e.*, root-inducing plasmid (Ri) (250 kbp). Ri plasmid harbors a self-reverting sequence of DNA, *i.e.*, transfer-DNA (T-DNA) of approximately 25 kbp, which contains root-inducing loci such as Rol A, B, C and D, which are responsible for the formation and promotion of root growth from infected plant parts. Rol B is, however, considered to be the most important because insertion of Rol B promotes induction and proliferation of hairy roots from *Nicotiana tabaccum* leaf segments. Whereas, other root loci only promote the growth of roots that were already induced by Rol B and could not induce new roots on their own [70]. Bacterial Rol B locus codes for β-glucosidase which hydrolyses inactive indoxyl-β-glucosides. Due to the structural similarity between indoxyl-β-glucosides and indole-3-acety--β-glucosides in transformed plants, β-glucosidase hydrolyses indole-3-acety--β-glucosides and releases active auxins from inactive conjugates. This leads to an increase in intracellular auxin concentration and subsequently promotes root growth and development [71]. Transformed hairy roots show high proliferation rates even in growth regulator free medium as compared to untransformed roots. Due to minimal nutritional requirements and faster growth rates, hairy roots can be employed to scale up biomolecules production in bioreactors [72]. They produce an array of bioactive constituents which are produced by the plant in normal conditions but in relatively lower amounts. There are many studies regarding the elicitation of biologically active molecules using hairy root cultures in *C. corylifoilum* and *Psoralea corylifolia* were carried out, where, independently, Ri T-DNA integration as well as differences in their secondary metabolite profile were observed [73]. Furthermore, when hairy root cultures were supplemented with NaCl for 5 weeks, which consequently, enhanced the daidzein production twice compared to control, however, a reduction in its growth has also been observed after 14 days of culture [74]. It has also been observed that salt stress stimulated the production of daidzein with increased oxidative stress. Simultaneously, it has also been recorded that intracellular Ca^{2+} concentration increased with enhanced activities of calcineurin and Ca^{2+}-dependent protein kinases but downregulated the activity of Ca^{2+} – calmodulin-dependent protein kinase. Hairy roots of *Psoralea corylifolia* have responded well when treated with acetyl salicylic acid and jasmonic acid for 10 weeks, where jasmonic acid has been shown to be optimum for the increased production of 7-O-glucoside of daidzein [75]. Hairy roots have been cultured on a modified MS medium, wherein NH_4^+ and NO_3^- increased the biomass and productivity, and enhancement in the daidzein content and decline in genistein production with an increased level of sucrose were recorded. However, low concentrations of PO_4^{3-} have elevated the level of daidzein and genistein in the hairy roots of *P. corylifolia*.

In Vitro Elicitation Using Biotic & Abiotic Elicitors

Elicitation is one of the most efficient biotechnological techniques for ameliorating the yield of biomolecules. Elicitation can be achieved by employing various elicitors in *in vitro* plant cell and tissue cultures or *in vivo* plants. Elicitors are substances which have the ability to improve or induce the biosynthesis of certain compounds which play an important function in the adaptations of plants under stressful conditions. On the basis of origin, elicitors may be classified as biotic (biological origin), abiotic (non–biological origin) and signalling molecules (phytohormones). In plants, elicitors serve as ligands or secondary messengers that bind to receptors on the plasma membrane in order to trigger signal transduction that affects physiological and biochemical processes. In the presence of an elicitor, the intracellular transduction system is activated, the NADPH cascade is activated, reactive oxygen species are produced, defense genes are expressed, GTP-binding proteins are expressed, intracellular cAMP levels and calcium levels are high, and other secondary messengers with mitogen-activated protein kinases are generated [76, 77]. Biotic elicitors such as fungal extract, including chitosan and yeast extract, have been employed to enhance the quantity of psoralen in cell suspension cultures of *P. corylifolia* [78]. Supplemented chitosan, proline and salicylic acid to the cotyledons of green seed derived callus of *C. corylifolium* for the increased synthesis of psoralen, daidzein and genistein content.

Cell Suspension Cultures/Bioreactors

Plant cell suspension cultures have been studied by employing biotechnological approaches and offer a potential and alternative approach for the synthesis of biomolecules of interest. Cell suspension culture can be considered as free cells or small groups of totipotent cells derived from callus of various parts of medicinally important plants inoculated in a liquid medium that can be exploited for the production of high-value natural products which can later be utilized by therapeutical and pharmaceutical industries [79]. Over the last 20 years, significant progress has been recorded in plant molecular biology; consequently, cell suspension culture has emerged as an alternative, reliable and sustainable production tool for plant-derived therapeutics. This technique is simple and cost-effective and has been extensively utilized for large-scale yields of various bioactive compounds.

Gene Cloning of Key Enzymes and Overexpression

The aim of improved production of bioactive compounds can be accomplished by metabolic engineering, which involved cloning of genes/enzymes in the biosynthetic pathway, their characterization, evaluation of gene expression and regulation of metabolism [80]. The cloning and characterization of psoralen synthase gene from *Psoralea corylifolia* was done by preparing cDNA from RNA isolated from leaf tissues using the forward primer RBL2 (5'-CGC AAG CAT CAC TCA GAG AC-3') and the reverse primer RBR (5'-GAG CTG GGA ATG AGT ATT GAC ATG G -3') RBR1 (5' -CAA ACA TGT GGT CTG GCA AC-3'). They have also developed transgenic tobacco plants to study the expression of the gene, and have demonstrated that *PcIFS (Psoralea corylifolia Isolflavone synthase)* expression results in genistein production in a non-leguminous plant that does not normally biosynthesize these molecules. *Isolflavone synthase* (IFS) is a member of the CYP450 protein family, which plays an important role in the 2-hydroxylation of flavanone precursors as well as 1, 2-aryl migration reaction. There is high homology between PcIFS and IFS from other plants, and PcIFS encoding a polypeptide of 520 amino acids contains conserved amino acid residues necessary for the reaction catalyzed by CYP93C to cause aryl migration [81]. In a study reported more than 95% of IFS amino acid sequences were identical across non-leguminous plants, suggesting an enzyme function is influenced by a specific sequence requirement. Based on the phylogenetic evaluation, PcIFS and PmIFS appear to be closely related. The phylogenetic analysis and conservation of activity-specific motifs support that PcIFS encodes functional IFS from *P. corylifolia*. Genes of the isoflavonoid biosynthetic pathway are transcriptionally controlled by tissue specific and environmental factors [82 - 84]. PcIFS expression and isoflavone synthesis have been observed in different parts of *Psoralea*. On the other hand, IFS genes in *Glycine max* have been majorly expressed in the seeds and roots [85]. PcIFS is widely expressed and isoflavones accumulate in different plant parts, indicating that this gene is actively expressed. There are several reports which showed that elicitors and pathogens can induce the production of isoflavonoids in plants. Alfalfa leaves have been found to express IFS and other related genes involved in isoflavone biosynthesis [86]. In the case of soybean seedling, IFS expression has been stimulated by defense signals salicylic acid SA) and jasmonic acid (JA). Therefore, the upregulation of PcIFS by SA, methyl jasmonate (MJ) and wound indicates the conserved function of IFS in response to defense among various plant species. JA and wound-induced IFS expression results in a significant increment in the genistein and daidzein contents. Thus, it can be concluded that defense signals have the ability to induce the expression of PcIFS, whereas other genes of the isoflavonoid pathway may show differential expression. Similarly, differential expression of genes involved in the flavonoid pathway was observed when suspension cultures of *Medicago truncatula* treated with yeast extract and MJ [87]. As well as the expression analysis of other genes of the flavonoid pathway, a comprehensive analysis of flavonoids might reveal the molecular events responsible for the variation in isoflavonoid levels in different tissues and their response to different environmental stimuli. Based on the above reports, it has been suggested that significant levels of isoflavones may only be produced in the tissues where the activity of the phenylpropanoid pathway is occurred [88]. For a better understanding of isoflavonoid biosynthesis in *C. corylifolium*, genes involved in the phenylpropanoid pathway need to be cloned and characterized, and their metabolism must be profiled using high-throughput metabolic analysis.

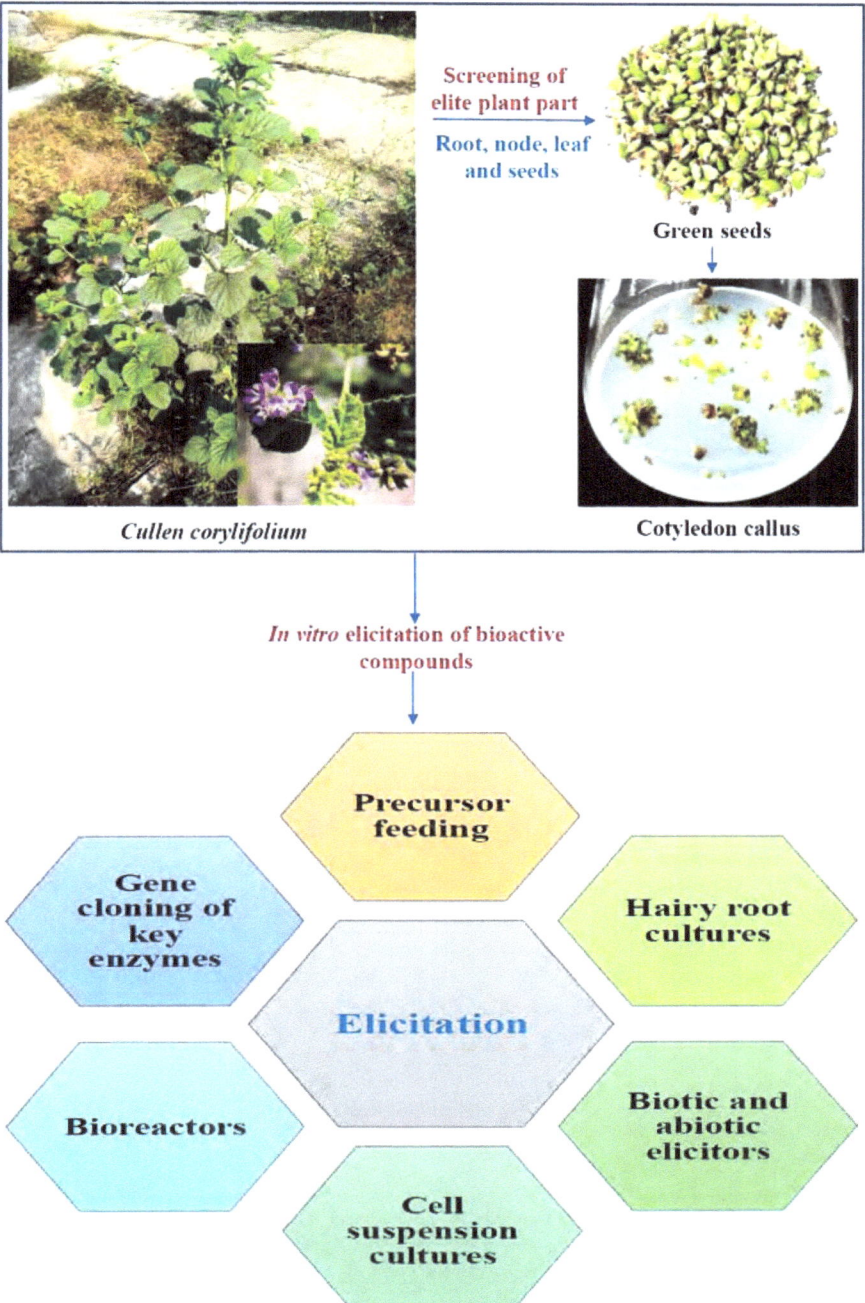

Fig. (3). Induction of callus from cotyledons of green seeds and *in vitro* elicitation of bioactive compounds using various strategies.

Consequently, such and similar reports will not only provide a deeper understanding of the biosynthesis of isoflavonoid in *C. corylifolium*, but will also provide molecular insight into its potential biotechnological improvement and other non-leguminous plants.

CONCLUSION

C. corylifolium is a traditional medicinal herb extensively utilized in Ayurvedic medicine to cure many skin ailments such as leprosy, psoriasis and leucoderma due to the presence of a variety of therapeutically valuable isoflavonoids and furanocoumarins. Marker compounds of this herb, such as psoralen, genistein and daidzein, have proved their efficacy as anticancerous agents. Incidentally, the demand for these marker compounds does not meet the need of therapeutic and pharmaceutical industries; thus, the synthesis and elicitation of above said biomolecules is a topic of concern. The current chapter underlined the various strategies for the isolation as well as scaling up of biomolecules, *viz.*, psoralen, genistein and daidzein from the green seeds of *C. corylifolium*. Interestingly, bioreactors can be employed for the large-scale production of psoralen, daidzein and genistein using the optimized concentration of elicitors, *e.g.*, salicylic acid and chitosan. Besides, further transcriptomics and metabolomics studies can also be conducted in order to find out the other possible genes or pathways involved in their synthesis in *C. corylifolium,* as reviewed [89]. There are few or no studies pertaining to gene expression studies during the *in vitro* elicitation, which can further be the area of investigation for this medicinally important plant.

REFERENCES

[1] Buchanan BB, Gruissem W, Jones RL. Biochemistry and molecular biology of plants. Hoboken, NJ: Wiley-Blackwell: Biochemistry and Molecular Biology of Plants 2015.

[2] Mulabagal V, Tsay HS. Plant cell cultures-an alternative and efficient source for the production of biologically important secondary metabolites. Int J Appl Sci Eng 2004; 2: 29-48.

[3] Baenas N, García-Viguera C, Moreno D. Elicitation: A tool for enriching the bioactive composition of foods. Molecules 2014; 19(9): 13541-63.
[http://dx.doi.org/10.3390/molecules190913541] [PMID: 25255755]

[4] Singh T, Sharma U, Agrawal V. Isolation and optimization of plumbagin production in root callus of Plumbago zeylanica L. augmented with chitosan and yeast extract. Ind Crops Prod 2020; 151: 112446.
[http://dx.doi.org/10.1016/j.indcrop.2020.112446]

[5] Singh T, Yadav R, Agrawal V. Effective protocol for isolation and marked enhancement of psoralen, daidzein and genistein in the cotyledon callus cultures of *Cullen corylifolium* (L.) Medik. Ind Crop Prod 2020; 143: 111905.

[6] Vasconsuelo A, Boland R. Molecular aspects of the early stages of elicitation of secondary metabolites in plants. Plant Sci 2007; 172(5): 861-75.
[http://dx.doi.org/10.1016/j.plantsci.2007.01.006]

[7] Goel MK, Mehrotra S, Kukreja AK. Elicitor-induced cellular and molecular events are responsible for productivity enhancement in hairy root cultures: An insight study. Appl Biochem Biotechnol 2011;

165(5-6): 1342-55.
[http://dx.doi.org/10.1007/s12010-011-9351-7] [PMID: 21909631]

[8] Chopra B, Dhingra AK, Dhar KL. Psoralea corylifolia L. *Buguchi* — Folklore to modern evidence: Review. Fitoterapia 2013; 90: 44-56.
[http://dx.doi.org/10.1016/j.fitote.2013.06.016] [PMID: 23831482]

[9] Christenhusz MJM, Byng JW. The number of known plants species in the world and its annual increase. Phytotaxa 2016; 261(3): 201-17.
[http://dx.doi.org/10.11646/phytotaxa.261.3.1]

[10] Sah P, Agarwal D, Garg SP. Isolation and identification of furocoumarins from the seeds of *Psoralea corylifolia* linn. Indian J Pharm Sci 2006; 68(6): 768-71.
[http://dx.doi.org/10.4103/0250-474X.31012]

[11] Qiao CF, Han QB, Song JZ, *et al.* Quality assessment of *fructus psoraleae*. Chem Pharm Bull 2006; 54(6): 887-90.
[http://dx.doi.org/10.1248/cpb.54.887] [PMID: 16755065]

[12] Khushboo PS, Jadhav VM, Kadam VJ, Sathe NS. Psoralea corylifolia Linn.-"Kushtanashini". Pharmacogn Rev 2010; 4(7): 69-76.
[http://dx.doi.org/10.4103/0973-7847.65331] [PMID: 22228944]

[13] Yang YM, Hyun JW, Sung MS, *et al.* The cytotoxicity of psoralidin from *Psoralea corylifolia*. Planta Med 1996; 62(4): 353-4.
[http://dx.doi.org/10.1055/s-2006-957901] [PMID: 8792669]

[14] Limper C, Wang Y, Ruhl S, *et al.* Compounds isolated from *Psoralea corylifolia* seeds inhibit protein kinase activity and induce apoptotic cell death in mammalian cells. J Pharm Pharmacol 2013; 65(9): 1393-408.
[http://dx.doi.org/10.1111/jphp.12107] [PMID: 23927478]

[15] Khatune NA, Ekramul Islam M, Ekramul Haque M, Khondkar P, Mukhlesur Rahman M. Antibacterial compounds from the seeds of *Psoralea corylifolia*. Fitoterapia 2004; 75(2): 228-30.
[http://dx.doi.org/10.1016/j.fitote.2003.12.018] [PMID: 15030932]

[16] Parast BM, Chetri SK, Sharma K, Agrawal V. *In vitro* isolation, elicitation of psoralen in callus cultures of *psoralea corylifolia* and cloning of psoralen synthase gene. Plant Physiol Biochem 2011; 49(10): 1138-46.
[http://dx.doi.org/10.1016/j.plaphy.2011.03.017] [PMID: 21524916]

[17] Chai MY. A new bioactive coumestan from the seeds of *psoralea corylifolia*. J Asian Nat Prod Res 2020; 22(3): 295-301.
[http://dx.doi.org/10.1080/10286020.2018.1563073] [PMID: 30678490]

[18] Yu O, Shi J, Hession AO, Maxwell CA, McGonigle B, Odell JT. Metabolic engineering to increase isoflavone biosynthesis in soybean seed. Phytochemistry 2003; 63(7): 753-63.
[http://dx.doi.org/10.1016/S0031-9422(03)00345-5] [PMID: 12877915]

[19] Munakata R, Olry A, Karamat F, *et al.* Molecular evolution of parsnip (*Pastinaca sativa*) membrane-bound prenyltransferases for linear and/or angular furanocoumarin biosynthesis. New Phytol 2016; 211(1): 332-44.
[http://dx.doi.org/10.1111/nph.13899] [PMID: 26918393]

[20] Karamat F, Olry A, Munakata R, *et al.* A coumarin-specific prenyltransferase catalyzes the crucial biosynthetic reaction for furanocoumarin formation in parsley. Plant J 2014; 77(4): 627-38.
[http://dx.doi.org/10.1111/tpj.12409] [PMID: 24354545]

[21] Bourgaud F, Hehn A, Larbat R, *et al.* Biosynthesis of coumarins in plants: A major pathway still to be unravelled for cytochrome P450 enzymes. Phytochem Rev 2006; 5(2-3): 293-308.
[http://dx.doi.org/10.1007/s11101-006-9040-2]

[22] Jian X, Zhao Y, Wang Z, *et al.* Two CYP71AJ enzymes function as psoralen synthase and angelicin

synthase in the biosynthesis of furanocoumarins in *peucedanum praeruptorum* dunn. Plant Mol Biol 2020; 104(3): 327-37.
[http://dx.doi.org/10.1007/s11103-020-01045-4] [PMID: 32761540]

[23] Dastmalchi M, Dhaubhadel S. Proteomic insights into synthesis of isoflavonoids in soybean seeds. Proteomics 2015; 15(10): 1646-57.
[http://dx.doi.org/10.1002/pmic.201400444] [PMID: 25757747]

[24] Jung W, Yu O, Lau SMC, *et al.* Identification and expression of isoflavone synthase, the key enzyme for biosynthesis of isoflavones in legumes. Nat Biotechnol 2000; 18(2): 208-12.
[http://dx.doi.org/10.1038/72671] [PMID: 10657130]

[25] Yu O, Jung W, Shi J, *et al.* Production of the isoflavones genistein and daidzein in non-legume dicot and monocot tissues. Plant Physiol 2000; 124(2): 781-94.
[http://dx.doi.org/10.1104/pp.124.2.781] [PMID: 11027726]

[26] Deng Y, Lu S. Biosynthesis and regulation of phenylpropanoids in plants. Crit Rev Plant Sci 2017; 36(4): 257-90.
[http://dx.doi.org/10.1080/07352689.2017.1402852]

[27] Page M, West L, Northcote P, Battershill C, Kelly M. Spatial and temporal variability of cytotoxic metabolites in populations of the new zealand sponge *Mycale hentscheli*. J Chem Ecol 2005; 31(5): 1161-74.
[http://dx.doi.org/10.1007/s10886-005-4254-0] [PMID: 16124239]

[28] Chetri SP, Sharma K, Agrawal V. Genetic diversity analysis and screening of high psoralen yielding chemotype of *psoralea corylifolia* from different regions of india employing HPLC and RAPD marker. Int J Plant Res 2013; 26: 88-95.

[29] Sharma U, Agrawal V. *In vitro* shoot regeneration and enhanced synthesis of plumbagin in root callus of *Plumbago zeylanica* L.—an important medicinal herb. *In Vitro*. Cell Dev Biol Plant 2018; 54(4): 423-35.
[http://dx.doi.org/10.1007/s11627-018-9889-y]

[30] Shamloo M, Babawale EA, Furtado A, Henry RJ, Eck PK, Jones PJH. Effects of genotype and temperature on accumulation of plant secondary metabolites in canadian and australian wheat grown under controlled environments. Sci Rep 2017; 7(1): 9133.
[http://dx.doi.org/10.1038/s41598-017-09681-5] [PMID: 28831148]

[31] Popović Z, Milošević DK, Stefanović M, *et al.* Variability of six secondary metabolites in plant parts and developmental stages in natural populations of rare gentiana pneumonanthe. Plant Biosystems-An International Journal Dealing with all Aspects of Plant Biology 2020; 155: 816-22.

[32] Yang L, Wen KS, Ruan X, Zhao YX, Wei F, Wang Q. Response of plant secondary metabolites to environmental factors. Molecules 2018; 23(4): 762.
[http://dx.doi.org/10.3390/molecules23040762] [PMID: 29584636]

[33] Parast BM, Rasouli M, Manafi M, Agrawal V. Quantification of psoralen in plant parts of *psoralea corylifolia* growing *in vivo* and *in vitro* and enhancement of psoralen by organic elicitors. Anal Chem Lett 2012; 2(4): 227-34.
[http://dx.doi.org/10.1080/22297928.2012.10648273]

[34] Azwanida NN. A review on the extraction methods use in medicinal plants, principle, strength and limitation. Med Aromat Plants 2015; 4: 1-6.

[35] Ingle KP, Deshmukh AG, Padole DA, Dudhare MS, Moharil MP, Khelurkar VC. Phytochemicals: Extraction methods, identification and detection of bioactive compounds from plant extracts. J Pharmacogn Phytochem 2017; 6: 32-6.

[36] Sasidharan S, Chen Y, Saravanan D, Sundram KM, Yoga Latha L. Extraction, isolation and characterization of bioactive compounds from plants' extracts. Afr J Tradit Complement Altern Med 2011; 8(1): 1-10.

[PMID: 22238476]

[37] Miura H, Nishida H, Iinuma M. Effect of crude fractions of *psoralea corylifolia* seed extract on bone calcification. Planta Med 1996; 62(2): 150-3.
[http://dx.doi.org/10.1055/s-2006-957839] [PMID: 8657749]

[38] Iwamura J, Dohi T, Tanaka H, Odani T, Kubo M. Cytotoxicity of corylifoliae fructus. ii. cytotoxicity of bakuchiol and the analogues. yakugaku zasshi. J Pharm Soc Japan 1989; 109: 962-5.

[39] Sun NJ, Woo SH, Cassady JM, Snapka RM. DNA polymerase and topoisomerase II inhibitors from *Psoralea corylifolia*. J Nat Prod 2003; 66(5): 734-4.
[http://dx.doi.org/10.1021/np030135t] [PMID: 9544566]

[40] Yin S, Fan CQ, Wang Y, Dong L, Yue JM. Antibacterial prenylflavone derivatives from *Psoralea corylifolia*, and their structure–activity relationship study. Bioorg Med Chem 2004; 12(16): 4387-92.
[http://dx.doi.org/10.1016/j.bmc.2004.06.014] [PMID: 15265490]

[41] Shinde AN, Malpathak N, Fulzele DP. Determination of isoflavone content and antioxidant activity in *psoralea corylifolia* L. callus cultures. Food Chem 2010; 118(1): 128-32.
[http://dx.doi.org/10.1016/j.foodchem.2009.04.093]

[42] Lim SH, Ha TY, Ahn J, Kim S. Estrogenic activities of *Psoralea corylifolia* L. seed extracts and main constituents. Phytomedicine 2011; 18(5): 425-30.
[http://dx.doi.org/10.1016/j.phymed.2011.02.002] [PMID: 21382704]

[43] Behloul N, Wu G. Genistein: A promising therapeutic agent for obesity and diabetes treatment. Eur J Pharmacol 2013; 698(1-3): 31-8.
[http://dx.doi.org/10.1016/j.ejphar.2012.11.013] [PMID: 23178528]

[44] Jeong D, Watari K, Shirouzu T, *et al*. Studies on lymphangiogenesis inhibitors from Korean and Japanese crude drugs. Biol Pharm Bull 2013; 36(1): 152-7.
[http://dx.doi.org/10.1248/bpb.b12-00871] [PMID: 23302649]

[45] Kim KA, Shim SH, Ahn HR, Jung SH. Protective effects of the compounds isolated from the seed of *Psoralea corylifolia* on oxidative stress-induced retinal damage. Toxicol Appl Pharmacol 2013; 269(2): 109-20.
[http://dx.doi.org/10.1016/j.taap.2013.03.017] [PMID: 23545180]

[46] Liu X, Nam JW, Song YS, *et al*. Psoralidin, a coumestan analogue, as a novel potent estrogen receptor signaling molecule isolated from *Psoralea corylifolia*. Bioorg Med Chem Lett 2014; 24(5): 1403-6.
[http://dx.doi.org/10.1016/j.bmcl.2014.01.029] [PMID: 24507928]

[47] Shan L, Yang S, Zhang G, *et al*. Comparison of the inhibitory potential of bavachalcone and corylin against UDP-glucuronosyltransferases. Evid Based Complement Alternat Med 2014; 2014: 1-6.
[http://dx.doi.org/10.1155/2014/958937] [PMID: 24829606]

[48] Teschke R, Wolff A, Frenzel C, Schulze J. Review article: Herbal hepatotoxicity : An update on traditional Chinese medicine preparations. Aliment Pharmacol Ther 2014; 40(1): 32-50.
[http://dx.doi.org/10.1111/apt.12798] [PMID: 24844799]

[49] Li YG, Hou J, Li SY, *et al*. Fructus Psoraleae contains natural compounds with potent inhibitory effects towards human carboxylesterase 2. Fitoterapia 2015; 101: 99-106.
[http://dx.doi.org/10.1016/j.fitote.2015.01.004] [PMID: 25596095]

[50] Siva G, Sivakumar S, Prem Kumar G, *et al*. Optimization of elicitation condition with jasmonic acid, characterization and antimicrobial activity of psoralen from direct regenerated plants of *psoralea corylifolia* L. Biocatal Agric Biotechnol 2015; 4(4): 624-31.
[http://dx.doi.org/10.1016/j.bcab.2015.10.012]

[51] Song K, Ling F, Huang A, *et al*. *In vitro* and *in vivo* assessment of the effect of antiprotozoal compounds isolated from *Psoralea corylifolia* against *Ichthyophthirius multifiliis* in fish. Int J Parasitol Drugs Drug Resist 2015; 5(2): 58-64.
[http://dx.doi.org/10.1016/j.ijpddr.2015.04.001] [PMID: 26042195]

[52]   Won TH, Song IH, Kim KH, *et al.* Bioactive metabolites from the fruits of *Psoralea corylifolia*. J Nat Prod 2015; 78(4): 666-73.
[http://dx.doi.org/10.1021/np500834d] [PMID: 25710081]

[53]   Zhang X, Zhao W, Wang Y, Lu J, Chen X. The chemical constituents and bioactivities of *Psoralea corylifolia* Linn.: A review. Am J Chin Med 2016; 44(1): 35-60.
[http://dx.doi.org/10.1142/S0192415X16500038] [PMID: 26916913]

[54]   Ramachandra Rao S, Ravishankar GA. Plant cell cultures: Chemical factories of secondary metabolites. Biotechnol Adv 2002; 20(2): 101-53.
[http://dx.doi.org/10.1016/S0734-9750(02)00007-1] [PMID: 14538059]

[55]   Oksman KM, Inze D. Plant cell factories in the post-genomic era: New ways to produce designer secondary metabolites. Trend Pl Sci 2004.

[56]   Alfermann AW, Petersen M. Natural product formation by plant cell biotechnology. Plant Cell Tissue Organ Cult 1995; 43(2): 199-205.
[http://dx.doi.org/10.1007/BF00052176]

[57]   Chetri SK, Kapoor H, Agrawal V. Marked enhancement of sennoside bioactive compounds through precursor feeding in *cassia angustifolia* vahl and cloning of isochorismate synthase gene involved in its biosynthesis. Plant Cell Tissue Organ Cult 2016; 124(2): 431-46.
[http://dx.doi.org/10.1007/s11240-015-0905-1]

[58]   Kapoor H, Yadav N, Chopra M, Mahapatra S, Agrawal V. Strong anti-tumorous potential of *nardostachys jatamansi* rhizome extract on glioblastoma and *in silico* analysis of its molecular drug targets. Curr Cancer Drug Targets 2016; 17(1): 74-88.
[http://dx.doi.org/10.2174/1570163813666161019143740] [PMID: 27774879]

[59]   Ramirez-Estrada K, Vidal-Limon H, Hidalgo D, *et al.* Elicitation, an effective strategy for the biotechnological production of bioactive high-added value compounds in plant cell factories. Molecules 2016; 21(2): 182.
[http://dx.doi.org/10.3390/molecules21020182] [PMID: 26848649]

[60]   Ye M, Ning L, Zhan J, Guo H, Guo D. Biotransformation of cinobufagin by cell suspension cultures of *catharanthus roseus* and *Platycodon grandiflorum*. J Mol Catal, B Enzym 2003; 22(1-2): 89-95.
[http://dx.doi.org/10.1016/S1381-1177(03)00011-0]

[61]   Xu J, Ge X, Dolan MC. Towards high-yield production of pharmaceutical proteins with plant cell suspension cultures. Biotechnol Adv 2011; 29(3): 278-99.
[http://dx.doi.org/10.1016/j.biotechadv.2011.01.002] [PMID: 21236330]

[62]   Xu XH, Zhang W, Cao XP, Xue S. Abietane diterpenoids synthesized by suspension-cultured cells of *Cephalotaxus fortunei*. Phytochem Lett 2011; 4(1): 52-5.
[http://dx.doi.org/10.1016/j.phytol.2010.12.003]

[63]   De Pádua RM, Meitinger N, Filho JDS, *et al.* Biotransformation of 21-O-acetyl-deoxycorticosterone by cell suspension cultures of Digitalis lanata strain *W.1.4*. Steroids 2012; 77(13): 1373-80.
[http://dx.doi.org/10.1016/j.steroids.2012.07.016] [PMID: 22917633]

[64]   Zhang X, Ye M, Dong Y, *et al.* Biotransformation of bufadienolides by cell suspension cultures of *saussurea involucrata*. Phytochemistry 2011; 72(14-15): 1779-85.
[http://dx.doi.org/10.1016/j.phytochem.2011.05.004] [PMID: 21636103]

[65]   Yue W, Ming Q, Lin B, *et al.* Medicinal plant cell suspension cultures: Pharmaceutical applications and high-yielding strategies for the desired secondary metabolites. Crit Rev Biotechnol 2016; 36(2): 215-32.
[http://dx.doi.org/10.3109/07388551.2014.923986] [PMID: 24963701]

[66]   Qu J, Zhang W, Yu X. A combination of elicitation and precursor feeding leads to increased anthocyanin synthesis in cell suspension cultures of *Vitis vinifera*. Plant Cell Tissue Organ Cult 2011; 107(2): 261-9.

[http://dx.doi.org/10.1007/s11240-011-9977-8]

[67] Hari G, Vadlapudi K, Vijendra PD, Rajashekar J, Sannabommaji T, Basappa G. A combination of elicitor and precursor enhances psoralen production in Psoralea corylifolia Linn. suspension cultures. Ind Crop Prod 2018; 124: 685-91.

[68] Shinde AN, Malpathak N, Fulzele DP. Enhanced production of phytoestrogenic isoflavones from hairy root cultures of *Psoralea corylifolia* L. Using elicitation and precursor feeding. Biotechnol Bioprocess Eng; BBE 2009; 14(3): 288-94.
[http://dx.doi.org/10.1007/s12257-008-0238-6]

[69] Shinde AN, Malpathak N, Fulzele DP. Optimized production of isoflavones in cell cultures of *Psoralea corylifolia* L. Using elicitation and precursor feeding. Biotechnol Bioprocess Eng; BBE 2009; 14(5): 612-8.
[http://dx.doi.org/10.1007/s12257-008-0316-9]

[70] Aoki S, Syono K. Synergistic function of rolB, rolC, ORF13 and ORF14 of TL-DNA of *Agrobacterium rhizogenes* in hairy root induction in *Nicotiana tabacum*. Plant Cell Physiol 1999; 40(2): 252-6.
[http://dx.doi.org/10.1093/oxfordjournals.pcp.a029535]

[71] Estruch JJ, Schell J, Spena A. The protein encoded by the rolB plant oncogene hydrolyses indole glucosides. EMBO J 1991; 10(11): 3125-8.
[http://dx.doi.org/10.1002/j.1460-2075.1991.tb04873.x] [PMID: 1915286]

[72] Patra N, Srivastava AK. Enhanced production of artemisinin by hairy root cultivation of *Artemisia annua* in a modified stirred tank reactor. Appl Biochem Biotechnol 2014; 174(6): 2209-22.
[http://dx.doi.org/10.1007/s12010-014-1176-8] [PMID: 25172060]

[73] Abhyankar G, Reddy VD, Giri CC, *et al.* Amplified fragment length polymorphism and metabolomic profiles of hairy roots of *Psoralea corylifolia* L. Phytochemistry 2005; 66(20): 2441-57.
[http://dx.doi.org/10.1016/j.phytochem.2005.08.003] [PMID: 16169025]

[74] Manikonda PK, Abhyanikarn G, Rao KV, Reddy VD, Subramanyam C. Salt stress enhances daidzein production in hairy root cultures of *Psoralea corylifolia* L. (Fabaceae). Proc AP Akad Sci. 35-49.

[75] Zaheer M, Reddy VD, Giri CC. Enhanced daidzin production from jasmonic and acetyl salicylic acid elicited hairy root cultures of *Psoralea corylifolia* L. *Fabaceae*. Nat Prod Res 2016; 30(13): 1542-7.
[http://dx.doi.org/10.1080/14786419.2015.1054823] [PMID: 26156378]

[76] Ferrari S. Biological elicitors of plant secondary metabolites: Mode of action and use in the production of nutraceutics. In: Giardi MT, Rea G, Berra B, Eds. Adv Exp Med Biol Springer,. Boston, MA: Bio-Farms for Nutraceuticals 2010.

[77] Zhang B, Zheng LP, Wang JW. Nitric oxide elicitation for secondary metabolite production in cultured plant cells. Appl Microbiol Biotechnol 2012; 93(2): 455-66.
[http://dx.doi.org/10.1007/s00253-011-3658-8] [PMID: 22089384]

[78] Ahmed SA, Baig MMV. Biotic elicitor enhanced production of psoralen in suspension cultures of *Psoralea corylifolia* L. Saudi J Biol Sci 2014; 21(5): 499-504.
[http://dx.doi.org/10.1016/j.sjbs.2013.12.008] [PMID: 25313287]

[79] Moscatiello R, Baldan B, Navazio L. Plant cell suspension cultures. Plant Mineral Nutrients. Totowa, NJ: Humana Press 2013; pp. 77-93.
[http://dx.doi.org/10.1007/978-1-62703-152-3_5]

[80] Hughes EH, Shanks JV. Metabolic engineering of plants for alkaloid production. Metab Eng 2002; 4(1): 41-8.
[http://dx.doi.org/10.1006/mben.2001.0205] [PMID: 11800573]

[81] Sawada Y, Kinoshita K, Akashi T, Aoki T, Ayabe S. Key amino acid residues required for aryl migration catalysed by the cytochrome p450 2-hydroxyisoflavanone synthase. Plant J 2002; 31(5): 555-64.

[http://dx.doi.org/10.1046/j.1365-313X.2002.01378.x] [PMID: 12207646]

[82] Dixon RA, Paiva NL. Stress-induced phenylpropanoid metabolism. Plant Cell 1995; 7(7): 1085-97.
[http://dx.doi.org/10.2307/3870059] [PMID: 12242399]

[83] Weisshaar B, Jenkins GI. Phenylpropanoid biosynthesis and its regulation. Curr Opin Plant Biol 1998; 1(3): 251-7.
[http://dx.doi.org/10.1016/S1369-5266(98)80113-1] [PMID: 10066590]

[84] Bernards MA, Susag LM, Bedgar DL, Anterola AM, Lewis NG. Induced phenylpropanoid metabolism during suberization and lignification: A comparative analysis. J Plant Physiol 2000; 157(6): 601-7.
[http://dx.doi.org/10.1016/S0176-1617(00)80002-4] [PMID: 11858251]

[85] Subramanian S, Hu X, Lu G, Odelland JT, Yu O. The promoters of two isoflavone synthase genes respond differentially to nodulation and defense signals in transgenic soybean roots. Plant Mol Biol 2004; 54(5): 623-39.
[http://dx.doi.org/10.1023/B:PLAN.0000040814.28507.35] [PMID: 15356384]

[86] He XZ, Dixon RA. Genetic manipulation of isoflavone 7-O-methyltransferase enhances biosynthesis of 4'-O-methylated isoflavonoid phytoalexins and disease resistance in alfalfa. Plant Cell 2000; 12(9): 1689-702.
[PMID: 11006341]

[87] Naoumkina M, Farag MA, Sumner LW, Tang Y, Liu CJ, Dixon RA. Different mechanisms for phytoalexin induction by pathogen and wound signals in *Medicago truncatula*. Proc Natl Acad Sci 2007; 104(46): 17909-15.
[http://dx.doi.org/10.1073/pnas.0708697104] [PMID: 17971436]

[88] Misra P, Pandey A, Tewari SK, Nath P, Trivedi PK. Characterization of isoflavone synthase gene from *Psoralea corylifolia*: A medicinal plant. Plant Cell Rep 2010; 29(7): 747-55.
[http://dx.doi.org/10.1007/s00299-010-0861-5] [PMID: 20437049]

[89] Lu X, Tang K, Li P. Plant metabolic engineering strategies for the production of pharmaceutical terpenoids. Front Pl Sci 2016; 7: 1647.

CHAPTER 12

Genetic Improvement of Pelargonium, an Important Aromatic Plant, through Biotechnological Approaches

Pooja Singh[1]**, Syed Saema**[2] **and Laiq ur Rahman**[1,*]

[1] *Plant Biotechnology Division, Central Institute of Medicinal and Aromatic Plants, P.O. CIMAP, Picnic Spot Road, Lucknow, India*

[2] *Environmental Science Department, Integral University, Lucknow, India*

Abstract: Pelargonium is one of the most recognized aromatic herbs due to its wide distribution around several countries and its perfumery and aromatherapy properties. The present chapter aims at exploring the current scientific study on the various species of Pelargonium along with its significance. The essential oil of Pelargonium contains more than 120 monoterpenes and sesquiterpenes obtained from the steam distillation of herbaceous parts. Citronellol, geraniol, rhodinol, 6, 9 –guaidiene, and 10-epi-γ eudesmol are the principal components responsible for its oil quality. Traditionally, propagation of pelargonium is done through cuttings from its mother plant material. However, the tissue culture approach is one of the reliable techniques for propagation and conservation, not influenced by environmental conditions. More likely, tissue culture approaches used are somatic embryogenesis, callus culture, direct regeneration, meristem culture, and hairy root culture. Transcriptome analysis has also been carried out in *Pelargonium graveolens* to understand the metabolic pathway. In order to accomplish the maximum oil production and better geranium varieties through genetic engineering, *Agrobacterium* mediated transformation systems have been developed. These standardised genetic transformation procedures were used to over-express, silencing, and heterologous expression of desired genes in Pelargonium to understand the outcome and succeed with enhanced essential oil production with better quality for the ultimate benefit.

Keywords: Biosynthesis, Essential oil, Genetic transformation, Pelargonium, Tissue culture, Terpene.

INTRODUCTION

Pelargonium (Geraniaceae family) is one of the most recognized aromatic and medicinal herbs due to its wide distribution around several countries. Pelargonium

---

[*] **Corresponding author Laiq ur Rahman:** Plant Biotechnology Division, Central Institute of Medicinal and Aromatic Plants, P.O. CIMAP, Picnic Spot Road, Lucknow, India; E-mail: l.rahman@cimap.res.in

Mohammad Anis & Mehrun Nisha Khanam (Eds.)
All rights reserved-© 2024 Bentham Science Publishers

has been renowned for its perfumery and aromatherapy properties. Besides being used in cosmetics, it is highly denuded to cure a number of diseases due to antibacterial, antifungal, antioxidant, anti-inflammatory and anticancer activities [1 - 5]. The aerial parts are used in folk medicine as a food and tea drinks additive and for relieving some gastrointestinal, topical, dental, and cardiovascular disorders and are effective in preventing haemorrhoids. This genus comprises 750 species, showing variation in floral morphology, chemical constituent and life forms [6, 7]. The essential oil of Pelargonium is obtained from steam distillation of herbaceous parts; they contain a mixture of more than 120 monoterpenes, sesquiterpenes and other low molecular weight aromatic compounds [8]. The most abundant monoterpenoids of geranium oil are citronellol, geraniol, caryophyllene oxide, menthone, linalool, β-bourbonene, iso-menthone, and geranyl formate (Fig. **1**) [9, 10]. However, the chemical composition of the oil is variable in different species, which could be due to differences in cultivars used, climate conditions, origin, time of harvest, fertilizers used, *etc*. Some of the Pelargonium species are rich in different aroma compounds mentioned in Table **1**. The commercial value of geranium oil depends on its quality, determined by the total rhodinol content and ratio of citronellol and geraniol [11]. Other two constituents of geranium oil, *viz*. 6, 9 –guaidiene and 10-epi-γ eudesmol, provide olfactory value, which are commercially utilized in the perfumery, cosmetic and aromatherapy industries worldwide. This wide area of importance of geranium essential oil enlarged the market demand (600 tons/per) at the estimated cost of $ 225/kg [12]. Propagation of Pelargonium was done by cuttings (10 to 15 cm in length) from top young shoots of mother plant material in sandy soil. Only a few species are capable of producing seeds, while most Pelargonium species do not produce or produce non-viable seeds and hence the occurrence of Pelargonium crops throughout the year is restricted [13 - 15]. However, alternative approaches for propagation and conservation are adapted by researchers through tissue culture, such as clonal propagation that involves somatic embryogenesis, callus culture, direct regeneration, and meristem culture. Most of the conventional breeding programmes cannot be applied for the genetic improvement of geranium owing to the vegetative mode of propagation; biotechnological approaches are more likely to be successful. *Agrobacterium* mediated transformation with relevant genes could provide avenues to make better geranium varieties for commercial cultivation [16].

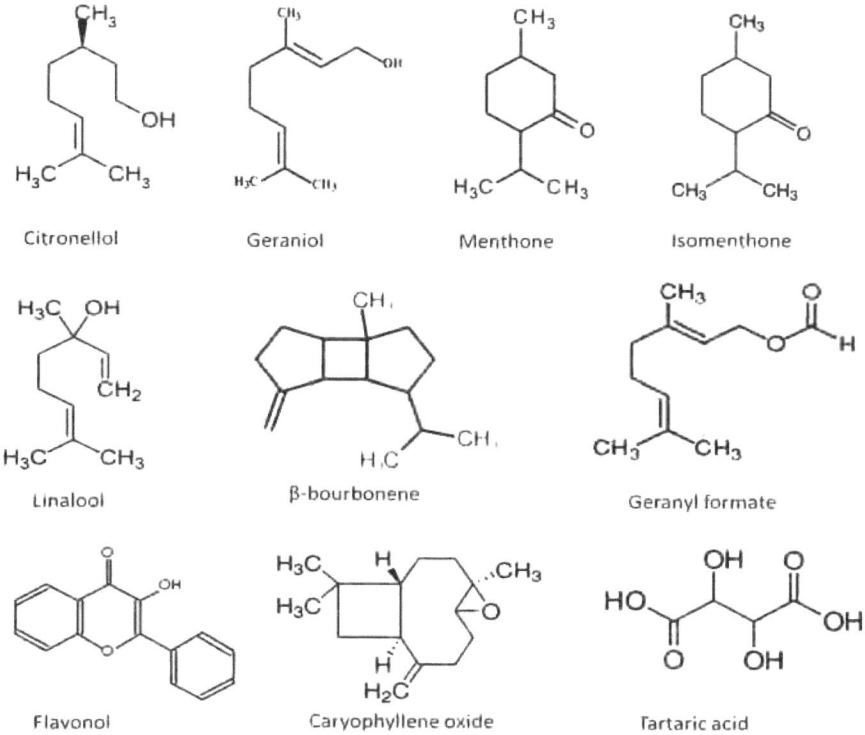

Fig. (1). Structure of some important compounds in *Pelargonium* species.

Table 1. Major oil constituents of some reported *Pelargonium* species.

Pelargonium Species	Major Constituent	References
P. graveolens	Citronellol (27%), Geraniol (25%)	[51]
P. graveolens (South African)	Isomenthone (65.8-83.3%).	[7]
P. radens	Isomenthone (81.5%)	[7]
P. tomentosum	Isomenthone (61-62%), menthone (25-27%)	[7]
P. vitifolium	Citronellic acid (77-85%)	[62]
P. citronellum	Geranic acid (36%), Neral (17.4%) and Geranial (27.2%)	[7]
P. papilionaceum	Citronellic acid (96.2%)	[7]
P. grossulariodes	Isomenthone (13%), Citronellol (12%) geraniol (16%), methyl eugenol (11%)	[8]
P. capitatum	Citronellol (76.6%)	[7]

TISSUE CULTURE STUDIES IN PELARGONIUM

Plant tissue culture is one of the reliable techniques for plant propagation and conservation, which is not influenced by environmental conditions. This method is used to propagate plants rapidly, by which many endangered medicinal, ornamental, and vegetable plants are propagated routinely [17, 18]. Traditionally, Pelargonium propagated through cutting. However, this plant is susceptible to several diseases that cause extensive damage to the crop at all stages of growth and development (Tables 2-4) and could not be available throughout the year. Tissue culture offers an effective and alternate way for rapid and safe propagation. Pelargonium was, for the first time, regenerated *in vitro* [19] through the callus culture. *In vitro,* standardized protocols have been developed for many species of Pelargonium through the induction of different regeneration modes such as meristem tip culture, organogenesis, somatic embryogenesis, callus cultures, suspension culture, and direct regeneration.

An efficient protocol of regeneration in *Pelargonium graveolens* was established from the leaf and nodal explants in MS medium supplemented with different concentrations of auxin and cytokinins [20]. Direct regeneration was achieved from the leaf explants in MS medium supplemented with 0.5 mg/l kinetin and 1.0 mg/l NAA, while callus induced from nodal explants had 10 mg/l kinetin and 1.0 mg/l NAA. Shoots regenerated from callus (0.5 mg/l BAP+0.1 mg/l NAA) were rooted in half-strength MS medium supplemented with 1.0 mg/l IBA. Rooted plantlet was acclimatised and successfully established in the soil in a greenhouse with 90% survival. They developed two types of calli clones represented by highly dented leaves which were similar to the parent type, and the other were low-dented round leaves. The low-dented leaf somaclones showed genetic variability and improved characteristics of *P. graveolens* regarding growth and morphology. After two years of the field trial, they reported that the less dented round leaf achieved better biomass and total oil yield than the wild-type parents which had been released for commercial cultivation. These *in vitro* regeneration protocols were further used for comparison of three cultivar of *P. graveolens* on the basis of chemical and molecular fingerprinting [21].

The somoclones of *P. graveolens* from callus culture and found both positive and negative impact than the wild-type parent plants [22]. Besides the use of phytohormone, auxin, and cytokine, they also used the adenine di sulphate that enhances the callus formation. Callus cultures were also obtained by using leaf, petioles, stem and nodal explants of *Pelargonium graveolens* in a B5 medium containing a different concentration of BA and NAA [23]. They conclude that maximum callus induced from leaf explants cultured on a medium containing 0.5 mg/l BA and 2.0 mg/l NAA. A medium containing higher auxin and low concentration of cytokinins yielded better shoots through the callus induced from leaf explants of *Pelargonium graveolens* [24]. They achieved maximum shoot formation on a medium containing higher auxin and low concentration of cytokinins. Regeneration through the callus is also reported in other species of Pelargonium [25].

Table 2. Disease and symptoms of *Pelargonium* species caused by a fungal pathogen.

Name of Pathogen	Disease	Symptom	References
Botrytis cinerea	Botrytis blight	Various shapes lesions develop at leaves, stem and cutting; flowers turn brown and drop prematurely.	[42]
Colletotrichum acutatum	Anthracnose disease	Invasions are sunken necrotic lesions on leaves, stems and flowers, as well as crown and stem rots.	[63]
Botryosphaeria ribis	Botryosphaeria stem rot	Reduces leaf and petiole size, leaf yellowing and bunching of leaves and cessation of plant growth resulting in plant mortality.	[64]
Sclerotinia sclerotiorum	Cottony stem rot	Produces black structures known as sclerotia and white fuzzy growths of mycelium on the plant it infects.	[65]
Pythium species	Cutting rot, Damping-off	At the base of cutting, brown, water-soaked lesions develop.	[66]
Peronospora conglomerata	Downy mildew	Downy mildews produce grayish, fuzzy looking spores and mycelium on the lower leaf surfaces and beneath areas of upper leaf discoloration. However due to the wide host range of this pathogen, symptoms vary significantly from host to host and even between cultivars.	Article published by Beckerman (https://www.extension.purdue.edu/extmedia)

(Table 2) cont.....

Name of Pathogen	Disease	Symptom	References
Fusarium oxysporum	Fusarium wilt	Seedlings lack vigor. Older leaves turn yellow. A brown rotted area develops at or near the soil line, over which a white mold (Fusarium) may form.	https://ipm.illinois.edu/diseases/rpds/650
Alternaria alternata	Leaf mold	Brown spot developed at lower leaf surface.	[67]
Sphaerotheca macularis	Powdery mildew	Irregular chlorotic or necrotic lesions develop on the leaf surface, followed by the typical white, powdery appearance. Other symptoms include scab-like lesions, witches'-brooms, twisting and distortion of newly emerging shoots, premature leaf coloration and drop, slowed or stunted growth, and leaf rolling.	[68]
Pucciniapelargonii zonalis	Pelargonium rust	Yellow spots are developed at the upper surface of the leaf, and rusty coloured pustules are on the underside of the leaf. Leaf drop prematurely.	[69]
Verticillium alboatrum, Verticillium dahliae	Verticillium wilt	Middle and upper leaves collapse, dry, and fall. The vascular tissue of affected stems is browned. Symptoms are readily confused with those of bacterial blight.	https://pnwhandbooks.org/plantdisease/disease/geranium-pelargonium-spp.

Table 3. Disease and symptoms of *Pelargonium* species caused by a bacterial pathogen.

Name of Pathogen	Disease	Symptom	References
Xanthomonas campestris	Bacterial blight	On leaf: Brown spot and V-shaped lesion occurs; some leaves develop distinctly darkened veins and wilt at leaf margins. When infection is systemic, vascular tissue of stems or branches with wilted leaves usually has brown discoloration	[61]
Rhodococcus fascians	Bacterial fasciation	Short, thick, fleshy, aborted stems that form at the base of the main stem at or below soil level are pale green or green-yellow. The rest of the plant appears healthy.	[70]
Pseudomonas cichorii, Pseudomonas syringae	Bacterial leaf spot	Elliptical, water-soaked 1/4 - 1/2 inch spots form on leaves. Spots become dark brown to black and irregularly shaped. A yellow halo may or may not surround each spot.	www.greenhousemag.com/article/gm0313-managing-bacterial-leaf-spots/
Ralstonia solanacearum	Southern wilt	Wilting of lower leaves followed by yellowing and necrosis. The root of an infected plant is generally black, and finally, death of the plant occurs.	[71]

Table 4. Disease and symptoms of *Pelargonium* species caused by a viral pathogen.

Name of Pathogen	Disease	Symptom	References
Cucumber mosaic virus	Leaf breaking and mosaic	Foliage appears in the venal area; the plant becomes dwarf with small leaves and short internodes.	[72]
Pelargonium leaf-curl virus (PLCV)	*Pelargonium* Leaf Curl or Crinkle	The leaves of infected plants have round to star-shaped spots or irregular yellow ones. The leaves become crinkled, puckered, and may split. As they expand, severely infected leaves may turn yellow and drop off.	[73]
Pelargonium chlorotic ring pattern virus	Pelargonium ring pattern	Symptomless	[74]

(Table 4) cont.....

Name of Pathogen	Disease	Symptom	References
Tobacco ring spot viruses	Pelargonium ring spot	Plants infected by this virus produced small and malformed leaves with small blister-like areas and irregular yellow rings.	[75]
Pelargonium ring pattern virus	*Pelargonium* Ring Pattern	Symptoms similar to Pelargonium ring spot.	[74]

Direct regeneration has an advantage over indirect regeneration for clonal propagation as callus-induced regenerations of plantlets have shown soma clonal variation, while plants regenerated through direct regeneration are genetically similar to the parent plants [26]. The direct regeneration protocol could be utilized for the clonal propagation of elite varieties. The direct regeneration from leaf and petiole explants of *P. graveolens* (CIM BIO 171) using adenine di sulphate along with a different combination of BAP and NAA has been reported. It has been suggested that direct regeneration from leaf and petiole explants have the advantage over indirect regeneration, which provide clonal propagation of plants with similar characteristic to the mother plant.

Somatic embryogenesis leads to the production of a bipolar structure containing a root and shoot axis and a closed, independent vascular system. The plant regenerated from the somatic embryo originates from single sub-epidermal parenchyma cells following characteristic embryological stages, *viz.* globular, heart torpedo and cotyledonous stages [27]. Several reports are available in which somatic embryogenesis has the potential to propagate Pelargonium species, *viz. Pelargonium hortorum, Pelargonium domesticum,* and *Pelargonium peltatum* [28]. However, to date, somatic embryogenesis has not been reported in *Pelargonium graveolens*.

Production of artificial seeds from somatic embryogenesis has been reported in *Pelargonium hortorum* and *Pelargonium domesticum* [29]. It has been reported that in several plant species, including Pelargonium, somatic embryogenesis is induced by thidiazuron (TDZ) phytohormone [30]. TDZ-induced somatic embryogenesis was the first time described in Zonal Pelargonium explants [31]. Besides TDZ, it has also been reported that Arabinogalactan-proteins (AGPs) also induced somatic embryogenesis in *Pelargonium sidoides*.

In addition to TDZ, other compounds such as forchlorfenuron and smoke-saturated water in TDZ-enriched media enhanced somatic embryogenesis in *Pelargonium hortorum* [32, 33].

A suspension culture is a single cell or a few of the aggregate cells that multiply and grow in an agitated liquid medium initiated by the inoculation of a fresh fragment of friable calli [34]. Little work has been done on cell suspension culture of *Pelargonium graveolens*. The cell suspension culture of *Pelargonium graveolens* through the transfer of leaf-generated callus in B5 liquid medium is supplemented with BA and NAA phytohormone. They report the essential oil production from suspension culture of *Pelargonium graveolens* through standardization of carbon (sucrose) and nitrogen (nitrate) sources. Shoot regeneration was also achieved by transforming the leaves and internodal explants of *Pelargonium graveolens* with *Agrobacterium rhizogenes*. The hairy root-derived plant showed the same pattern of monoterpenoids as that of the wild-type mother plant, with the exception of two plants which contain a higher concentration of geraniol and geranyl esters.

AGROBACTERIUM MEDIATED GENETIC TRANSFORMATION IN PELARAGONIUM

To modulate the expression of any gene in a plant system, alteration of the biosynthetic pathway of that particular gene is essential, which could be possible through genetic engineering. In the plant system, genetic transformation is the most suitable approach which could be possible by protoplast fusion, particle bombardment, and *agrobacterium*-mediated genetic transformation. *Agrobacterium* mediated genetic transformation is the most suitable method for the insertion of a foreign gene [35, 36]. Due to increasing demands of phytochemicals, drugs, and increasing environmental stress, the medicinal and aromatic plant also required transformation to enhance the quantity and quality of the desired plant. This approach of gene transfer is applied by researchers in several medicinal plants such as *Ocimum* sp. *Artemisia*, *Atropa,etc.* , to enhance the performance of the respective plant regarding the accumulation of metabolites [37 - 39]. In order to modulate and improve the quality of essential oil and to enhance survivorship, a genetic transformation protocol through *Agrobacterium* was reported in many species. The cotyledon and hypocotyls of *Pelargonium hortorum* were infected with *Agrobacterium tumefaciens* and developed the transgenic plants containing nptII, hpt II and gus genes showing normal phenotype [40]. These reports provide the foundation to develop the transformation in scented geranium [41]. From the leaf and petiole explants of Pelargonium, the somatic embryo was generated and developed the transformation system using kanamycin as a selection marker. In subsequent studies, this protocol was used to produce transgenic plants containing a gene encoding for the antimicrobial protein Ace- AMP1. The incorporation of this gene into *Pelargonium frensham* conferred an increased resistance to infection by *Botrytis cinerea* [42]. Transformation through *A. tumefaciens* containing *nptII* and *uidA* genes has also been established in *P. hortorum* and *P. capitatum* [43]. In order to induce the vir gene in *Agrobacterium tumefaciens* Ti plasmid and enhance the transformation efficiency in *Pelargonium domesticum,* acetosyringone was used [44]. Three species of transgenic Pelargonium were produced with *Agrobacterium rhizogenes* (A4 strain) mediated transformation that shows enhanced leaf and internodes, a source of essential oil than the control plant [45]. The development of branch roots and reduction of internodes was also recorded. The phenotype effect of the *rolC* gene in regal Pelargonium has been studied through *Agrobacterium rhizogenes* mediated transformation under the control of CaMV 35S promoters [46]. Among the six transformants developed, three plants exhibited a dwarf phenotype, and all transgenic showed a reduction in flowering period. The transformed leaf, petiole, and internode explants of *Pelargonium graveolens* with *Agrobacterium rhizogenes* to induce the hairy root were established. From these roots, the shoot was derived by optimization of phytohormone to achieve the transgenic *Pelargonium graveolens* expressing Ri plasmid. The qualitative analysis of the two transgenic lines reveals the enhanced concentration of geraniol and geranyl esters which have improved the quality of the oil. With the standardization of several parameters such as acetosyringone concentration, cocultivation time and bacterial cell density, the development of transformation protocol was carried out in *P. graveolens* using *A. tumefaciens* which provides an efficient and easy way for desired gene insertion. With the advancement of transformation protocol development, transgenic *P. graveolens* was developed with heterologous expression of the ACC deaminase gene, with the enhancement of total biomass. This transgenic plant also shows tolerance toward salt and drought stress *via* the reduction of stress ethylene. Expression of the *RoDELLA* gene under the control of CaMV 35S promoter in *Pelargonium domesticum* showed reduced plant height and increased node and branch number [47]. However, overexpression of the *RoDELLA* gene has negatively affected *Pelargonium domesticum by* delayed flowering and defects in root formation. *Agrobacterium*-mediated transformation of the *TrMYB1* gene in *Pelargonium crispum* shows the enhanced anthocyanins content and hence alters leaf color phenotype [48]. Overexpression of *mannitol dehydrogenase* gene in zonal geranium enhanced tolerance level against *B. cinerea*

Terpene Biosynthesis Pathway

Terpene is the major constituent of geranium essential oil, the key components are citronellol, geraniol, linalool, and their esters, produced by the glandular trichome of leaf tissue. Terpene is derived from the five-carbon (C5) element isopentenyl diphosphate (IDP) and dimethylallyl diphosphate (DMADP), also called the isoprene unit. They are synthesized by two biosynthetic pathways, *viz.* classical mevalonate pathway (MVA) and the 2-C-methyl-D-erythritol-4-phosphate (MEP) pathway. MVA pathway functions in eukaryotes, archaea, and some bacteria cytosol, while the MEP pathway is found in eubacteria, algae, cyanobacteria and plant chloroplast. Seven enzymatic reactions involved the MVA pathway to convert acetyl-CoA, a precursor, to IPP and DMAPP (Fig. 2). In the MEP pathway, eight enzymatic reactions convert pyruvate and glyceraldehydes-3-phasphate to IPP and DMAPP (Fig. 2). Geranyl pyrophosphate synthases (GPPS) catalyzes the condensation of IPP and DMAPP to produce geranyl pyrophosphate (GPP- C_{10}), a precursor of monoterpenes. Farnesyl pyrophosphate synthases catalyze the conversion of GPP to FPP (C_{15}), a precursor of sesquiterpenes and triterpenes. Geranylgeranyl pyrophosphate (GGPP-C 20) is produced by the enzymatic reaction of GGPP synthase (GGPPS), while farnesyl geranyl pyrophosphate (FGPP-C25) is synthesized by the catalysis of FGPP synthase. GGPP is the precursor of diterpenes and tetraterpene, while FGPP is the precursor of sesquiterpene catalyzed by terpene synthase enzyme. Condensation reactions occur during the synthesis of GPP, FPP, GGPP and FGPP through prenyltransferase, a group of enzymes adding a five-carbon component, *i.e.*, isoprene unit to DMAPP or IPP, in a head-to-to-tail fashion.

For understanding the metabolic pathway and engineering, transcriptome analysis of rose-scented geranium was carried out and reported 78,943 unique contigs [6]. This study identifies the biosynthetic pathway genes of terpenes, tartaric acid, ascorbic acid, phytohormone, and transcription factors. From 6.8% of the expressed transcripts, 6,040 simple sequence repeat motifs were identified, among which 50% SSR was trinucliotide. Besides rose-scented geranium, the raw sequence of 13 species of Pelargonium is also available in the SRA database.

Importance of Pelargonium

Different groups of indigenous populations used Pelargonium for medicinal purposes in multiple ways. β-citronellol and geraniol were the most abundant component, showing antioxidant activity, preventing lipid peroxidation, and testicular oxidative damage [49, 50]. The antimicrobial properties of Pelargonium were also identified against different bacterial species. The essential oil of *P. graveolens* acts against both gram-positive and gram-negative bacteria, due to high contents of oxygenated monoterpenes [51]. It was also a study that geraniol, geranyl acetate, and citronellol in essential oil damage the membrane integrity of fungi and show anti-candida activity [52]. Nitric oxide induced the neuroinflammation and neuronal cell death leading to neurodegenerative diseases. However, the essential oil of *P. graveolense* inhibited NO production, as well as the expression of the proinflammatory enzymes cyclooxygenase-2 (COX-2) and induced nitric oxide synthase (iNOS) in primary cultures of activated microglial cells [53]. Intraperitoneal administration of geranium oil, suppressed the acute and chronic inflammatory response by inhibiting neutrophil accumulation [54].

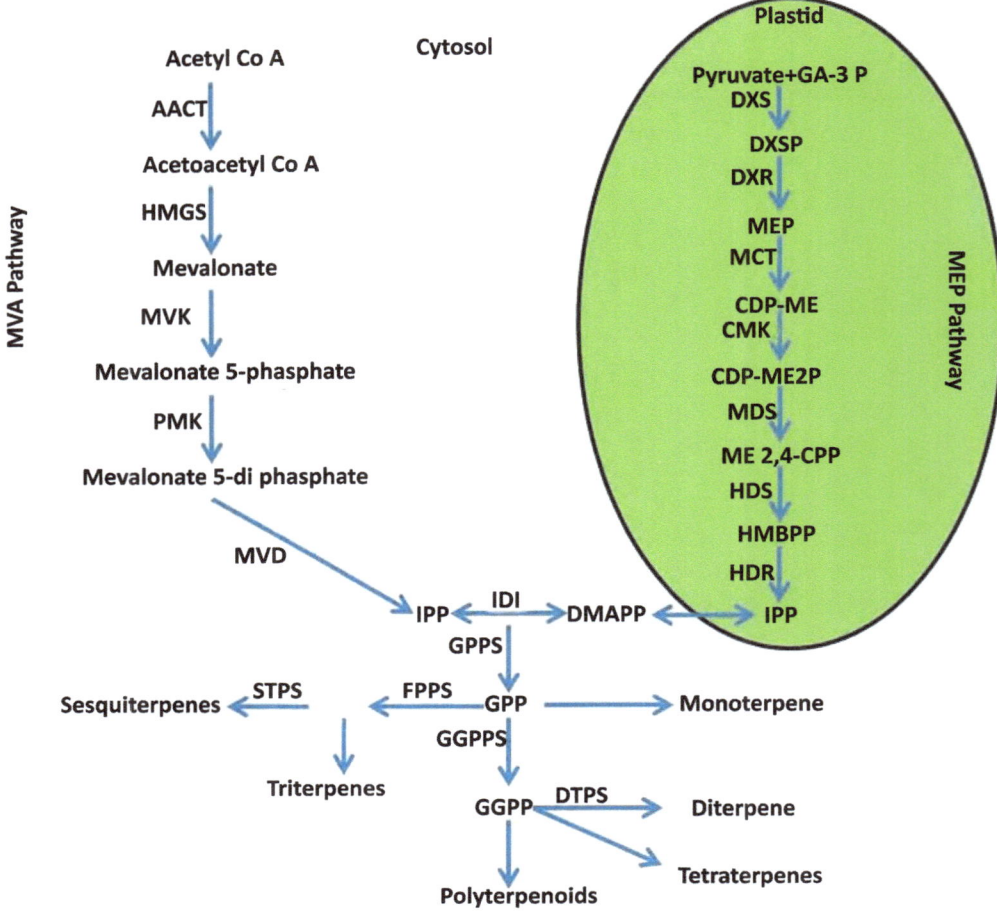

Fig. (2). Terpene Biosynthesis Pathway.

Like the roots of other medicinal plants, Pelargonium roots are also the most bioactive part, containing tannin, and are responsible for treating syphilis [55, 56]. Umckaloabo is the root ethanolic extract of two Pelargonium species used to treat chronic infections in Germany since 1980 as herbal medicine [57]. Methanolic extracts of *P. endlicherianum* root, containing tannin acting on *Haemonchus contortus* larvae, showed an anthelmintic effect against the eggs [58]. Insect repellent properties were also recorded for *P. graveolens*, *P. capitatum*, *P. odoratissimum,* and *P. radula* due to the presence of α-pinene, limonene, carvone and β-myrcene [59, 60].

Concerns with Pelargonium species

The growth and productivity of *Pelargonium* species are affected by a number of biotic and abiotic factors. The biotic stresses are induced by a vast number of plant pathogens, including fungi, bacteria, viruses, and nematodes. Their effects range from mild symptoms to catastrophes, in which large areas were planted and cuttings were destroyed. The fungal diseases, including rust (*Puccinia pelargonii-zonalis*), botrytis blight (*Botrytis cinerea*), root rot (*Pythium* spp.) and other fungal pathogens, affect the Pelargonium species detailed in Table **2**. Among these fungal diseases, botrytis blight is the most common, caused by *Botrytis cinerea,* which spreads through the air. This fungus grows in cool, moist conditions, and attacks free living plant tissue.

Several bacterial diseases also affect the Pelargonium species (Table **3**). Among them, the most destructive disease of Pelargonium is bacterial blight caused by *Xanthomonas campestris* pv. *Pelargonii*, which attacks only geraniums, resulting in 10 to 15% annual losses [61]. It affects the water-conducting cells of the plant, which results in wilting of leaves. In some cases, leaf spots and stem rot features also appear. Other less common bacterial species which cause disease in Pelargonium species are mentioned in Table **3**.

In modern Pelargonium species, viruses are not a major problem. However, several viruses and nematodes were also reported, which can infect and cause a range of symptoms such as yellow spots, rings, petal distortions, leaf curl or crinkle, dark green patterns along veins, white or yellow splotchy patterns on the leaves, yellow veins, and cupping of leaves in *pelargonium* species (Table **4**). Unlike animal viruses, plant viruses require a vector to transport the infection from the infected plants to healthy plants. Little knowledge about the vector for *Pelargonium* viruses is available, but it has been supported that they are transmitted mechanically through contact between plants and soil. It has been reported that some viruses, such as tomato ring spot virus and tobacco ring spot virus, are transmitted by nematodes.

CONCLUSION

Pelargonium species offer many promising prospects for cosmetics, perfumery, and, apparently, a potential herbal therapy for many ailments. This review summarized the important chemical constituents, tissue culture studies, genetic transformations, and diseases associated with *Pelargonium* species. The significance of this crop leads researchers to use the best possible way to enhance *Pelargonium* production to fulfil the market demands. In order to complete the requirement, information related to propagating *in vitro* regeneration systems through various tissue culture techniques and modification, and a biosynthetic pathway, is essential. This chapter describes all analysed combinations of tissue culture techniques such as callus culture, suspension culture, meristem tip culture, somatic embryogenesis, hairy root culture, and direct regeneration systems that make this crop available throughout the year without influencing environmental stresses. Although in synthetic seed technology, more detailed research is required mainly for improvement in the germination frequency of synthetic seeds and subsequent plantlet growth in soil, it can be used on a commercial scale. Genetic transformation and transcriptome studies made it possible to modulate the pathway by desired gene transfer that enhanced the quality and quantity of *Pelargonium* species. Disease control strategies through transgenic technology are very limited, and need to be increased in continuation with the transformation protocols developed. Such a combination of tissue culture techniques and genetic transformation provides a much better way to open immense perspectives to improve commercially important traits such as cultivar resistance to diseases caused by bacteria and fungi and yield of high biomass with better oil quality.

REFERENCES

[1] Boukhatem MN, Kameli A, Saidi F. Essential oil of algerian rose-scented geranium (*Pelargonium graveolens*): Chemical composition and antimicrobial activity against food spoilage pathogens. Food Control 2013; 34(1): 208-13.
[http://dx.doi.org/10.1016/j.foodcont.2013.03.045]

[2] Bigos M, Wasiela M, Kalemba D, Sienkiewicz M. Antimicrobial activity of geranium oil against clinical strains of *staphylococcus aureus*. Molecules 2012; 17(9): 10276-91.
[http://dx.doi.org/10.3390/molecules170910276] [PMID: 22929626]

[3] Shokri H, Khosravi AR, Mansouri M, Ziglari T. Effects of Zataria multiflora and Geranium essential oils on growth-inhibiting of some toxigenic fungi. Iran J Vet Res 2011; 12(3): 247-51.

[4] Fayed SA. Antioxidant and anticancer activities of *citrus reticulatepetitgrain mandarin* and *pelargonium graveolensgeranium* essential oils. Res J Agric Biol Sci 2009; 5: 740-7.

[5] Dorman HJD, Surai P, Deans SG. *In vitro* Antioxidant activity of a number of plant essential oils and phytoconstituents. J Essent Oil Res 2000; 12(2): 241-8.
[http://dx.doi.org/10.1080/10412905.2000.9699508]

[6] Narnoliya LK, Kaushal G, Singh SP, Sangwan RS. *De novo* transcriptome analysis of rose-scented geranium provides insights into the metabolic specificity of terpene and tartaric acid biosynthesis. BMC Genomics 2017; 18(1): 74.
[http://dx.doi.org/10.1186/s12864-016-3437-0] [PMID: 28086783]

[7] Lalli JYY, Viljoen AM, Başer KHSC, Demirci B, Özek T. The essential oil composition and chemotaxonomical appraisal of south african pelargoniums (geraniaceae). J Essent Oil Res 2006; 18(sup1): 89-105.
[http://dx.doi.org/10.1080/10412905.2006.12067128]

[8] Lis-Balchin M. Geranium essential oil: Standardisation, iso; adulteration and its detection using gc, enantiomeric columns, toxicity and bioactivity. CRC Press eBook 2002.

[9] Sharopov FS, Zhang H, Setzer WN. Composition of geranium (*pelargonium graveolens*) essential oil from tajikistan. Am J Essential Oils Nat Prod 2014; 2(2): 13-6.

[10] Shawl AS, Kumar T, Chishiti N, Shabir S. Cultivation of rose scented geranium (*Pelargonium sp.*) as a cash crop in kashmir valley. Asian J Plant Sci 2006; 5(4): 673-5.

[http://dx.doi.org/10.3923/ajps.2006.673.675]

[11] Babu KGD, Kaul VK. Variation in essential oil composition of rose-scented geranium (*Pelargonium* sp.) distilled by different distillation techniques. Flavour Fragrance J 2005; 20(2): 222-31.
[http://dx.doi.org/10.1002/ffj.1414]

[12] Singh P, Pandey SS, Dubey BK, *et al.* Salt and drought stress tolerance with increased biomass in transgenic *pelargonium graveolens* through heterologous expression of *acc deaminase* gene from *achromobacter xylosoxidans*. Plant Cell Tissue Organ Cult 2021; 147(2): 297-311.
[http://dx.doi.org/10.1007/s11240-021-02124-0]

[13] Weiss EA. Essential oil crops. center of agriculture and biosciences (CAB). New York, UK: International 1997; pp. 24-50.

[14] Sedibe MM, Khetsha ZP, Malebo N. Salinity effects on external and internal morphology of rose geranium (*Pelargonium graveolens*L.) leaf. Life Sci J 2013; 10(11): 99-103.

[15] Singh P, Khan S, Kumar S, Rahman L. Establishment of an efficient *Agrobacterium*-mediated genetic transformation system in *Pelargonium graveolens*: An important aromatic plant. Plant Cell Tissue Organ Cult. Springer 2016; 147.(3)
[http://dx.doi.org/10.1007/s11240-016-1153-1158]

[16] Williamson JD, Desai A, Krasnyanski SF, *et al.* Overexpression of mannitol dehydrogenase in zonal geranium confers increased resistance to the mannitol secreting fungal pathogen Botrytis cinerea. Plant Cell Tissue Organ Cult 2013; 115(3): 367-75.
[http://dx.doi.org/10.1007/s11240-013-0368-1]

[17] Cui Y, Deng Y, Zheng K, *et al.* An efficient micropropagation protocol for an endangered ornamental tree species (magnolia sirindhorniae noot. & chalermglin) and assessment of genetic uniformity through DNA markers. Sci Rep 2019; 9(1): 9634.
[http://dx.doi.org/10.1038/s41598-019-46050-w] [PMID: 31270420]

[18] Sidhu Y. *In vitro* micropropagation of medicinal plants by tissue culture. Plymouth Student Scientist 2010; 4: 432-49.

[19] Stephamiak B, Zenkteler M. Regeneration of whole plants of geranium from petiole cultured *in vitro*. Acta Soc Bot Pol 1982; 51: 161-72.

[20] Saxena G, Banerjee S, Laiq-ur-Rahman , Verma PC, Mallavarapu GR, Kumar S. Rose-scented geranium *pelargonium sp.* generated by agrobacterium rhizogenes mediated ri-insertion for improved essential oil quality. Plant Cell Tissue Organ Cult 2007; 90(2): 215-23.
[http://dx.doi.org/10.1007/s11240-007-9261-0]

[21] Tembe RP, Deodhar MA. Chemical and molecular fingerprinting of different cultivars of pelargonium graveolens *l' herit. viz.* reunion, bourbon and egyptian. Biotechnology 2010; 9(4): 485-91.
[http://dx.doi.org/10.3923/biotech.2010.485.491]

[22] Gupta R, Banerjee S, Mallavarapu GR, *et al.* Development of a superior somaclone of rose-scented geranium and a protocol for inducing variants. HortScience 2002; 37(4): 632-6.
[http://dx.doi.org/10.21273/HORTSCI.37.4.632]

[23] Aly U, Hanafy M. Geranium oil production in suspension culture of pelargonium graveolens. Plant Biotechnol J 2010; 24-8.

[24] Benazir JF, Suganthi R, Chandrika P, Mathithumilan B. *In vitro* regeneration and transformation studies on *Pelargonium graveolens geranium* : An important medicinal and aromatic plant. J Med Chem Plant Res 2013; 7(38): 2815-22.

[25] Zuraida AR, Mohd SMA, Ayu Nazreena O, Zamri Z. Improved micropropagation of biopesticidal plant, *Pelargonium radulavia* direct shoot regeneration. Am J Res Commun 2013.

[26] Kumar S, Bhat V. High-frequency direct plant regeneration *via* multiple shoot induction in the apomictic forage grass Cenchrus ciliaris L. *In Vitro* Cell Deve. Biol Plant 2012; 48(2): 241-8.

[27]   Maheswaran G, Williams EG. Uniformity of plants regenerated by direct somatic embryogenesis from zygotic embryos of *trifolium repens*. Ann Bot 1987; 59(1): 93-7.
[http://dx.doi.org/10.1093/oxfordjournals.aob.a087290]

[28]   Madden JI, Jones CS, Auer CA. Modes of regeneration in pelargonium hortorum *geraniaceae* and three closely related species. In Vitro Cell and Develop Bio Plant. 2005; 41: pp. (1)37-46.

[29]   Marsolais AA, Wilson DPM, Tsujita MJ, Senaratna T. Somatic embryogenesis and artificial seed production in zonal (*pelargonium* × *hortorum*) and regal (*pelargonium* × *domesticum)* geranium. Can J Bot 1991; 69(6): 1188-93.
[http://dx.doi.org/10.1139/b91-152]

[30]   Duchow S, Dahlke RI, Geske T, Blaschek W, Classen B. Arabinogalactan-proteins stimulate somatic embryogenesis and plant propagation of pelargonium sidoides. Carbohydr Polym 2016; 152: 149-55.
[http://dx.doi.org/10.1016/j.carbpol.2016.07.015] [PMID: 27516259]

[31]   Visser C, Qureshi JA, Gill R, Saxena PK. Morphoregulatory role of thidiazuron : Substitution of auxin and cytokinin requirement for the induction of somatic embryogenesis in geranium hypocotyl cultures. Plant Physiol 1992; 99(4): 1704-7.
[http://dx.doi.org/10.1104/pp.99.4.1704] [PMID: 16669097]

[32]   Senaratna T, Dixon K, Bunn E, Touchell D. Smoke-saturated water promotes somatic embryogenesis in geranium. Plant Growth Regul 1999; 28(2): 95-9.
[http://dx.doi.org/10.1023/A:1006213400737]

[33]   Singh RP, Murthy BNS, Saxena PK. *In vitro* morphogenetic competence of diploid zonal geranium (pelargonium x hortorum bailey cv. scarlet orbit improved) cotyledonary tissue induced with phenylurea compounds. Physiol Mol Biol Plants 1996; 2: 53-8.

[34]   Kong EYY, Biddle J, Foale M, Adkins SW. Cell suspension culture: A potential *in vitro* culture method for clonal propagation of coconut plantlets *via* somatic embryogenesis. Ind Crops Prod 2020; 147: 112125.
[http://dx.doi.org/10.1016/j.indcrop.2020.112125]

[35]   Narusaka Y, Narusaka M, Yamasaki S, Iwabuchi M. Methods to transfer foreign genes to plants book chapter. RIBS 2012; 173-88.

[36]   Rivera AL, Gómez-Lim M, Fernández F, Loske AM. Physical methods for genetic plant transformation. Phys Life Rev 2012; 9(3): 308-45.
[http://dx.doi.org/10.1016/j.plrev.2012.06.002] [PMID: 22704230]

[37]   Simon J, Deschamps C. Agrobacterium tumefaciens -mediated transformation of *Ocimum basilicum* and O. citriodorum. Plant Cell Rep 2002; 21(4): 359-64.
[http://dx.doi.org/10.1007/s00299-002-0526-0]

[38]   Dilshad E, Cusido RM, Estrada KR, Bonfill M, Mirza B. Genetic transformation of artemisia carvifolia buch with rol genes enhances artemisinin accumulation. PLoS One 2015; 10(10): e0140266.
[http://dx.doi.org/10.1371/journal.pone.0140266] [PMID: 26444558]

[39]   Moharrami F, Hosseini B, Sharafi A, Farjaminezhad M. Enhanced production of hyoscyamine and scopolamine from genetically transformed root culture of *hyoscyamus reticulatus* l. elicited by iron oxide nanoparticles. In Vitro Cell Dev Biol Plant 2017; 53(2): 104-11.
[http://dx.doi.org/10.1007/s11627-017-9802-0] [PMID: 28553065]

[40]   Robichon MP, Renou JP, Jalouzot R. Genetic transformation of Pelargonium X hortorum. Plant Cell Rep 1995; 15(1-2): 63-7.
[http://dx.doi.org/10.1007/BF01690255] [PMID: 24185656]

[41]   KrishnaRaj S, Bi Y-M, Saxena PK. Somatic embryogenesis and agrobacterium -mediated transformation system for scented geraniums (pelargonium sp. 'frensham'). Plants 1997; 201(4): 434-40.
[http://dx.doi.org/10.1007/s004250050086]

[42] Bi YM, Cammue BPA, Goodwin PH, KrishnaRaj S, Saxena PK. Resistance to botrytis cinerea in scented geranium transformed with a gene encoding the antimicrobial protein ace -AMP1. Plant Cell Rep 1999; 18(10): 835-40.
[http://dx.doi.org/10.1007/s002990050670]

[43] Hassanein A, Chevreau E, Dorion N. Highly efficient transformation of zonal (pelargonium x hortorum) and scented (p. capitatum) geraniums *via* agrobacterium tumefaciens using leaf discs. Plant Sci 2005; 169(3): 532-41.
[http://dx.doi.org/10.1016/j.plantsci.2005.04.014]

[44] Boase MR, Bradley JM, Borst NK. An improved method for transformation of regal pelargonium (pelargonium xdomesticum dubonnet) by *agrobacterium tumefaciens*. Plant Sci 1998; 139(1): 59-69.
[http://dx.doi.org/10.1016/S0168-9452(98)00177-0]

[45] Pellegrineschi A, Davolio-Mariani O. *Agrobacterium rhizogenes*-mediated transformation of scented geranium. Plant Cell Tissue Organ Cult 1996; 47(1): 79-86.
[http://dx.doi.org/10.1007/BF02318969]

[46] Boase MR, Winefield CS, Lill TA, Bendall MJ. Transgenic regal pelargoniums that express the rolc gene from agrobacterium rhizogenes exhibit a dwarf floral and vegetative phenotype. *In Vitro*. Cell Dev Biol Plant 2004; 40(1): 46-50.
[http://dx.doi.org/10.1079/IVP2003476]

[47] Hamama L, Naouar A, Gala R, *et al.* Overexpression of RoDELLA impacts the height, branching, and flowering behaviour of Pelargonium × domesticum transgenic plants. Plant Cell Rep 2012; 31(11): 2015-29.
[http://dx.doi.org/10.1007/s00299-012-1313-1] [PMID: 22898902]

[48] Kanemaki A, Otani M, Takano M, *et al.* Ectopic expression of the R2R3-MYB gene from Tricyrtis sp. results in leaf color alteration in transgenic *Pelargonium crispum*. Sci Hortic 2018; 240: 411-6.
[http://dx.doi.org/10.1016/j.scienta.2018.06.029]

[49] Rana V, Rana JP, Juyal M, Amparo B. Chemical constituents of essential oil of *Pelargonium graveolens* leaves. Int J Aromather 2002; 12(4): 216-8.
[http://dx.doi.org/10.1016/S0962-4562(03)00003-1]

[50] Ben Slima A, Ali MB, Barkallah M, *et al.* Antioxidant properties of pelargonium graveolens l'her essential oil on the reproductive damage induced by deltamethrin in mice as compared to alpha-tocopherol. Lipids Health Dis 2013; 12(1): 30.
[http://dx.doi.org/10.1186/1476-511X-12-30] [PMID: 23496944]

[51] Hsouna AB, Hamdi N. Phytochemical composition and antimicrobial activities of the essential oils and organic extracts from *pelargonium graveolens* growing in Tunisia. Lipids Health Dis 2012; 11(1): 167.
[http://dx.doi.org/10.1186/1476-511X-11-167] [PMID: 23216669]

[52] Zore GB, Thakre AD, Rathod V, Karuppayil SM. Evaluation of anti-candida potential of geranium oil constituents against clinical isolates of *candida albicans* differentially sensitive to fluconazole: inhibition of growth, dimorphism and sensitization. Mycoses 2009.
[http://dx.doi.org/10.1111/j.1439-0507.01852] [PMID: 20337938]

[53] Elmann A, Mordechay S, Rindner M, Ravid U. Anti-neuroinflammatory effects of geranium oil in microglial cells. J Funct Foods 2010; 2(1): 17-22.
[http://dx.doi.org/10.1016/j.jff.2009.12.001]

[54] Maruyama N, Ishibashi H, Hu W, *et al.* Suppression of carrageenan- and collagen II-induced inflammation in mice by geranium oil. Mediators Inflamm 2006; 2006(3): 1-7.
[http://dx.doi.org/10.1155/MI/2006/62537] [PMID: 16951493]

[55] Helfer M, Koppensteiner H, Schneider M, *et al.* The root extract of the medicinal plant *Pelargonium sidoides* is a potent HIV-1 attachment inhibitor. PLoS One 2014; 9(1): e87487.

[http://dx.doi.org/10.1371/journal.pone.0087487] [PMID: 24489923]

[56] Asgarpanah J, Ramezanloo F. An overview of phytopharmacology of Pelargonium graveolens. Indian J tradi know 2015; 14(4): 558-63.

[57] Saraswathi J, Venkatesh K, Baburao N, Hilal MH, Rani AR. Phytopharmacological importance of *Pelargonium* species. J Med Plants Res 2011; 5(13): 2587-98.

[58] Kozan E, Küpeli Akkol E, Süntar I. Potential anthelmintic activity of *Pelargonium endlicherianum* Fenzl. J Ethnopharmacol 2016; 187: 183-6.
[http://dx.doi.org/10.1016/j.jep.2016.04.044] [PMID: 27130640]

[59] Kolodziej H, Kayser O, Radtke OA, Kiderlen AF, Koch E. Pharmacological profile of extracts of *Pelargonium sidoides* and their constituents. Phytomedicine 2003; 10 (4): 18-24.
[http://dx.doi.org/10.1078/1433-187X-00307] [PMID: 12807338]

[60] Simmonds MSJ. Interactions between arthropod pests and pelargoniums. geranium and pelargonium: History of nomenclature, usage and cultivation. UK: CRC Press 2002; pp. 291-8.

[61] Nameth ST, Daughtrey ML, Moorman GW, Sulzinski MA. Bacterial blight of geranium: A history of diagnostic challenges. Plant Dis 1999; 83(3): 204-12.
[http://dx.doi.org/10.1094/PDIS.1999.83.3.204] [PMID: 30845495]

[62] Demarne FE, Van der Walt JJA. Composition of the essential oil of *pelargonium citronellum* (*geraniaceae*). J Essent Oil Res 1993; 5(3): 233-8.
[http://dx.doi.org/10.1080/10412905.1993.9698214]

[63] Gautam AK. The genera colletotrichum: An incitant of numerous new plant diseases in India. Journal on New Biol Rep 2014; 3(1): 9-21.

[64] Rao BRR, Kaul PN, Mallavarapu GR, Ramesh S. First observation of little leaf disease and its impact on the yield and composition of the essential oil of rose-scented geranium *Pelargonium sp.* Flavour Fragr 2000; J15: 137-40.

[65] Kareem FA. Induced resistance in bean plants against root rot and alternaria leaf spot diseases using biotic and abiotic inducers under field conditions. Res J Agric Biol Sci 2007; 3(6): 767-74.

[66] Filonow AB. Biological control of pythium damping-off and root rot of greenhouse-grown geraniums and poinsettias. Proc Okla Acad Sci 1999; 79: 29-32.

[67] Furukawa T, Kishi K. Alternaria leaf spot on three species of *pelargonium* caused by *alternaria alternata* in Japan. J Gen Plant Pathol 2001; 67(4): 268-72.
[http://dx.doi.org/10.1007/PL00013028]

[68] Elad Y, Malathrakis NE, Dik AJ. Biological control of *botrytis*-incited diseases and powdery mildews in greenhouse crops. Crop Prot 1996; 15(3): 229-40.
[http://dx.doi.org/10.1016/0261-2194(95)00129-8]

[69] Rytter JL, Lukezic FL, Craig R, Moorman GW. Biological control of geranium rust by *bacillus subtilis*. APS 1989; 79(3): 367-70.

[70] Sule S. Bacterial fasciation of *pelargonium hortorum* in hungary. Acta Phytopathologica Academiae Scientiarum Hungaricae 1976; 11: 223-30.

[71] Harmon PF, Harmon CL, Norman D, Momol T. Southern wilt of geranium. university of florida, cooperative extension service. Florida, U. S.: Institute of Food and Agricultural Sciences 2005; p. 206.

[72] Verma N, Mahinghara BK, Ram R, Zaidi AA. Coat protein sequence shows thatcucumber mosaic virus isolate from geraniums *pelargonium spp.* belongs to subgroup II. J Biosci 2006; 31(1): 47-54.
[http://dx.doi.org/10.1007/BF02705234] [PMID: 16595874]

[73] Alemzadeh E, Ghorbani A. Occurrence of *pelargonium* leaf curl virus and moroccan pepper virus on natural hosts. Australas Plant Dis Notes 2016; 11(1): 8.
[http://dx.doi.org/10.1007/s13314-016-0194-5]

[74] Ruiza L, Castanoa A, Borjab M, Hernándeza C. Pelargonium chlorotic ring pattern virus: First report in Spain. Plant Patholo. Springer 2008; 57: p. 396.
[http://dx.doi.org/10.1111/j.1365-3059.2007.01750.x]

[75] Ryden K. Pelargonium ring-spot: A virus disease caused by tomato ring-spot virus in sweden. J Phytopathology 1972; 73(2): 178-8.

SUBJECT INDEX

A

Acaciella angustissima 32
Acid(s) 7, 8, 11, 18, 20, 23, 25, 26, 29, 30, 31, 32, 33, 34, 35, 38, 71, 72, 73, 76, 77, 79, 80, 83, 101, 102, 103, 104, 106, 107, 108, 109, 110, 111, 112, 122, 123, 124, 125, 129, 135, 137, 138, 139, 140, 141, 142, 143, 144, 145, 146, 168, 171, 187, 188, 201, 202, 204, 209, 225, 246, 261, 262, 265, 267, 269, 271, 272, 292, 295, 304, 305, 312
   ascorbic 135, 137, 138, 140, 141, 142, 312
   butyric 83
   caffeic 20, 35, 124, 187
   chicoric 187
   citronellum Geranic 304
   coumaric 124, 187
   dicarboxylic 122, 123
   dichlorophenoxyacetic 34
   fatty 122, 123, 129, 171, 271
   ferulic 20, 35, 124
   gallic 32, 35, 38, 124, 209, 272
   gentisic 124
   gibberellic 109, 112
   glacial acetic 8
   glucuronic 124
   glutamic 272
   glycyrrhizic 30, 34
   indole acetic (IAA) 26, 30, 31, 71, 72, 76, 77, 106, 108, 109, 110, 111, 112, 143, 144
   indole butyric (IBA) 23, 26, 31, 72, 73, 77, 79, 102, 103, 104, 106, 109, 110, 111, 143, 145, 146, 262, 265, 269
   lactic 18, 125
   linoleic 204
   malic 123, 124, 125
   mevalonic 18
   myristic 101
   naphthalene acetic (NAA) 33, 71, 72, 79, 102, 103, 106, 107, 110, 111, 112, 143, 144, 145, 146, 168, 187, 261, 267, 305
   oleic 122
   organic 122, 123, 124, 125, 139
   oxalic 122, 123, 201, 202
   palmitic 129
   pcoumaric 31, 35
   phenolic 31
   rosmarinic (RA) 25, 32, 187, 188
   salicyclic 135
   salicylic 29, 33, 35, 79, 80, 138, 142, 225, 292, 295
   shikimic 18, 272
   sinapic 35, 124
   syringic 124
   tannic 38
   tartaric 312
   trans-cinnamic 18
   turpethinic 246
   uronic 123
   ursolic 188
   vanillic 35, 124, 129
Activities 3, 17, 19, 29, 99, 121, 162, 170, 171, 174, 175, 176, 205, 206, 208, 271
   anti-angiogenic 162, 176
   anti-diabetic 99, 205
   anti-hyperglycemic 205
   antidiabetic 99, 174, 271
   antihelminthic 171
   antimutagenic 175
   antimycoplasmic 121
   cholinesterase 208
   enzymatic 29
   immunomodulatory 174, 175
   metabolic 3, 17, 19, 170
   nephroprotective 174, 206
Anthracnose disease 306
Anthropogenic pressures 196, 214, 227
Anti-cancer 65, 175
   agent 65
   property 175

Anti-inflammatory activity 65, 174
Anti-oxidant properties 121
Antibacterial 98, 129, 130, 136, 172, 204, 246
   activity 129, 130, 172, 204, 246
   agents 98
   drugs 129
    properties 136
Antibiotic(s) 119, 132
   conventional 119
   streptomycin 132
Antibodies, monoclonal 227
Anticancer activities 175, 205, 303
Anticancerous agents 282, 284, 295
Antimicrobial 36, 65, 122, 125, 129, 136, 148, 149, 171, 172, 312
   activity 65, 129, 149, 172
   agents 36, 125, 148
   properties 122, 129, 136, 149, 171, 312
Antioxidant activity 100, 127, 128, 140, 141, 173, 175, 204, 271, 312
Aspergillus 30, 133, 134, 173
   flavus 133, 134
   niger 30, 133, 134, 173

B

Bacteria, multidrug-resistant 132
Biochemical pathways 223
Biomolecules, hydrophobic 287
Biotechnological techniques 66

C

Cambial meristematic cell (CMCs) 188
Cardiovascular effects 212
Casein hydrolysate (CH) 170
Coating, modified atmosphere packaging 139
Completely randomized design (CRD) 248
Conditions, inflammatory 101

D

Diseases 95, 96, 99, 101, 105, 119, 174, 209, 271, 273, 312, 314
   cardiovascular 174, 273
   eye 101, 105
   fungal 314
   heart 209, 271
   infectious 119
   mouth 96
   neurodegenerative 312
   tumor 99
   upper respiratory 95
Disorders 95, 175, 176, 203, 206, 259, 271, 303
   cardiovascular 303
   gastric 175
   inflammatory 206
   neurological 95
Drought stress 311
Drugs 1, 2, 3, 4, 6, 12, 16, 17, 18, 19, 93, 98, 161, 175, 188, 210, 212, 283
   anticancer 98, 210
   immunomodulator 175
   natural products-based 283
   plant-based 1, 2, 4
   plant-derived 93
   therapeutic 188
Dyspepsia 98

E

Effects 15, 65, 122, 127, 174, 181, 202, 209, 212
   cardio-vasculoprotective 209
   cytotoxicity 181
   harmonizing 202
   immunity-enhancing 174
   synergistic 65, 122
   therapeutic 15, 127
   toxicological 212
Elicitation, salinity stress 32

Embryogenesis 259, 265, 270, 273
   micropropagation and somatic 259, 273
   zygotic 265, 270
Embryos 22, 27, 185, 186, 218, 220, 259, 264, 265, 266, 267
   androgenic 264
   cotyledonary-shaped 267
   cotyledonary stage 220
   mature 186
   parthenogenetic 264
   triangular globular 266
Enzyme(s) 18, 68, 81, 121, 138, 140, 141, 174, 312
   activity 81, 138
   chorismate mutase 18
   HMG-CoA synthase 68
   phenylalanine ammonia-lyase 141
   polyphenol oxidase 140
   polyphenolic biosynthesis 141
   serum 174
   superoxide dismutase 121
   terpene synthase 312

F

Fruit(s) 136, 138, 139, 140, 141, 142
   avocado 138
   coated 139, 142
   cucumber 136
   gel-coated 138, 139
   mango 139, 140, 141, 142
   raspberry 139, 141
   ripening 138
   strawberries 141

G

Gene cloning 289
Genetic transformation 33, 226, 264, 302, 310, 314, 315
Genome information 36

Germination-regulating compounds (GRCs) 112
Glutathione peroxidase 121
Glycoproteins 188
Glycosides 1, 8, 9, 10, 11, 16, 27, 99, 100, 123, 127, 128, 223
   cardiac 99, 100, 223
GPP synthase 70
Gum inflammation 202

H

Health disorders 161
Hemicelluloses 122
Hemorrhoids 202
Hepatocellular liver cirrhosis 206
Hepatoprotective agent 206
Herbal 3, 4, 6, 12, 93, 200, 202, 203, 228, 283, 313
   industries 4, 6, 228
   medicine (HM) 3, 6, 12, 93, 200, 202, 203, 283, 313
   phytomedicine 203
High
   performance liquid chromatography (HPLC) 21, 286, 287
   resolution mass spectrometry (HRMS) 287
HPLC analysis 187
Hydrolyzed casein (HC) 34

I

Infection 94, 132, 203
   respiratory 203
   therapy 132
   throat 94
Infrared spectroscopy 287
Inhibition, mycelial growth 136
Inhibitory 112, 135, 205
   activity 135
   effects 112, 205

Inter-simple sequence repeats (ISSR) 182, 183, 263
International board for plant genetic resources (IBPGR) 214

K

Klebsiella pneumoniae 130

L

Lactobacillus fermentation 125
Lesions, necrotic 306, 307
Leucoderma 108, 171, 282, 295
Lipid peroxidation 127, 129, 138, 141, 205
   inhibiting 205
   reduced 141
Liquid suspension 38
Listeria monocytogenes 131
Liver 101, 205, 206
   disorders 101
   fibrosis 206
   injury 205

M

Mass 95, 96, 97, 209, 287
   micropropagation 95, 96, 97
   spectrometry 209, 287
Medicinal herbs 93, 197, 198, 203, 215, 302
Metabolic 17, 18, 19, 35, 36, 37, 121, 189, 225, 302, 312
   pathways 17, 18, 19, 35, 36, 37, 189, 225, 302, 312
   processes 121
Metabolism pathway 27
Metabolites 4, 38, 64, 76, 218
   bioactive 4, 38, 76
   desired 38, 64, 218
Microbial 6, 39, 135, 207, 212, 263
   activity 135
   attacks 207, 212
   contamination 6, 263
   fermentation 39
   populations 135
Micrococcus flavus 131
Micropropagation 92, 162, 163, 260, 261
   methods 163
   technique 92, 162, 260, 261
Mitogen-activated protein kinases (MAPKs) 283, 291
Murine interleukins 227
Mycelial growth 135

N

Nanoparticles for secondary metabolite 38

O

Oil 3, 18, 19, 37, 94, 97, 100, 135, 141, 180, 181, 246, 302, 303, 310, 311, 312
   essential 18, 19, 37, 100, 141, 180, 302, 303, 310, 311, 312
   extracts 246
   volatile 3
Oleuropein glucoside 101
Oxidative stress 25, 29, 127, 206

P

Pain, stomach 94, 202
Pathogenic 29, 133
   fungi 133
   invasion 29
Pathogens 136, 149, 306, 314
   fungal 136, 306, 314
   resistance 149
Pathways 18, 28, 29, 35, 36, 39, 68, 81, 132, 211, 218, 223, 228, 271, 282, 283, 284, 285, 290, 293, 295, 311, 315
   catabolic 271

glyphosate 18
isoflavonoid 293
isoprenoid 18, 35, 81
mevalonic acid 35
phenylpropanoid 18, 211, 282, 284, 285, 290, 293
ribosomal 132
secondary metabolite 28, 223, 228
signal transduction 29
Pelargonium 308, 314
  leaf-curl virus (PLCV) 308
  viruses 314
Phenolic 31, 123, 124, 128, 129, 204, 211, 263, 272
  agent 204
  compounds 31, 123, 124, 128, 129, 211, 263, 272
Phenylalanine ammonia lyase 18, 264, 284, 290
Physiological 19, 137
  disorders 137
  function of secondary metabolites 19
Phytochemical composition 119, 135
Plant(s) 3, 20, 27, 82, 84, 102, 111, 181, 202, 216, 219, 224, 226, 263, 290, 310, 311, 312
  based medicines 3
  chloroplast 312
  henna 226
  herbaceous 181
  hormones 102, 111, 219
  mature 202
  metabolism 226
  metabolites 3, 224
  micropropagated 84, 263
  transform 82
  transformed 290
  transgenic 20, 27, 82, 216, 310, 311
  transplanted 263
Plant cell 27, 39, 186, 223, 224
  culture systems 223, 224
  culture technology 27

immobilization 39, 186
Plant growth 29, 30, 31, 32, 33, 34, 75, 103, 108, 109, 112, 143, 164, 165, 217, 249, 254, 263, 265, 270
  hormones 29, 34, 108, 109, 265
  regulators (PGRs) 30, 31, 32, 33, 75, 103, 112, 143, 164, 165, 217, 249, 254, 263, 270
Plasmid 225, 226, 290, 310
  root-inducing 226, 290
Polymerase chain reaction (PCR) 78
Production 29, 162, 169, 172, 225
  biopharmaceutical 169
  ginsenoside 225
  metabolite 29
  silk 162, 172
Products 4, 16, 181, 196, 221, 223, 226
  industrial 4, 196
  meat 181
  plant-derived 16
  renewable 223
  secondary phytochemical 221
  therapeutic 226
Propagation 26, 64, 66, 94, 95, 98, 101, 103, 109, 110, 214, 215, 218, 220, 246, 302, 303
  commercial 220
  cutting 109
  mass 94, 218
  methods 66, 109
  vegetative 26, 109, 246
Properties 6, 98, 100, 103, 121, 122, 135, 138, 161, 171, 173, 181, 202, 205, 206, 209, 272, 282, 284, 302, 303
  anti-angiogenic 161
  anti-helminthic 284
  anti-hepatotoxic 98
  anticancer 205, 206
  anticholinergic 103
  antioxidative 272
  antiplatelet 181
  antiviral 100, 122

aromatherapy 302, 303
hygroscopic 138
leucodermatic 171
Protein(s) 18, 29, 36, 37, 121, 123, 201, 227, 267, 270, 283
   green fluorescent 227
   kinases 270
Proteus 130, 131, 132
   mirabilis 130, 131
   vulgaris 131, 132
Pseudomonas aeruginosa 130, 131, 132
Psoriasis 101, 223, 282, 284, 295
PTC techniques 21

R

Radiation therapy 271
Random amplified polymorphic DNA (RAPD) 182, 183, 263
Ripening 138, 139
   index 139
   process 138
RNA 36, 37, 143, 292
   antisense 37
   interference method 37
   messenger 36
   synthesis 143

S

Saccharomyces cerevisiae 30
Salmonella typhii 131
Sapodilla fruits, gel-coated 139
Scavenging activities 174
Secondary metabolite(s) 18, 221, 226, 259, 271
   biosynthesis 18, 221, 259, 271
   producing 226
Shoot 34, 76, 102, 104, 107, 108, 109, 111, 183, 249, 250, 252, 261, 310
   organogenesis 34, 76, 102, 109
   regeneration 104, 107, 108, 111, 183, 249, 250, 252, 261, 310
Side effects, toxic 212
Signal transduction 29, 291
Single-gene breeding 27
Sitoindosides 65
Skin 94, 101, 120, 245, 246, 282
   diseases 94, 101, 245, 282
   disorders 246
   irritations 120
Sleep problems 95
Soil, sterilized 102
Soilrite, sterilized 185
Staphylococcus 130, 131, 132, 173
   aureus 130, 131, 132, 173
   pyogenes 130, 131
Stress 17, 27, 29, 36, 84, 93, 126, 209, 227, 263, 273, 290, 315
   abiotic 36, 263, 273
   ecosystem 93
   environmental 315
   oxygen 227
   seawater 126
Synthesis 15, 18, 19, 25, 28, 29, 30, 34, 35, 37, 68, 132, 217, 223, 282, 283, 292, 293, 295
   bacterial protein 132
   isoflavone 293
   isoprenoid 68
   pathways 217
Synthetic antioxidants 127
Systems 3, 4, 21, 64, 65, 93, 106, 132, 176, 207, 208, 220, 245
   biotechnological production 21
   enzymatic anti-oxidative defense 208
   mass production 106
   plant-based therapeutic 93
   traditional 65, 245
   traditional herbal 207
   traditional medicinal 176

T

Traditional medicine 176
Taxadiene synthase 37
Tea 263, 264, 268
   plants, micropropagated 263
   somatic embryogenesis 264, 268
Techniques 15, 21, 22, 24, 25, 26, 27, 28, 31, 32, 33, 34, 35, 37, 38, 102, 214, 216, 217, 224, 288
  callus formation 24
  immobilization 15, 38, 224
  micro-propagation 102
  sustainable 288
Terpene biosynthesis pathway 311, 313
Thin layer chromatography (TLC) 287
Tissue culture 113, 224, 259, 273
   methods 113, 259, 273
   technology 224
Tools 4, 85
  bioinformatics 85
  biotechnological 4
Traditional Chinese medicine (TCM) 200, 202, 203
Transcription factors (TFs) 24, 36, 219, 269, 270, 312
Transformation 78, 223, 226, 310
  rhizogenes-mediated 226
Transgenic 27, 227, 315, 311
   breeding 27
   technology 315
   tomato 227
Transpiration, reducing 138
Transport 37, 212, 266, 314
  polar 266
Transverse leaf sections 148

V

Vascular 176, 206
   endothelium growth factor (VEGF) 176
   inflammatory process 206

W

Water 138, 309
   smoke-saturated 309
   transmission 138
Withanolide biosynthetic pathway 70, 81